Automated Deduction – A Basis for Applications

APPLIED LOGIC SERIES

VOLUME 9

Managing Editor

Dov M. Gabbay, *Department of Computing, Imperial College, London, U.K.*

Co-Editor

Jon Barwise, *Department of Philosophy, Indiana University, Bloomington, IN, U.S.A.*

Editorial Assistant

Jane Spurr, *Department of Computing, Imperial College, London, U.K.*

SCOPE OF THE SERIES

Logic is applied in an increasingly wide variety of disciplines, from the traditional subjects of philosophy and mathematics to the more recent disciplines of cognitive science, computer science, artificial intelligence, and linguistics, leading to new vigor in this ancient subject. Kluwer, through its Applied Logic Series, seeks to provide a home for outstanding books and research monographs in applied logic, and in doing so demonstrates the underlying unity and applicability of logic.

Automated Deduction – A Basis for Applications

Volume II

Systems and Implementation Techniques

edited by

WOLFGANG BIBEL
Darmstadt University of Technology,
Germany

and

PETER H. SCHMITT
University of Karlsruhe,
Germany

KLUWER ACADEMIC PUBLISHERS
DORDRECHT / BOSTON / LONDON

A C.I.P. Catalogue record for this book is available from the Library of Congress.

ISBN 0-7923-5129-0 (Vol. I)
ISBN 0-7923-5130-4 (Vol. II)
ISBN 0-7923-5131-2 (Vol. III)
ISBN 0-7923-5132-0 (Set)

Published by Kluwer Academic Publishers,
P.O. Box 17, 3300 AA Dordrecht, The Netherlands.

Sold and distributed in North, Central and South America
by Kluwer Academic Publishers,
101 Philip Drive, Norwell, MA 02061, U.S.A.

In all other countries, sold and distributed
by Kluwer Academic Publishers,
P.O. Box 322, 3300 AH Dordrecht, The Netherlands.

Printed on acid-free paper

Printed in the Netherlands.

TABLE OF CONTENTS

VOLUME II:

SYSTEMS AND IMPLEMENTATION TECHNIQUES

VOLUME I:

FOUNDATIONS. CALCULI AND METHODS

VOLUME III:

APPLICATIONS

AFFILIATIONS OF THE AUTHORS

Wolfgang Ahrendt
University of Karlsruhe
Institute for Logic, Complexity
and Deduction Systems
ahrendt@ira.uka.de

Jürgen Avenhaus
University of Kaiserslautern
Department of Computer Science
avenhaus@informatik.uni-kl.de

Matthias Baaz
Vienna University of Technology
Institute of Algebra
baaz@logic.at

Michael Balser
University of Ulm
Faculty of Computer Science
Department for Software Engineering
and Compiler Construction (PM)
89069 Ulm, Germany
balser@informatik.uni-ulm.de

Bernhard Beckert
University of Karlsruhe
Institute for Logic, Complexity
and Deduction Systems
76128 Karlsruhe, Germany
beckert@ira.uka.de

Holger Benl
Ludwig-Maximilians-University Munich
Mathematical Institute
80333 München, Germany
benl@rz.mathematik.uni-muenchen.de

Ulrich Berger
Ludwig-Maximilians-University Munich
Mathematical Institute
80333 München, Germany
berger@rz.mathematik.uni-muenchen.de

Sven-Erik Bornscheuer
Dresden University of Technology
Department of Computer Science
Artificial Intelligence Institute
Knowledge Representation
and Reasoning Group
seb@inf.tu-dresden.de

Reinhard Bündgen
IBM Germany Development GmbH
71003 Böblingen, Germany
buendgen@de.ibm.com

Ingo Dahn
Humboldt University Berlin
Institute of Pure Mathematics
dahn@mathematik.hu-berlin.de

Jörg Denzinger
University of Kaiserslautern
Department of Computer Science
denzinge@informatik.uni-kl.de

Uwe Egly
Vienna University of Technology
Institute of Information Systems
A-1040 Wien, Austria
uwe@kr.tuwien.ac.at

Detlef Fehrer
University of the Saarland
Department of Computer Science
Saarbrücken, Germany
fehrer@ags.uni-sb.de

Matthias Fuchs
University of Kaiserslautern
Department of Computer Science
fuchs@informatik.uni-kl.de

Manfred Göbel
International Computer Science Institute
Berkeley, CA 94704-1198, USA
manfredg@icsi.berkeley.edu

Peter Graf
SAP AG
peter.graf@sap-ag.de

Reiner Hähnle
University of Karlsruhe
Institute for Logic, Complexity
and Deduction Systems
76128 Karlsruhe, Germany
reiner@ira.uka.de

Friedrich von Henke
University of Ulm
89069 Ulm, Germany
vhenke@informatik.uni-ulm.de

Steffen Hölldobler
Dresden University of Technology
Department of Computer Science
Artificial Intelligence Institute
Knowledge Representation and Reasoning Group
sh@inf.tu-dresden.de

Yvonne Kalinke
Dresden University of Technology
Department of Computer Science
Artificial Intelligence Institute
Knowledge Representation and Reasoning Group
yve@inf.tu-dresden.de

Thomas Kolbe
Darmstadt University of Technology
Department of Computer Science
kolbe@informatik.tu-darmstadt.de

Wolfgang W. Küchlin
University of Tübingen
Faculty of Informatics
Wilhelm-Schickard Institute
kuechlin@informatik.uni-tuebingen.de

Alexander Leitsch
Vienna University of Technology
Institute of Computer Languages
leitsch@logic.at

Marko Luther
University of Ulm
Artificial Intelligence
89069 Ulm, Germany
luther@ki.informatik.uni-ulm.de

Wolfram Menzel
University of Karlsruhe
Institute for Logic, Complexity
and Deduction Systems
menzel@ira.uka.de

Gerd Neugebauer
University of Koblenz-Landau
Department of Computer Science
gerd@uni-koblenz.de

Tobias Nipkow
Technical University of Munich
Institute for Informatics
80290 München, Germany
nipkow@informatik.tu-muenchen.de

Uwe Petermann
Leipzig University of Applied Sciences
Department of Computer Science
uwe@imn.htwk-leipzig.de

Wolfgang Reif
University of Ulm
Faculty of Computer Science
Department for Software Engineering
and Compiler Construction (PM)
89069 Ulm, Germany
reif@informatik.uni-ulm.de

Gerhard Schellhorn
University of Ulm
Faculty of Computer Science
Department for Software Engineering
and Compiler Construction (PM)
89069 Ulm, Germany
schellhorn@informatik.uni-ulm.de

Peter H. Schmitt
University of Karlsruhe
Institute for Logic, Complexity
and Deduction Systems
76128 Karlsruhe, Germany
Email: pschmitt@ira.uka.de

Johann Schumann
Technical University Munich
Department of Informatics
schumann@informatik.tu-muenchen.de

Helmut Schwichtenberg
Ludwig-Maximilians-University Munich
Mathematical Institute
80333 München, Germany
schwicht@rz.mathematik.uni-muenchen.de

Monika Seisenberger
Ludwig-Maximilians-University Munich
Mathematical Institute
80333 München, Germany
seisenb@rz.mathematik.uni-muenchen.de

Jörg H. Siekmann
DFKI and University of the Saarland
66123 Saarbrücken, Germany
siekmann@dfki.uni-sb.de

Kurt Stenzel
University of Ulm
Faculty of Computer Science
Department for Software Engineering
and Compiler Construction (PM)
89069 Ulm, Germany
stenzel@informatik.uni-ulm.de

Martin Strecker
University of Ulm
Artificial Intelligence
89069 Ulm, Germany
strecker@ki.informatik.uni-ulm.de

Antje Strohmaier
TLC GmbH
Antje.Strohmaier@bku.db.de

Christian Suttner
Technical University of Munich
Institute for Informatics
Automated Reasoning
D-80290 München, Germany
suttner@informatik.tu-muenchen.de

Christoph Walther
Darmstadt University of Technology
Department of Computer Science
Chr.Walther@informatik.tu-darmstadt.de

Andreas Weber
University of Tübingen
Faculty of Informatics
Wilhelm-Schickard Institute
weber@informatik.uni-tuebingen.de

Andreas Wolf
Technical University of Munich
Institute for Informatics
Automated Reasoning
80290 München, Germany
wolfa@informatik.tu-muenchen.de

Wolfgang Zuber
Ludwig-Maximilians-University Munich
Mathematical Institute
80333 München, Germany
zuber@rz.mathematik.uni-muenchen.de

Part 1

Interactive Theorem Proving

Editor: Wolfgang Reif

INTRODUCTION

1. BASIC CONCEPTS OF INTERACTIVE THEOREM PROVING

Interactive Theorem Proving ultimately aims at the construction of powerful reasoning tools that let us (computer scientists) prove things we cannot prove without the tools, and the tools cannot prove without us. Interaction typically is needed, for example, to direct and control the reasoning, to speculate or generalize strategic lemmas, and sometimes simply because the conjecture to be proved does not hold. In software verification, for example, correct versions of specifications and programs typically are obtained only after a number of failed proof attempts and subsequent error corrections.

Different interactive theorem provers may actually look quite different: They may support different logics (first- or higher-order, logics of programs, type theory etc.), may be generic or special-purpose tools, or may be targeted to different applications. Nevertheless, they share common concepts and paradigms (e.g. architectural design, tactics, tactical reasoning etc.). The aim of this chapter is to describe the common concepts, design principles, and basic requirements of interactive theorem provers, and to explore the band-width of variations.

Having a 'person in the loop', strongly influences the design of the proof tool:

- proofs must remain comprehensible,
- proof rules must be high-level and human-oriented,
- persistent proof presentation and visualization becomes very important.

In (Paulson, 1997) this is called the *transparency* requirement. Many interactive theorem provers offer expressive specification languages to ease problem formulation, and use sequent calculi or natural deduction to ease the reading and understanding of machine proofs.

Transparency is not the only desideratum. The proof tool must be powerful enough to carry out large parts of the proof on its own. We expect the tool, for example, to fill in technical details, to simplify goals, and to discharge automatically as many proof obligations as possible. Automation may be attained both by appropriate proof strategies within the interactive system, and by external reasoners such as model checkers, automated theorem

3

W. Bibel, P. H. Schmitt (eds.), Automated Deduction. A basis for applications. Vol. II
© 1998 *Kluwer Academic Publishers. Printed in the Netherlands*

provers, decision procedures or computer algebra systems. In the latter case, however, care has to be taken to maintain transparency. Possible indicators for the *power* of an interactive theorem prover are, for example, the overall time to solve a problem, and the total number of required interactions. Other indicators may be defined to measure the prover's goal directedness (ratio of required proof steps and total proof steps including failed proof attempts).

Many interactive theorem provers follow the LCF paradigm of proof tactics and tactical reasoning. The basic principles are recalled in Section 2. This paradigm offers a.o. three attractive features:

— *soundness by construction.* If the basic logic is sound, arbitrary combinations of tactics preserve the soundness.
— *safe customizing.* The user may program problem specific high-level tactics, that are tailor-made for the application under consideration. The proof search then may rely mainly on these new tactics.
— *free combination of reasoning styles.* System designers may easily integrate different reasoning styles such as top-down, bottom-up, or middle-out.

In general, interactive theorem provers support structured theory formation. Since formal models of large soft- or hardware systems are large, it is very important not only to organize the formal theories in a structured way but also to make use of the structure in the proofs. Most interactive theorem provers therefore support deduction in structured theories, whereas most automatic theorem provers do not. Dealing with large applications sometimes means hundreds of proofs and lemmas, including dead ends, unlucky proof decisions, and failed proof attempts. Working out the entire theory and all proofs of a comprehensive application with an interactive theorem prover is not only a mathematical but to a large extent a *(proof-) engineering* challenge. Therefore, many interactive provers give the users access to different kinds of *proof engineering capabilities* such as

— *organizational support* for keeping track of proof obligations, proofs, lemmas, theories and dependencies,
— *proof manipulation support* for inspecting proofs, withdrawing proof steps, replaying proofs or parts of it, and
— *library support* for reusing frequently required theories and lemmas.

Some provers offer more advanced proof engineering techniques, like proof reuse and -adaption, proof generalization, proof transformation, learning from failure, or proof by analogy. The common goal of all these techniques is to take computers closer to a human's problem solving repertoire, a goal which was formulated e.g. in (Constable et al., 1985).

Historically, the first generation of interactive theorem provers was developed in the second half of the 1960s, e.g. AUTOMATH (de Bruijn, 1980), FOL (Weyhrauch, 1981), and Stanford LCF. In the decade from 1975 to 1985 the systems PL/CV (Constable et al., 1982), λ-PRL (Constable et al., 1985), Edinburgh LCF (Gordon et al., 1979), Cambridge LCF (Paulson, 1987), and Nqthm (Boyer and Moore, 1979) emerged. These tools are the forerunners of (or at least strongly influenced) almost all later developments. In the second half of the 1980s the first versions of HOL (Gordon, 1988), Nuprl (Constable et al., 1986), KIV (Reif et al., 1997), Isabelle (Paulson, 1994), EVES (Craigen et al., 1992), IMPS (Farmer et al., 1996), and Coq (Bertot and Bertot, 1996) showed up. All these provers have been steadily enhanced up to now. In the 1990s the spectrum has been enriched by a number of systems, e.g. LP (Guttag and Horning, 1993), PVS (Owre et al., 1992), ACL2 (Kaufmann and Moore, 1997), Alf (Magnusson and Nordström, 1994), LEGO (Luo and Pollack, 1992), TYPELAB (Chapter II.1.3), and MINLOG (Chapter II.1.2).

2. ARCHITECTURES

2.1. *The LCF approach*

LCF is certainly the most influential interactive theorem prover/proof checker. (Paulson, 1987) presents a cleaned up version of LCF and (Gordon, 1998) a very readable account of the history of LCF.

Although LCF supports a very specific logical system, its architecture is quite generic because it is based merely on functional programming and data abstraction:

- Theorems are an abstract type.
- The constructors for theorems correspond exactly to the inference rules of the logic.

As a trivial example, let us build a system for Hilbert-style minimal propositional logic. Given some concrete type *form* of propositional formulae (based on some type *name* of variable names and the only connective '→'), we can define the abstract type *thm* of theorems as follows. The representing type is *form*. Given $f : form$ we write $\lfloor f \rfloor$ for the corresponding theorem. Note that $\lfloor . \rfloor$ is not available outside the abstract type! The only functions for building theorems are $S : name \to name \to name \to form$, $K : name \to name \to form$ and $MP : form \to form \to form$ which are defined such that

$$S\,P\,Q\,R = \lfloor (P \to Q \to R) \to (P \to Q) \to (P \to R) \rfloor$$
$$K\,P\,Q = \lfloor P \to (Q \to P) \rfloor$$
$$MP \lfloor \Phi \to \Psi \rfloor \lfloor \Phi \rfloor = \lfloor \Psi \rfloor$$

Since *thm* is defined as an abstract type, the semantics of the programming language ensures that all theorems are 'correct by construction', i.e. they are derived using only *S*, *K* and *MP*.

This is a very clear architecture for forward proofs. Backward proofs can be simulated by *tactics*. A tactic is simply a function which takes the formula to be proved and returns a list of formulae (the subgoals) together with a function from a list of theorems to a theorem. The latter function is called a *validation*, and should produce the overall theorem given theorems for the subgoal: if *tac* is a tactic and *tac* $\Phi = ([\Psi_1,...,\Psi_n], v)$ then $v \lfloor\Psi_1\rfloor \ ... \ \lfloor\Psi_n\rfloor = \lfloor\Phi\rfloor$ should hold. For example, a tactic for inverting *MP* can be defined as follows: $MP\_tac \ \Phi \ \Psi = ([\Phi \rightarrow \Psi, \Phi], MP)$. The first parameter Φ is necessary because given only the conclusion Ψ of *modus ponens*, there are many possible premises which lead to it. Note that some theorem provers (e.g. Isabelle) overcome this inconvenience by supporting unknowns that can be instantiated by unification later in the proof process.

Tactics are always accompanied by a (usually state-based) mechanism that keeps track of the subgoals and applies the validation functions once all the corresponding subgoals have been proved. Usually there is also a collection of predefined combinators for tactics such as THEN (sequential composition), ORELSE (choice), and REPEAT (iteration).

The important point about the LCF architecture is that both new forward inference rules and new tactics are user-definable in some general-purpose programming language. Thus the system can be customized to an arbitrary degree without endangering its soundness.

2.2. *Variations*

The simplicity of the LCF approach has made it quite popular but also means that it has been adapted in various ways to fit the requirements of particular applications. Below we give a short overview of some of the more important variations.

Meta-languages

LCF is based on a functional programming language with data abstraction. This language is usually called the *meta-language*. In fact, the designers of LCF first designed the programming language ML because at the time there was no language around that suited their purpose. Most interactive theorem provers have followed the lead of a functional meta-language (an exception is the work on embedding theorem provers in λProlog (Nadathur and Miller, 1988)). However, another popular choice is Lisp. Since Lisp does not offer data abstraction, the user cannot manipulate theorems directly in the meta-language (because he could fake a theorem). Instead, an isolating layer in the form of a restricted programming language or a special interface has to be inserted between the user and Lisp. Even if the meta-language does offer data abstraction, some systems, e.g. Coq and KIV, provide a separate interface to shield the user from a direct exposure to the meta-language. Systems implemented in an imperative language, such as LP, usually do not provide a proper meta-language but merely a fixed command language.

Tactics

Not all systems offer them, or if they do, the user may not be able to extend them. This is largely a function of the meta-language: if the user interacts with the system directly via a programming language, user-definable tactics are usually available. Otherwise there may only be a fixed number of inbuilt tactics and maybe a few simple combinators.

Although the LCF-model of tactics is quite general, there are alternatives. In particular Isabelle offers an interesting variation: tactics are functions from proof states to (possibly infinite) lists of new proof states. This is the functional programmer's way of coding alternatives and is ideally suited to express different kinds of search strategies with just a few combinators.

Theorems versus proofs

LCF knows only about theorems. The proof of a theorem is a computation and not a value of the meta-language. Many systems treat proofs as concrete data objects which can be displayed, manipulated and recorded. It is not very hard to extend an LCF-style system with a notion of proof term: simply define a datatype with one constructor for each primitive rule of the logic and a function *exec* which recursively traverses a proof term and extracts the corresponding theorem. In our example above, the type of proof trees has the three constructors *Sp*, *Kp* and *MPp*, and *exec* is defined in the obvious way:

$$exec(Sp\ P\ Q\ R) = S\ P\ Q\ R$$
$$exec(Kp\ P\ Q) = K\ P\ Q$$
$$exec(MPp\ t_1\ t_2) = MP(exec\ t_1)(exec\ t_2)$$

Based on this principle, Wong (Wong, 1995) extended HOL, a direct descendent of LCF, with proof terms. However, proof terms are not an integral part of HOL. This is in contrast to, for example, KIV, where proof terms are an essential feature: they do not just provide a record of the proof but can also be browsed interactively and are the key to reuse of proofs after changes.

A more abstract approach to proof terms is embodied in the Curry-Howard isomorphism which equates formulae with types and proofs with λ-terms.

Checking a proof is thus the same as type-checking a λ-term. Systems such as Coq, ELF (Pfenning, 1991) and TYPELAB follow this approach.

Soundness versus efficiency
Very few interactive theorem provers are completely sound. For systems in the LCF-tradition, unsoundness should only be due to an incorrect implementation of a basic rule of inference. This should be easy to detect, assuming that the basic rules of the system are simple. However, the desire to reduce every proof to a composition of basic inference rules can lead to very slow proofs: frequently, it is much easier to check that a theorem is true than to construct a formal derivation. In particular, it can be very tricky to write the required validation functions for complex tactics. Therefore many systems take short cuts: for efficiency reasons, important derived inference rules are implemented directly in the kernel of the system without reducing them to the basic rules. A typical example is rewriting, which is much easier to implement directly than by going via the rules of equational logic. However, each proof procedure that is included in the core for efficiency reasons is a threat to soundness. An extreme example is PVS, which features a tightly integrated collection of decision procedures which are not based on formal proofs.

Genericity
Theorem provers differ widely in the logical systems they support, and most systems support quite specific logics. The reason is that more customized systems tend to be more effective. In the 80s it was noticed that a) much of the basic infrastructure for interactive theorem proving is independent of the specific logic and b) there are meta-logics which allow formal presentations of object-logics. Isabelle was the first system to implement these insights. Given a presentation of some logical system in Isabelle's meta-logic, Isabelle becomes an interactive theorem prover for this object logic. The resulting system is customizable because Isabelle follows the LCF tradition. Although generic systems are a great asset because one can quickly prototype a proof checker for a new logic, turning this first prototype into a fully fledged theorem prover still requires years of work in populating the libraries with useful theories and theorems and implementing specialized proof procedures.

Isabelle is probably still the most flexible generic theorem prover. Other similar efforts are either strongly biased towards executability (λProlog and ELF) or have adopted the meta-logic as their sole object-logic (e.g. Coq and LEGO).

3. Interactive Theorem Provers Discussed in This Part

The remaining chapters of this part describe three interactive systems, and an approach to integrate interactive and automated theorem proving.

Chapter II.1.1 is on KIV, an interactive system for formal software modeling and verification. The focus is on structured specification development, and on interactive, and automated proof construction for algebraic specifications. An application of KIV in compiler correctness can be found in Chapter III.2.7.

Chapter II.1.2 highlights MINLOG, an interactive prover following the proofs-as-programs paradigm. It is based on a first-order logic over a simply typed λ-calculus, and exploits proof theoretic constructions to verify, extract, or optimize programs.

Chapter II.1.3 is on TYPELAB, a specification and verification environment based on type theory. The goals of MINLOG and TYPELAB are quite similar. However, TYPELAB relies on the Calculus of Constructions. The chapter focuses on theoretical foundations, in particular on a calculus with explicit substitutions.

Finally, Chapter II.1.4 reports on a systematic approach to integrate automated and interactive theorem proving for software verification. The technical basis is an integration of the systems KIV and $_3TAP$ (Beckert et al., 1996). $_3TAP$ is an automated tableau prover for full first-order logic with equality not requiring normal forms. The chapter identifies the major problems of an integration (e.g. different logics, large theories, equality handling), and presents solutions for them. Experimental data can be found in the chapter as well as in Chapter III.2.9.

References

Beckert, B., Hähnle, R., Oel, P., and Sulzmann, M. (1996). The tableau-based theorem prover $_3TAP$, version 4.0. In McRobbie, M., editor, *Proc. 13th CADE, New Brunswick/NJ, USA*, LNCS 1104, pages 303–307. Springer.

Bertot, J. and Bertot, Y. (1996). Ctcoq : a system presentation. In *13th International Conference on Automated Deduction, CADE-13*, Springer LNCS.

Boyer, R. S. and Moore, J. S. (1979). *A Computational Logic*. Academic Press, New York.

Constable, R. L., Allen, S. F., Bromley, H. M., Cleaveland, W. R., Cremer, J. F., Harper, R. W., Howe, D. J., Knoblock, T. B., Mendler, N. P., Panagaden, P., Sasaki, J. T., and Smith, S. F. (1986). *Implementing Mathematics with the Nuprl Proof Development System*. Prentice Hall.

Constable, R. L., Johnson, S. D., and Eichenlaub, C. D. (1982). *An introduction to the PL/CV2 programming logic.* Springer LNCS 135.

Constable, R. L., Knoblock, T. B., and Bates, J. L. (1985). Writing programs that construct proofs. *Journal of Automated Reasoning,* 1(3):285 – 326.

Craigen, D., Kromodimoeljo, S., Meisels, I., Pase, B., and Saaltink, M. (1992). The EVES system. In *Proceedings of the International Lecture Series on "Functional Programming, Concurrency, Simulation and Automated Reasoning" (FPCSAR).* McMaster University.

de Bruijn, N. G. (1980). A survey of the project AUTOMATH. In Hindley, J. R. and Seldin, J. P., editors, *Essays on Combinatory Logic, Lambda Calculus and Formalism,* pages 580–606. Academic Press, London.

Farmer, W. M., Guttman, J. D., and Fábrega, F. J. T. (1996). IMPS: an updated description. In *13th International Conference on Automated Deduction, CADE-13,* Springer LNCS.

Gordon, M. (1988). HOL: A Proof Generating System for Higher-order Logic. In Birtwistle, G. and Subrahmanyam, P., editors, *VLSI Specification and Synthesis.* Kluwer Academic Publishers.

Gordon, M., Milner, R., and Wadsworth, C. (1979). *Edinburgh LCF: A Mechanised Logic of Computation.* Springer LNCS 78.

Gordon, M. J. (1998). From LCF to HOL: a short history. In Plotkin, G., Stirling, C., and Tofte, M., editors, *Essays in Honour of Robin Milner.* MIT Press.

Guttag, J. V. and Horning, J. (1993). *Larch: Languages and Tools for Formal Specifications.* Springer New-York.

Kaufmann, M. and Moore, J. S. (1997). An industrial strength theorem prover for a logic based on common lisp. *IEEE Transactions on Software Engineering,* 23(4).

Luo, Z. and Pollack, R. (1992). LEGO proof development system: User's manual. Technical Report ECS-LFCS-92-211, LFCS, Computer Science Dept., University of Edinburgh, The King's Buildings, Edinburgh EH9 3JZ. Updated version.

Magnusson, L. and Nordström, B. (1994). The ALF proof editor and its proof engine. In Barendregt, H. and Nipkow, T., editors, *Types for Proofs and Programs,* pages 213–237. Springer LNCS 806.

Nadathur, G. and Miller, D. (1988). An overview of λProlog. In Kowalski, R. A. and Bowen, K. A., editors, *Proc. 5th Int. Logic Programming Conference,* pages 810–827. MIT Press.

Owre, S., Rushby, J. M., and Shankar, N. (1992). PVS: A Prototype Verification System. In Kapur, D., editor, *Automated Deduction - CADE-11. Proceedings,* Springer LNAI 607. Saratoga Springs, NY, USA.

Paulson, L. C. (1987). *Logic and Computation, Interactive Proof with Cambridge LCF,* volume 2 of *Cambridge Tracts in Theoretical Computer Science.* Cambridge University Press.

Paulson, L. C. (1994). *Isabelle: A Generic Theorem Prover.* Springer LNCS 828.

Paulson, L. C. (Published 1997). Tool support for logics of programs. In Broy, M., editor, *Mathematical Methods in Program Development: Summer School Marktoberdorf 1996,* NATO ASI Series F, pages 461–498. Springer.

Pfenning, F. (1991). Logic programming in the LF Logical Framework. In Huet, G. and Plotkin, G., editors, *Logical Frameworks*. Cambridge University Press.

Reif, W., Schellhorn, G., and Stenzel, K. (1997). Proving System Correctness with KIV 3.0. In *14th International Conference on Automated Deduction. Proceedings*, pages 69 – 72. Townsville, Australia, Springer LNCS 1249.

Weyhrauch, R. (1981). Prolegomena to a theory of mechanized formal reasoning. In Webber, B. and Nilsson, N., editors, *Readings in Artifical Intelligence*, chapter 2, pages 173–191. Tioga, Palo Alto.

Wong, W. (1995). Recording and checking HOL proofs. In Schubert, E., Windley, P., and Alves-Foss, J., editors, *Higher Order Logic Theorem Proving and Its Applications*, pages 353–368. Springer LNCS 971.

W. REIF, G. SCHELLHORN, K. STENZEL AND M. BALSER

CHAPTER 1

STRUCTURED SPECIFICATIONS AND INTERACTIVE PROOFS WITH KIV

1. INTRODUCTION

The aim of this chapter is to describe the integrated specification- and theorem proving environment of KIV. KIV is an advanced tool for developing high assurance systems. It supports:

- hierarchical formal specification of software and system designs
- specification of safety/security models
- proving properties of specifications
- modular implementation of specification components
- modular verification of implementations
- incremental verification and error correction
- reuse of specifications, proofs, and verified components

KIV supports the entire design process from formal specifications to verified code. It supports functional as well as state-based modeling. KIV is ready for use, and has been tested in a number of industrial pilot applications. It can also be used as a pure specification environment with a proof component. Furthermore, KIV serves as an educational and experimental platform in formal methods courses. For details on KIV see (Reif, 1995), (Reif et al., 1995a).

In this chapter we focus on one particular aspect of KIV: its integrated specification- and theorem proving environment. The concepts are illustrated by a running example: an algebraic theory of directed graphs (Sect. 2). In Sect. 3 we prove a simple theorem, sketch the proof calculus, graphical proof construction, the proof engineering capabilities, and flexible proof automation. Sect. 4 is devoted to simplification, a key issue in interactive systems. Sect. 5 presents KIV's elaborate correctness management. It tracks the impact of error corrections and other modifications to specifications, programs, and lemmas. In Sect. 6 we discuss implementation correctness. Sect. 7 is devoted to related systems, and in Sect. 8 we draw some conclusions.

W. Bibel, P. H. Schmitt (eds.), Automated Deduction. A basis for applications. Vol. II
© 1998 *Kluwer Academic Publishers. Printed in the Netherlands*

2. Representation of Structured Designs

In KIV, formal specifications of software and system designs are represented explicitly as directed acyclic graphs called *development graphs*. Nodes correspond to specification components or program modules, and the edges indicate their interrelations (specification structure, refinement, implementation). After a short overview of the structuring operations in KIV we introduce the running example, and describe development graphs in more detail.

2.1. *Structured Algebraic Specifications*

The specification language and semantics for algebraic theories in KIV are similar to CASL (Mosses, 1996), PLUSS (Gaudel, 1992) or the first-order part of SPECTRUM (Broy et al., 1993). An elementary specification consists of a signature, a generation principle, and a set of axioms. The signature defines *sorts*, *constants*, *functions*, *predicates*, and *variables*. The optional generation principle restricts models to term-generated models and provides structural induction. Axioms are written in full first-order logic with equality. The semantics is loose, i.e. the set of all (term-generated) algebras that satisfy the axioms.

Structured specifications are built using the operations *union, enrichment, renaming, parameterization,* and *actualization*, and combinations thereof, similar to ASL (Sannella and Wirsing, 1983). A union $SP = SP_1 + SP_2$, is the union of the signatures and the axioms. An enrichment, $SP =$ **enrich** SP_1 **by** Δ, adds new signature symbols and new axioms to a specification SP_1. A renamed specification, $SP =$ **rename** SP_1 **by** M, renames some or all signature symbols of SP_1 using a mapping M. A parameterized specification, SP = **generic specification parameter** SP_1 **using** SP_2 **target** Δ, enriches SP_2 by Δ and parameterizes it with SP_1. The parameter specification can later be instantiated by an actual specification. In this manner, lists of naturals and lists of strings can be derived from a parameterized specification of lists of elements. An actualization, SP = **actualize** SP_1 **with** SP_2 **by** M, instantiates a parameterized specification SP_1 with an actual specification SP_2. The correspondence between parameter operations and actual operations is given by a mapping M. M also allows the renaming of target operations from SP_1.

2.2. *Example: Directed Graphs*

The structured specification for directed graphs is shown in Fig. 1. The toplevel specification *path* contains the definition of paths as lists of vertices of a graph. This specification enriches *digraph* (with the basic definitions of

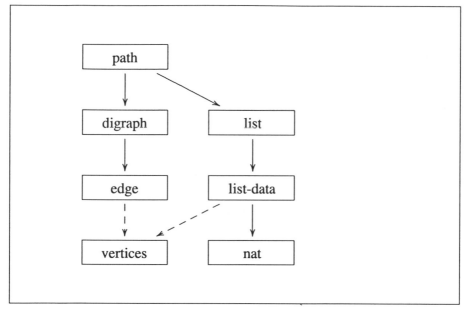

Figure 1.   Visualization of a development graph

directed graphs) and *list*. *list* is based on *list-data* (where elementary operations on lists are specified) and contains additional operations. *list-data* uses
natural numbers (specified in *nat*) and *vertices* for list elements. *digraph* is
based on *edge*, which is again based on *vertices*.

The specifications of *vertices* and *nat* can be found in Fig. 4. *edge* (Fig. 2)
is a generic specification with *vertices* as parameter (indicated by a dashed
line in the development graph). $a \Rightarrow b$ denotes an edge from a to b. In the
specification a special notation (introduced by the keyword **data**) is used to
define a free data type in a convenient manner similar to data type definitions
in ML (Harper et al., 1986). KIV automatically generates axioms for this kind
of data type definitions which are also listed in Fig. 2.

In *digraph* (Fig. 3), directed graphs are specified. Graphs are built up from
the empty graph $\emptyset_g$ by adding vertices with $+_v$ or edges with $+_e$. The specification also contains two predicates which test if a given vertex ($\in_v$) or edge
($\in_e$) is in the graph. The first axiom,

$$g_1 = g_2 \leftrightarrow (\forall\, a.\, a \in_v g_1 \leftrightarrow a \in_v g_2) \wedge (\forall\, e.\, e \in_e g_1 \leftrightarrow e \in_e g_2)$$

is an extensionality for graphs: Two graphs are equal if and only if they contain the same vertices and edges. Axioms 2–4 define $\in_v$. No vertex is in the
empty graph $\emptyset_g$, a vertex a is in a graph g with an added vertex b if and only
if either a equals b (i.e. this vertex was added) or a is already in g. The fourth

edge =
generic data specification
   parameter vertices
   edge = . $\Rightarrow$. (. .source : vertex, . .destination : vertex);
   variables e, e_0, e_1, e_2: edge;
end generic data specification

Generated axioms:
   edge **freely generated by** $\Rightarrow$;
   $(a \Rightarrow a_0)$.source = a,
   $(a \Rightarrow a_0)$.destination = a_0,
   $(a \Rightarrow a_0) = (a_1 \Rightarrow a_2) \leftrightarrow a = a_1 \wedge a_0 = a_2$,
   (e.source $\Rightarrow$ e.destination) = e

Figure 2.   The specification *edge*

digraph =
enrich edge **with**
   sorts graph;
   constants 0_g : graph;
   functions
        . $+_v$. : graph $\times$ vertex $\rightarrow$ graph;
        . $+_e$. : graph $\times$ edge $\rightarrow$ graph;
   predicates
        . $\in_v$. : vertex $\times$ graph;
        . $\in_e$. : edge $\times$ graph;
   variables g_2, g_1, g_0, g: graph; d: vertex;

axioms
   graph **generated by** 0_g, $+_v$, $+_e$;
   $g_1 = g_2 \leftrightarrow (\forall a.\ a \in_v g_1 \leftrightarrow a \in_v g_2) \wedge (\forall e.\ e \in_e g_1 \leftrightarrow e \in_e g_2)$,
   $\neg\, a \in_v 0_g$,
   $a \in_v g +_v b \leftrightarrow a = b \vee a \in_v g$,
   $a \in_v g +_e (b \Rightarrow c) \leftrightarrow a \in_v g$,
   $\neg\, e \in_e 0_g$,
   $e \in_e g +_v a \leftrightarrow e \in_e g$,
      $(a \Rightarrow b) \in_e g +_e (c \Rightarrow d)$
   $\leftrightarrow a = c \wedge b = d \wedge a \in_v g \wedge b \in_v g \vee (a \Rightarrow b) \in_e g$

Figure 3.   The specification *digraph*

nat =
specification
sorts nat;
constants 0 : nat;
functions . +1, . −1 : nat → nat; . + . : nat × nat → nat;
predicates . < . : nat × nat;
variables n, m, k: nat;
axioms
   nat **freely generated by** 0, +1;
   $n + 1 - 1 = n$, $m + 0 = m$, $m + n + 1 = (m + n) + 1$,
   $\neg\, m < 0$, $0 < n + 1$, $m + 1 < n + 1 \leftrightarrow m < n$,
   $\neg\, m < m$, $m < n \wedge n < k \rightarrow m < k$
end specification

---

vertices =
specification sorts vertex; **variables** a, b, c: vertex;
end specification

---

list-data =
generic data specification
parameter vertex **using** nat
list = nil | . ⊕ . (. .first : vertex, . .rest : list);
variables z, y, x: list; **size functions** # : list → nat ;
end generic data specification

Generated axioms:
   list **freely generated by** nil, ⊕;
   $(a \oplus x).\text{first} = a$, $(a \oplus x).\text{rest} = x$,
   $a \oplus x = a_0 \oplus x_0 \leftrightarrow a = a_0 \wedge x = x_0$, $\text{nil} \neq a \oplus x$,
   $x = \text{nil} \vee x = x.\text{first} \oplus x.\text{rest}$, $\#(\text{nil}) = 0$, $\#(a \oplus x) = \#(x) + 1$

---

list =
enrich list-data **with**
functions . ⊙ . : list × list → list; . .last : list → vertex;
predicates . ∈ . : vertex × list; dups : list;
axioms
   $\text{nil} \odot x = x$, $a \oplus x \odot y = a \oplus (x \odot y)$, $(x \odot a \oplus \text{nil}).\text{last} = a$,
   $a \in x \leftrightarrow (\exists\, y, z.\ x = y \odot a \oplus z)$,
   $\text{dups}(x) \leftrightarrow (\exists\, a, x_0, y, z.\ x = x_0 \odot a \oplus y \odot a \oplus z)$
end enrich

Figure 4.   The specifications *nat, vertices, list-data* and *list*

path =
enrich list + digraph **with**
   predicates
        path, spath   :   list × graph;
        $\exists_{path}, \exists_{spath}$:   vertex × vertex × graph;

axioms
   $\neg$ path(nil, g),
   path(a $\oplus$ nil, g) $\leftrightarrow$ a $\in_v$ g,
   path(a $\oplus$ b $\oplus$ x, g) $\leftrightarrow$ (a $\Rightarrow$ b) $\in_e$ g $\wedge$ path(b $\oplus$ x, g),
   spath(x, g) $\leftrightarrow$ path(x, g) $\wedge \neg$ dups(x),
   $\exists_{path}$(a, b, g) $\leftrightarrow$ ($\exists$ x. path(x, g) $\wedge$ x.first = a $\wedge$ x.last = b),
   $\exists_{spath}$(a, b, g) $\leftrightarrow$ ($\exists$ x. spath(x, g) $\wedge$ x.first = a $\wedge$ x.last = b)
end enrich

Figure 5.   The specification *path*

spath-exists: $\vdash \exists_{path}$(a, b, g) $\rightarrow \exists_{spath}$(a, b, g)

Figure 6.   A theorem to prove

axiom states that a vertex *a* is in a graph with an added edge if and only if it is already in *g*. This implies that adding edges never introduces new vertices.

In order to specify paths, we need lists of vertices (Fig. 4). *nil* denotes the empty list, $\oplus$ adds an element to the head of the list, *.first* and *.rest* select the first element and the rest of the list, respectively, $\odot$ concatenates two lists, *.last* selects the last element of a list, and *dups* is true if the list contains duplicates. # computes the length of a list.

The specification *path* (Fig. 5) finally is an enrichment of *list* and *digraph*. A list of vertices $(a_1, \ldots, a_n)$ is a *path* only if adjacent vertices are connected by an edge, i.e. $a_i \Rightarrow a_{i+1} \in_e g$ for $i = 1, \ldots, n-1$. The empty list is not a path, and the list with only one vertex is a path if this vertex is in *g*. Hence *path*(x, g) is true if *x* is a path in *g*. The predicate *spath*(x, g) is true if *x* is a simple path in *g*, i.e. *x* is a path in *g* and contains no duplicate vertices. The two predicates $\exists_{path}$ and $\exists_{spath}$ are true if a path (or a simple path, respectively) exists between two vertices in a graph. This concludes the specification.

Given the specification it is now possible to formulate theorems for directed graphs and paths and prove them with KIV. A typical textbook theorem states that if a path exists between two vertices, then also a simple path exists between them (Fig. 6). Section 3 shows how *spath-exists* is proved in KIV.

2.3. *Development Graphs*

Development graphs can be viewed and manipulated graphically using daVinci (Fröhlich and Werner, 1994). New specifications are added by creating new nodes and links, and by clicking on a node, a specification can be viewed, edited, modified, deleted etc. Selecting a node also gives access to the *theorem base* for the specification component. For each specification component one theorem base exists. Initially, the axioms of a component (written by the user or generated by the system) and proof obligations are stored there, later on it is extended by user-defined lemmas. In the remainder, we will call axioms, proof obligations, and lemmas *theorems*.

The proof of a theorem, e.g. *spath-exists*, is done relative to the specification component and the theorem base where the theorem is defined. In our example this is the component *path*. Since the structuring operations guarantee that every theorem that holds in a subspecification of *SP* also holds in *SP* itself (modulo renaming), the proof may as well rely on any theorem of any subspecification.

KIV has a very tight integration of design, proving, and support for modifications, and maintains detailed status information in the development graph and the theorem bases. The *correctness management* ensures that after modifications to specifications or theorems only those proofs are invalidated that are really affected by the modification. All other proofs will remain valid. (See Sect. 5 for a detailed discussion.) In order to achieve this goal, the theorem base does not only contain theorems, but stores additional information for each theorem,

1. its name (e.g. ax-01, ax-02, spath-exists, ...)
2. its status: axiom, proof obligation, user-defined
3. proved/unproved? (this is irrelevant for axioms)
4. does a (perhaps partial) proof exist?
5. theorems used in the proof
6. special usage of the theorem (e.g. as simplifier rule)
7. a link to the file where the proof is stored

Furthermore, the theorem base contains information used by heuristics, as well as information about well-founded orderings for induction.

A lemma needed in a proof is added to the theorem base of the lowest component in the specification hierarchy where it is valid. This allows the reuse of the lemma in other proofs in other specifications as well. Fig. 7 lists some of the used lemmas for the example.

---

path

$\vdash$ path($x \odot a \oplus y \odot a \oplus z$, g) $\rightarrow$ path($x \odot a \oplus z$, g)

$\vdash$ path(x, g) $\land$ path(y, g) $\land$ x.last = y.first $\rightarrow$ path($x \odot y$.rest, g)

$x \neq$ nil $\vdash$ path($x \odot y$, g) $\rightarrow$ path(x, g)

$y \neq$ nil $\vdash$ path($x \odot y$, g) $\rightarrow$ path(y, g)

digraph

$\vdash \neg a \in_v g \rightarrow \neg (a \Rightarrow b) \in_e g$

$\vdash (a \Rightarrow b) \in_e g \rightarrow a \in_v g \land b \in_v g$

list

$\vdash (x \odot a \oplus y)$.last = $(a \oplus y)$.last

$\vdash (a \oplus x \odot b \oplus y)$.last = $(b \oplus y)$.last

list-data

$x \neq$ nil $\vdash a = x$.first $\land$ $y = x$.rest $\leftrightarrow$ $x = a \oplus y$

Figure 7.   Some theorems

3. THEOREM PROVING IN KIV

The example theorem

$$\vdash \exists_{path}(a, b, g) \rightarrow \exists_{spath}(a, b, g)$$

states that if there is some path from *a* to *b* in a graph *g* then there is also a "simple" path which does not contain duplicate nodes. It can be proved by induction over the length of the path from *a* to *b*. If the path contains duplicates, the main lemma

elim-cycle: $\vdash$ path($x \odot a \oplus y \odot a \oplus z$, g) $\rightarrow$ path($x \odot a \oplus z$, g)

allows to eliminate one cycle, leading to a shorter path to which the induction hypothesis can be applied.

3.1. *The example proof*

Of course, there are a lot of details which have to be filled in for this informal argument to become a machine acceptable proof. In KIV formal proofs are carried out in the classical sequent calculus. Starting out from an initial goal, a fixed, predefined set of user-oriented rules (sometimes also called *tactics*) is applied either by the system or by the user. Rules reduce goals to sufficient subgoals until axioms (or other lemmas) are reached.

Thereby a proof tree is built up. The final one for our example theorem is shown in Fig. 8. The conclusion (node 1) contains the initial goal, all open premises in the tree (light circles) are used lemmas. We will briefly explain how it evolves.

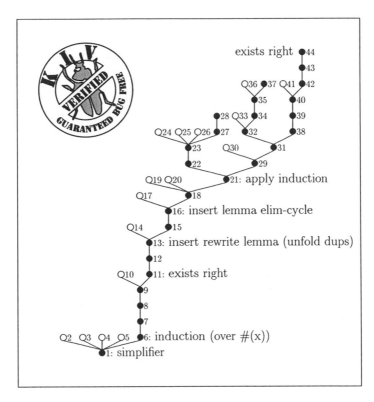

Figure 8. Proof tree for the theorem spath-exists

The first step in the proof is an application of the *simplifier* rule. Simplification includes simple propositional reasoning, rewriting (e.g. the definition of $\exists_{path}$ is unfolded; node 2 contains the corresponding axiom) and removal of quantifiers (the existential quantifier introduced by unfolding $\exists_{path}$ is dropped). The *simplifier* rule is the most complex rule available, see Sect. 4. The proof then proceeds with induction over $\#(x)$, the length of the existing path x. To this purpose the *induction* rule is applied to goal 6, resulting in the following goal (node 7)

path(x, g), Ind-Hyp
$\vdash x = nil, \exists\ x_0.\quad$ path$(x_0, g) \wedge \neg$ dups(x_0)
$\qquad\qquad \wedge\ x_0.first = x.first \wedge x_0.last = x.last)$

where Ind-Hyp abbreviates

$$\forall\, g_1, x_1. \quad \#(x_1) < \#(x) \wedge \text{path}(x_1,g_1)$$
$$\rightarrow x_1 = \text{nil} \vee \exists\, x_2. \quad \text{path}(x_2,g_1) \wedge \neg\, \text{dups}(x_2)$$
$$\wedge\, x_2.\text{first} = x_1.\text{first} \wedge x_2.\text{last} = x_1.\text{last}$$

The next important step is the instantiation of the existential quantifier in the succedent with $[x_0 \leftarrow x]$ with the *exists right* rule (node 11). This instance proves the case, in which path x is already duplicate free. Otherwise x contains duplicates. In goal 13, x has been replaced by $a \oplus y$, since it is non-empty

$$\text{dups}(a \oplus y), \text{path}(a \oplus y, g), \text{Ind-Hyp}$$
$$\vdash \exists\, x_0. \text{path}(x_0,g) \wedge \neg\, \text{dups}(x_0) \wedge x_0.\text{first} = a \wedge x_0.\text{last} = (a \oplus y).\text{last}$$

Next the definition of *dups* has to be unfolded (the axiom from specification *list* is contained in node 14). This is done with the rule *insert rewrite lemma*. This rule lets the user first select a subspecification (here *list*) from which the lemma is to be chosen, and then offers all relevant instances of lemmas. Then the main lemma *elim-cycle* is applied by *insert lemma* (node 16). Here a substitution has to be given for the free variables of the lemma. Then just as in the informal proof, the induction hypothesis is applied. This is done in goal 21 (also visible in Fig. 9) which reads as

$$a \oplus y = x_0 \odot a_0 \oplus (y_0 \odot a_0 \oplus z), x_0 \odot a_0 \oplus z \neq \text{nil}$$
$$\text{path}(a_0 \oplus z, g), \text{path}(x_0 \odot a_0 \oplus z, g), \text{path}(a \oplus y, g), \text{Ind-Hyp}$$
$$\vdash \exists\, x_0. x_0.\text{last} = (a \oplus y).\text{last} \wedge \text{path}(x_0,g) \wedge \neg\, \text{dups}(x_0) \wedge x_0.\text{first} = a$$

The fact that the element a_0 is contained twice in the path $a \oplus y$ is represented by the first equation. Therefore the induction hypothesis must be applied with the instance $[x_1 \leftarrow x_0 \odot a_0 \oplus z, g_1 \leftarrow g]$. This results in two subgoals. In the first goal (node 22) it has to be proved that the induction hypothesis was applied on a path shorter than the original. In the second it has to be shown that the path x_2 delivered by the induction hypothesis is indeed a simple path (node 29). This (after some technical steps) is finally done by instantiating the existential quantifier in the succedent with x_2 (node 44).

3.2. Proof Trees and Rules

Proof trees are explicitly represented as data structures in KIV. They can be permanently stored on disk and are visualized by an elaborate graphical interface, which conveys a lot of information to the user. Some examples:

— Red branches lead to open premises, while the edges of already closed branches are colored in green. This allows to track paths to open goals easily even in large proofs.

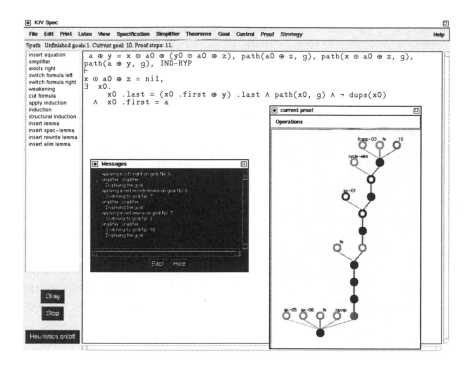

Figure 9. Proving with KIV

— Nodes are given different color depending on the tactic that was applied. This allows to spot important sequents immediately. E.g. in our example proof the use of the *induction* and *apply induction* rule is indicated by red nodes. For the same reason, nodes in which rules were applied interactively are emphasized.

— All sequents can be displayed by clicking a node in the proof tree. Switching to another open premise can also be done in this way.

Fig. 9 shows a snapshot of the graphical interface during the example proof (the current goal is goal 21). The proof tree is shown to the right. The current goal is visible in the large background frame. Applicable tactics for this goal are shown to the left. The dark window in the middle displays system messages.

KIV allows to withdraw wrong decisions in a proof by pruning the proof tree. The faulty decision is cut off the proof, and the user can correct his (or a heuristic's) decision. Pruning is more flexible than chronological backtracking to the last user interaction (which is also possible).

Proof trees are of invaluable help in case something goes wrong, which is the normal case rather than the exception. If an unprovable subgoal is reached, the source of the error can be found by tracing back the branch to the conclusion. In case the theorem must be changed or one of the lemmas used in its proof must be corrected, the now invalid proof tree can be reused to guide an automatic new proof. A very simple approach is proof replay. A more advanced method for program corrections computes the difference between the faulty and the corrected goal, analyses the old proof tree and guides a new proof in an intelligent way, (Reif and Stenzel, 1993), (Balser, 1997). Since the error may occur after several hours of work, this support for the correction of errors has a significant impact on the overall performance of the proof system. See (Reif and Stenzel, 1993) for a detailed example.

Back to the construction of the example proof. Applying a rule interactively to the currently selected goal is done by clicking an applicable rule offered by the proof strategy (see again Fig. 9). These rules have been designed to be as user-friendly as possible. Although a few of them implement basic rules of the sequent calculus (e.g. *cut formula*), most of them realize complex derived rules. If rules require additional parameters they can be selected via extra menus, which always suggest some reasonable defaults. E.g. the length of x is suggested as a possible induction ordering, when the *induction* rule is applied in our example. For all rules which require substitutions (quantifier rules, *apply induction* and all rules that introduce lemmas) proposals are computed by an elaborate program based on pattern matching, unification and simplification. In the example, all necessary substitutions are exactly the ones proposed by the system. Only for the induction hypothesis a selection of alternatives is computed.

3.3. *Automation and Heuristics*

Automation in KIV is attained by *heuristics* that control the application of tactics. To save interactions for non-inductive goals, KIV was recently integrated with an automated theorem prover. The integration is described in detail in Chapter II.1.4. The integration was not exploited in our example, but an application can be found in Chapter III.2.9. Here we describe KIV's heuristics.

KIV offers a number of heuristics to automate proofs. Our approach is not the one taken in most automated proof systems, where the proof strategy works with a fixed set of heuristics that can be modified only by the system developer. This approach has several disadvantages. First, the system developer must be sure that a new heuristic does not interfere with the existing ones. Second, the user has no possibility to tailor the heuristics to his special problem. Third, the user cannot ignore the heuristics. In case a proof fails, he or she must formulate lemmas to get around disturbing heuristics.

In our interactive proof environment, the user may select the heuristics individually depending on the estimated complexity of the proof, and decide how to combine them. This selection can be modified any time during the proof. The heuristics are tried in the order they were selected until a heuristic is applicable. If none is applicable, the user must help. Heuristics usually apply tactics to a goal, but can also influence other heuristics, add or remove them from the list of active heuristics, and, generally, may perform all actions the user might do.

A heuristic which is almost always selected is the simplifier. Two other frequently used heuristics deal with the elimination of selectors in favor of constructors and with the introduction of case splits using the *cut formula* rule. Both heuristics are parameterized with suitable lemmas and were used several times in our proof. Other heuristics deal with unfolding recursive definitions, the instantiation of quantifiers (the final substitution of the existential quantifier in node 44 was found automatically) and with induction (induction over the length of the path was done automatically). Altogether, the heuristics manage to find 24 of the 28 steps of our example proof automatically. The remaining 4 interactions were

1. instantiation of the existential quantifier with x (node 11)
2. unfolding the definition of *dups* (node 13)
3. application of the main lemma (node 16)
4. application of the induction hypothesis (node 21)

Working with a fixed set of tactics and a graphical interface is in contrast to the philosophy many other interactive theorem provers have followed for many years. These systems usually use commands of their programming language, to build up proofs. The advantage of this approach is its flexibility. At any time, the user can program new tactics exploiting the full power of the programming language. On the other hand, the user must first learn a programming language and a lot of commands to use the system. In KIV we have tried to achieve user-friendliness and flexibility. The simplifier is customizable by the various different forms of simplifier rules (see Sect. 4), standard heuristic sets are predefined, and the prover can be tailored to the problem by *problem specific heuristics*, which we introduce now.

Problem specific heuristics are relevant in large case studies on software development where proofs can take weeks or even months to be completed. During this time the same problem specific proof situations arise again and again. No general heuristic, however, will be able to cope with those situations. The user, on the other hand, will detect them after some time, but (according to our philosophy) cannot be expected to program his own heuristics.

Therefore, a mechanism is needed that allows to introduce new heuristics in a simple and, of course, correctness preserving way. Basically, this is done by specifying a proof situation and a rule to be applied. This specification is used by the system to detect the proof situation and to take the appropriate action. The specification is kept permanently, and will be used in all subsequent proofs. A proof situation is described by two lists of formula patterns. The patterns of the first list must match the formulas of the goal in any order, while no pattern of the second list may match a formula of the goal. This kind of description is refined by taking the proof history into account to allow for specifications like "apply a tactic only once in this branch of the proof", or "apply a tactic only once with the same instantiation".

As an example consider a large universally quantified formula $\forall x.\varphi$ that must be instantiated in many proofs (e.g. it must be proved that it is an invariant of all transitions of a state-based system). Then in many proofs one will find that the required instances will depend on some other critical formula $p(\tau)$. Assume the substitution with which the formula should be used is always $[x \leftarrow f(\tau)]$. Usually this instance will not close the goal immediately (then one could prove a lemma). Instead other subgoals are created that are specific to each proof. In this situation one would give a pattern for the two formulas $\forall x.\varphi$ and $p(\tau)$, and apply the rule for quantifier instantiation with the substitution $[x \leftarrow f(\tau)]$.

The benefits of problem specific heuristics increases with the number and size of proofs.

4. EFFICIENT SIMPLIFICATION

Theorem proving over a specification typically requires a large number of simplification steps, which are automated by the *simplifier* tactic of KIV. The various types of simplification steps, and the theorems used to guide these steps (the *simplifier rules*) are described in Sect. 4.1. The implementation technique, which allows to use a large number of simplifier rules efficiently, is explained in Sect. 4.2.

4.1. *Classification of Simplifier Rules*

Simplification steps can be classified into datatype independent ones (like simplification of $\varphi \wedge true$ to φ) and datatype dependent ones (like simplification of $x * 0$ to 0). For datatype independent simplification we found the following principles important.

- Apply all rules of the sequent calculus which remove connectives (like conjunction left) or quantifiers (like exists left).
- Never change the propositional structure of a formula unless a connective can be eliminated. For example, do not turn $\varphi \rightarrow \psi$ into $\neg \varphi \vee \psi$. Such changes immediately make formulas unreadable and destroy the intuitive understanding of the theorem. (Since the user formulated the theorem as an implication, he probably *has* an intuitive understanding of it.)
- Use local assumptions. To simplify $\varphi \rightarrow \psi$ simplify ψ under the assumption φ (with result ψ'), and then simplify $\neg \varphi$ under the assumption $\neg \psi'$ (with result $\neg \varphi'$). The result is $\varphi' \rightarrow \psi'$.
- Remove redundant quantifiers, like in $\forall x.\ x = \tau \rightarrow \varphi$ yielding φ_x^τ.
- Simplify as much as possible. Simplification is always repeated until a maximally simplified formula is obtained. In the example above, $\varphi \rightarrow \psi$ is simplified by repeatedly simplifying $\psi, \varphi, \psi', \varphi', \ldots$ until none of the two formulas changes any more.
- Use equations as local assumptions carefully. Unrestricted use as local rewrite rules can easily lead to non-termination, while not using them at all prevents trivial goals such as $\sigma = \tau \rightarrow f(\sigma) = f(\tau)$ to be proved automatically. The simplifier in KIV therefore carefully distinguishes equations according to various criteria (e.g. the information about constructors from **generated by** clauses is taken into account).

To capture the notion of datatype dependent simplification steps, theorems can be marked as *simplifier rules*. Then they are repeatedly applied as lemmas to simplify a proof goal (a sequent). KIV offers a variety of different simplifier rules, since we found it important that the user is able to state precisely, how and when the simplifier should try to apply a certain rule. Otherwise a lot of work can be caused by unsuccessful precondition checks or by unnecessary unfolding of definitions. The different types of simplifier rules and their effect when used by the simplifier are the following

- *Conditional rewrite rules.* These are theorems of the form

$$\varphi_1 \wedge \ldots \wedge \varphi_n \rightarrow \sigma = \tau$$

which are used to rewrite instances of σ to instances of τ. The preconditions $\varphi_1, \ldots, \varphi_n$ are checked before the application of the rewrite rule. The check may either be syntactical (by searching for instances of $\varphi_1, \ldots, \varphi_n$ in the sequent) or a proof (by recursively applying the simplifier).
- *Conditional replacement rules.* These are theorems of the form

$$\varphi_1 \wedge \ldots \wedge \varphi_n \rightarrow (\psi \leftrightarrow \chi)$$

where ψ is a literal and χ an arbitrary formula. They are used to replace instances of ψ by instances of χ (in case χ contains quantifiers, a bound renaming may be necessary). Preconditions $\varphi_1, \ldots, \varphi_n$ are treated as above. Several variants of replacement rules are available, depending on whether the rule should replace ψ by χ only in the antecedent/succedent, and on whether the rule is allowed to introduce new connectives.

— *Forward rules.* These rules are used to add information to goals (simplification here does not yield a "smaller" formula). Examples are transitivity rules (for orderings and equivalence relations) like

$x < y \wedge y < z \rightarrow x < z$

This rule adds $\sigma < \rho$ (an instance of $x < z$) to a goal, if $\sigma < \tau$ and $\tau < \rho$ (instances of the two preconditions) are present in the goal, and the rule has not yet been applied with this instance. Experience shows that forward reasoning works nicely together with simplification if the simplifier rules are formulated with some care.

— *AC rules.* Associativity and commutativity rules are not used directly for simplification. Instead they are used to find instances modulo AC (when applying rewrite rules, checking for preconditions, inserting equations etc.).

All simplifier rules are classified into *local* and *global* ones. Local simplifier rules are used in the specification in which they are defined, global simplifier rules are used in all specifications which use this specification. The reason for this distinction is illustrated by the extensionality axiom

$g_1 = g_2 \leftrightarrow (\forall a.\ a \in_v g_1 \leftrightarrow a \in_v g_2) \wedge (\forall e.\ e \in_e g_1 \leftrightarrow e \in_e g_2)$

from *digraph.* Used globally, application to $g = g_1 \cup g_2$ (where $\cup$ is the union of two graphs) would result in

$(\forall v.\ v \in_v g \leftrightarrow v \in_v g_1 \vee v \in_v g_2) \wedge (\forall e.\ e \in_e g \leftrightarrow e \in_e g_1 \vee e \in_e g_2)$

which is obviously not simplified. On the other hand, extensionality is used locally to automatically prove rewrite rules like

$g +_v a +_v a = g +_v a$ (*)

Therefore extensionality is used locally, whereas (*) is used globally. A typical example to achieve automation by the use of local simplifier rules, is the specification of sets (a specification from the KIV library). 60 of the 62 theorems defined over this specification (most are used as global simplifier rules) can be proved automatically.

We have the experience that after some work with a new datatype a stable set of simplifier rules is built up that does not change anymore and is reusable in all future projects.

4.2. *Efficient Rewriting*

In typical software projects, the number of simplifier rules ranges from several hundreds to more than thousand rules. Therefore an implementation technique is required that allows to efficiently determine the applicable simplifier rules on a goal. The technique used in KIV is based on (Kaplan, 1987) and consists of two steps. First, rewrite rules are sorted into a tree-like structure, called discrimination net. To be precise: compiled non-deterministic perfect AC discrimination net, see (Graf, 1996); for an overview on discrimination nets and other term indexing techniques see also Chapter II.2.5.

In the second step this discrimination net is compiled to functional programs (in our case compiled LISP programs). For each n-ary function symbol f, a LISP function with the same name is generated. This function, when called with already simplified terms $t_1, \ldots, t_n$ computes a maximally simplified version of the term $f(t_1, \ldots, t_n)$. Bottom-up rewriting is done by the following simple LISP program ('fct' and 'args' select the function symbol and the arguments of a term, 'apply' applies a function to a list of arguments, and 'mapcar' maps a function to each element of a list).

```
(defun rewrite (term)
  (if (is-a-variable term) term
    ;; else
    (apply (fct term)
           ;; simplify subterms
           (mapcar #'rewrite (args term)))))
```

The code shown above is somewhat simplified. The actual code generated by KIV additionally contains updates to the set of used simplifier rules, to be used by the correctness management. The code also becomes more complex when associative and commutative operators show up. Usually, the simplifier does not build up proof trees for efficiency reasons, but by enabling a suitable option, it can be forced to do so. This allows a precise analysis in case of unexpected behavior.

The compilation technique has the advantage, that determining a matching simplifier rule for a goal (and therefore rewriting as a whole) is much faster than using a generic retrieval routine from the discrimination net. On the other hand, every time the user starts to work with a new or changed set of simplifier rules, compiled LISP code has to be generated for the new discrimination net. This makes the compilation technique unsuitable for use in automated theorem provers, where the set of patterns, which should be matched (or unified), changes continuously during a proof.

Fortunately, in KIV changes to the simplifier rules are rare in comparison to the number of calls to the simplifier. Also, by using incremental compilation (each generated LISP function compiles itself only when first called) and by caching the LISP code for each function, unnecessary recompilation can be avoided (e.g. when switching between specifications with different sets of simplifier rules). The user will note a slight delay (usually below 2 seconds) only on the very first simplification step in the very first proof he does with KIV. In all following steps, compilation time is negligible.

Summarizing, with the compilation technique shown above, the system run time for typical KIV proofs was drastically reduced.

5. CORRECTNESS MANAGEMENT

KIV stores proofs and theorem bases permanently in different files, and does not enforce a bottom-up proof strategy. This means that theorems can be used in a proof before they are proved themselves. This happened e.g with *elim-cycle*, which was invented and used during the proof of *spath-exists*. Obviously, this poses some problems concerning overall correctness. The *correctness management* deals with these issues.

5.1. *Local Correctness*

Of course, cyclic dependencies between proofs must be avoided. It is illegal to use theorem *elim-cycle* in the proof of *spath-exists* and at the same time *spath-exists* in the proof of *elim-cycle*. To achieve this goal, it is necessary to keep track of each used lemma, rewrite rule, and simplification rule in a proof. KIV does this kind of book-keeping for each proof step, which is important since pruning the proof tree can reduce the number of used theorems.

Local correctness is concerned with used theorems from the same theorem base. Used theorems from subspecifications can never cause a cycle. Since *elim-cycle* is used in the proof of *spath-exists*, we say that *spath-exists* *depends* on *elim-cycle*. This notion leads to a *dependency graph*, which is a directed graph (and can be visualized in KIV). The local correctness management must ensure that the graph is always acyclic. When working with a theorem base, the initial dependency graph is computed based on the information stored for each theorem (Sect. 2.3). When beginning the proof for *elim-cycle*, the transitive closure of all theorems depending on it are *hidden* by inspecting the dependency graph (in our example just *spath-exists*). Hiding a theorem means that it is not available for the rule *insert lemma* and that it is temporarily removed from the simplifier rules (leading to partial recompilation). This approach is much more efficient than cycle checking before each application of a theorem, and, of course, much better than checking after a proof is finished. When the proof (finished or unfinished) is saved the dependency graph is updated accordingly. In this manner, cyclic dependencies are avoided.

Another task of the local correctness management is to deal with the deletion or modification of a theorem and the effects on other proofs in the same theorem base. Of course it is not possible to delete or modify an axiom or a proof obligation (an axiom can only be modified by modifying the specification). However, it is possible to modify a user-defined theorem even if it is used as a lemma in the proof of another theorem. The proofs for both theorems will become *invalid* which means that both theorems are considered unproved. However, the invalid proof is kept for reuse. If all theorems of a theorem base are proved, and their proofs are valid, then the theorem base is locally correct.

The modification of a specification normally leads to different axioms and proof obligations which are also handled by the local correctness management. The situation becomes more complicated if the signature or generation principles are modified. Then it is necessary to inspect each proof if it is affected by the modification. In principle, the second overall goal of the correctness management is the minimization of the effects of modifications, i.e. to invalidate as few proofs as possible.

5.2. *Global Correctness*

A theorem may be used in many proofs in different theorem bases. For example, the proof for *spath-exists* uses the theorem *elim-list* ($x \neq$ nil $\vdash$ a = x.first $\wedge$ y = x.rest $\leftrightarrow$ x = a $\oplus$ y) from *list-data*. Another example are rewrite rules, which typically are used in many proofs. What should be done with these proofs if the theorem is modified or deleted? Clearly they must be invalidated. However, KIV does not store the information where a theorem is used. The information that the proof for *spath-exists* uses *elim-list* from *list-data* is stored with the proof for *spath-exists* in *path*; the reverse information that *elim-list* is used by *spath-exists* is not stored in *list-data*. (This situation is depicted in the upper left corner of Fig. 10.) The reasons are efficiency considerations and the additional problems caused by redundant information. Again for efficiency considerations it is a bad idea to inspect every proof of every theorem base that is based on *list-data*. This could involve the inspection of hundreds of proofs.

KIV delays the invalidation of the proofs and uses the concept of *proved states* for theorem bases. A theorem base can *enter proved state* if it is locally correct (i.e. all theorems are proved and all proofs are valid) and all theorems

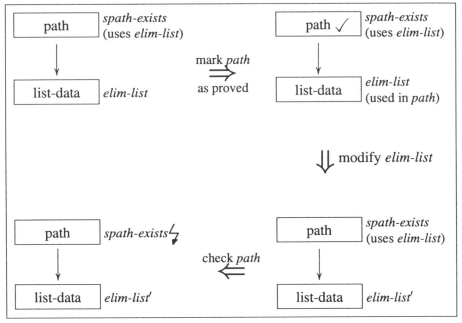

Figure 10.   Correctness Management

from other bases used in some proof exist (and are unchanged). Entering proved state (e.g. for *path*) is initiated by the user. Then the above mentioned checks are performed, and in every used base the list of theorems that are used in some proof in *path*, is stored (as depicted in the upper right corner in Fig. 10). For example, in *list-data* the information is stored that *elim-list* is used in *path*. It is not stored by which theorems of *path elim-list* is used. Finally, *path* is marked as proved.

If *elim-list* is modified (perhaps at first the precondition $x \neq nil$ was missing), *path* is removed from the proved state using the information stored in *list-data*. This leads to a situation where the proof for *spath-exists* is not marked as invalid but in fact is invalid (lower right corner in Fig. 10). Of course, trying to enter proved state again will fail. The user can run a check on *path* that marks *spath-exists* as invalid (lower left corner in Fig. 10). This approach allows complete control over the time when complex and expensive operations are performed. The main point is that for a theorem base in proved state everything is guaranteed correct. If all theorem bases are in proved state, everything is globally correct.

If a specification (e.g. *list-data*) is modified, the theorem bases of all specifications based on it (here *path* and *list*) are removed from proved state.

6. THEOREM PROVING WITH PROGRAMS

Implementing algebraic specifications in KIV is done by writing modules of imperative programs. Modules are added to the development graph as a special kind of node. Edges are drawn from the (export-) specification (the one to be implemented) to the module, and from the module to the (import-) specification (used in the programs). The related theorem base contains automatically generated proof obligations which guarantee the correctness of the module. Thus module verification smoothly fits into the framework described in the previous sections of this chapter. The only additional ingredients are rules and heuristics for program instructions to allow theorem proving with programs. The logic of programs used in KIV (Dynamic Logic, DL (Goldblatt, 1982), (Harel, 1984), (Heisel et al., 1989)) is an extension of first-order logic. Therefore, the rule sets for first-order goals and DL goals are semantically fully compatible. In this section we only sketch how to prove implementation correctness in KIV. For details we refer to (Reif et al., 1995a).

6.1. *Modules and Proof Obligations*

Let us consider a module which implements directed graphs by adjacency lists. The export specification is *path* from Fig. 5 which enriches *digraph*. The import interface is the specification *adjlists* of adjacency lists. Adjacency lists are specified as lists of lists of vertices, and therefore are an actualization of the generic specification *list* (see Fig. 4) with itself. The functions and predicates of *list* are renamed by adding *adj* as postfix. The module is intended to implement the (more abstract) data type *digraph* by procedures which act on the (more concrete) data type *adjlists*. Therefore the implementation consists of a procedure for every function and predicate of *path* and *digraph*. The procedures for $\in_v$ and *path* are shown in Fig. 11. The elementary operations used in the procedure bodies are the ones declared in the *adjlists*-specification. The procedures use value parameters (before the semicolon) and reference parameters, which are used only as result parameters. To verify correctness we (roughly) have to show that the procedures terminate and "behave" like the functions and predicates specified in *path* and *digraph*. Correct behavior can be characterized by a set of proof obligations which are expressed in Dynamic Logic (see (Reif, 1992)). In our example the axiom

$$\text{path}(a \oplus \text{nil}, g) \leftrightarrow a \in_v g,$$

of *path* (see Fig. 5) is transformed to the proof obligation

$$\text{restr}(xg) \vdash \langle \text{is\_path\#}(a \oplus \text{nil}, xg; r) \rangle\, r = \text{true} \leftrightarrow \langle \text{in\_v\#}(a, xg; r) \rangle\, r = \text{true}$$

```
in_v        implements    ∈ᵥ;
is_path     implements    path;

in_v(a, xg; var r)
if xg = nil_adj then r := false else
  if a = xg.firstadj.first then r := true else
    in_v(a, xg.restadj; r);

is_path(x, xg; var r)
if x = nil then r := false else
  if x.rest = nil then in_v(x.first, xg; r) else
    begin
        in_e(x.first ⊕ x.rest.first ⊕ nil, xg; r);
        if r = true then is_path(x.rest, xg; r) else skip
    end;
```

Figure 11.   Part of the implementation of directed graphs by adjacency lists

The proof obligation is obtained by replacing the predicates *path* and $\in_v$ with their implementations *is_path* and *in_v*. (As not all possible adjacency lists are correct representations of graphs, a predicate *restr(xg)* is added to apply certain restrictions on the input *xg*.) The given proof obligation states that under the assumption of *xg* being a correct representation of a graph, $a \oplus nil$ is a path in *xg*, if and only if *a* is a vertex of *xg*. The notation $\langle \alpha \rangle \, \varphi$ denotes that the program α terminates, and φ holds afterwards.

6.2. *Rules for Module Verification*

Module verification in KIV is based on symbolic execution of the procedures of the implementation, and induction over the data structures of the import specification. For induction the same rules as in Sect. 3 are applied. Additional rules for the symbolic execution of simple programs compute strongest postconditions, and thereby eliminate the instruction. They deal with compounds, assignments, variable declarations, conditionals, procedure calls, and loops. For example the rule schemes for handling assignments and conditionals are as follows (x' is a fresh variable).

$$\frac{x' = \tau, \Gamma \vdash \varphi_x^{x'}, \Delta}{\Gamma \vdash \langle x := \tau \rangle \, \varphi, \Delta} \text{ assign right} \qquad \frac{\varphi_x^{x'}, x' = \tau, \Gamma \vdash \Delta}{\langle x := \tau \rangle \, \varphi, \Gamma \vdash \Delta} \text{ assign left}$$

$$\frac{\Gamma, \varepsilon \vdash \langle \alpha \rangle \, \varphi, \Delta \quad \Gamma, \neg \varepsilon \vdash \langle \beta \rangle \, \varphi, \Delta}{\Gamma \vdash \langle \textbf{if } \varepsilon \textbf{ then } \alpha \textbf{ else } \beta \rangle \, \varphi, \Delta} \text{ if right}$$

assign right eliminates the assignment in favor of an equation. As programs may also occur in the antecedent of a sequent an additional rule *assign left* is provided. Applying *if right* on a conditional produces two premises dealing with the case of ε being true or false, respectively.

6.3. *Heuristics for Module Verification*

The application of rules for symbolic execution often is straightforward. Therefore heuristics for module verification lead to a comparatively high degree of automation. A simple heuristic *symbolic execution* eliminates compounds, assignments, conditionals with a decidable test, and local variables and is always used. Unfolding of procedure calls is more critical, since the repeated unfolding of recursively defined procedures generally does not terminate. As unfolding may cause case distinctions, the size of the resulting proof tree depends on the order in which different procedure calls are unfolded. Therefore different heuristics exist, which are more or less restrictive in unfolding a given call of a recursive procedure.

7. RELATED SYSTEMS

As an interactive theorem proving system KIV is in the tradition of the LCF-approach to tactical theorem proving, (Gordon et al., 1979) (for details we refer to (Heisel et al., 1990), (Reif et al., 1995b)). It is based on a typed functional language, called PPL ("proof programming language"). PPL is implemented with an abstract machine in LISP, and plays a role similar to ML (Harper et al., 1986) in the systems Isabelle (Paulson, 1994) and HOL (Gordon, 1988). For efficiency and scalability reasons, KIV deviates from the LCF-approach mainly in the context of efficient simplification (cf. Sect. 4) and flexible correctness management (cf. Sect. 5), which does not enforce a bottom-up proof strategy. KIV also deviates from the LCF-approach in that it does not use the programming language as a command language. Instead creating specifications, defining theorems, and theorem proving is done using a graphical interface.

KIV supports one particular target logic for algebraic specifications and imperative programs. In this respect it is similar to systems like ACL2 (Kaufmann and Moore, 1997), Nqthm (Boyer and Moore, 1979) and Eves (Craigen et al., 1992), which also provide specialized logics (for LISP programs and PASCAL-Programs respectively). Other systems, like Isabelle, HOL, IMPS (Farmer et al., 1996) and PVS (Owre et al., 1992) are designed as generic or general purpose theorem provers and use higher-order logic.

Explicitly represented, persisting proof trees distinguish KIV from most other interactive theorem provers. They keep only conclusion and premises of the proof and store proof scripts, i.e. listings of applied rules (e.g. ACL2, HOL, Isabelle and PVS). Exceptions are interactive provers which are based on constructive logics, like NUPRL (Constable et al., 1986) or Coq (Bertot and Bertot, 1996). These require proof trees for program extraction.

Browsing proof trees helps to detect errors, and to analyze where each program and formula part of an initial goal contributes to a proof. Such an analysis is done in KIV to reuse old proof attempts in order to guide new proofs automatically (Reif and Stenzel, 1993), (Balser, 1997). The analysis would not be possible without proof trees. Proof reuse in KIV goes beyond proof scripts in HOL, Isabelle, or PVS.

The use of heuristics for automating proof construction is a concept found in many interactive provers. E.g. NQTHM always uses a fixed, powerful set of predefined heuristics, some of which (especially the elimination heuristic, and the heuristics for unfolding procedures) have influenced the development of heuristics used in KIV. Other theorem provers, like PVS, offer predefined sets of heuristics as elementary proof strategies. A distinguishing feature of KIV is the flexible control of heuristics and the easy definition of problem-specific ones.

Using rewrite rules and a simplifier to automate first-order reasoning is also a standard in interactive theorem provers. KIV uses a predefined simplification strategy like NQTHM or PVS, which allows to use the simplifier automatically as a standard heuristic. KIV tries to achieve flexibility by offering a large set of different forms of simplifier rules (not just rewrite rules). In Isabelle and HOL flexibility is achieved by configuring different existing simplification strategies or by programming new ones.

8. CONCLUSION

This chapter presented the integrated specification and proof environment of KIV, an advanced modeling and verification tool for software and systems engineering. KIV shares several concepts with other state-of-the-art systems like Isabelle, PVS, ACL2, HOL, and others, but offers quite a number of distinguishing features. One of them is the handling of development graphs including an elaborate change- and correctness management. It keeps track of specifications, programs, proof obligations, proofs, and their interrelations, and computes the logical effects in case of modifications to them. KIV has a library of specifications, proofs, and verified components that can easily be included into a current development. Proving in KIV is the automated or interactive construction of an explicit proof tree through a graphical interface. The user has access to an arsenal of proof engineering facilities. Visual information in proof trees helps to follow the proof idea. These features (used in KIV for a couple of years) become now more and more popular. Automation is attained by an elaborate set of heuristics, efficient simplification, and a user friendly way to customize the prover to the problem. A recent development is the integration of external reasoners such as automated theorem provers. In software verification, proof attempts are more likely to reveal errors than to prove their absence. Therefore, in KIV strong emphasis was put on the tight integration of proving and error correction. Incremental verification by reuse of proofs in KIV goes beyond proof script replay mechanisms in other interactive systems. KIV has been applied in a number of industrial pilot applications in different areas. An application in compiler correctness can be found in Chapter III.2.7.

References

Balser, M. (1997). Wiederverwendung von Beweisen nach Modifikation. Diplomarbeit, Fakultät für Informatik, Universität Ulm, Germany. (in German).

Bertot, J. and Bertot, Y. (1996). Ctcoq : a system presentation. In *13th International Conference on Automated Deduction, CADE-13*, Springer LNCS.

Boyer, R. S. and Moore, J. S. (1979). *A Computational Logic*. Academic Press, New York.

Broy, M., Facchi, C., Grosu, R., Hettler, R., Hußmann, H., Nazareth, D., Regensburger, F., Slotosch, O., and Stolen, K. (1993). The Requirements and Design Specification Language SPECTRUM. An informal introduction. Version 1.0. Part I/II. Technical Report TUM-19311 and 19312, Technische Universität München, Fakultät für Informatik, TUM, 80290 München, Germany.

Constable, R. L., Allen, S. F., Bromley, H. M., Cleaveland, W. R., Cremer, J. F., Harper, R. W., Howe, D. J., Knoblock, T. B., Mendler, N. P., Panagaden, P., Sasaki, J. T., and Smith, S. F. (1986). *Implementing Mathematics with the Nuprl Proof Development System*. Prentice Hall.

Craigen, D., Kromodimoeljo, S., Meisels, I., Pase, B., and Saaltink, M. (1992). The EVES system. In *Proceedings of the International Lecture Series on "Functional Programming, Concurrency, Simulation and Automated Reasoning" (FPCSAR)*. McMaster University.

Farmer, W. M., Guttman, J. D., and Fábrega, F. J. T. (1996). IMPS: an updated description. In *13th International Conference on Automated Deduction, CADE-13*, Springer LNCS.

Fröhlich, M. and Werner, M. (1994). Demonstration of the interactive graph visualization system *daVinci*. In Tamassia, R. and Tollis, I., editors, *DIMACS Work-*

shop on Graph Drawing '94. Proceedings, Springer LNCS 894. Princeton (USA). http://www.informatik.uni-bremen.de/~davinci/.

Gaudel, M. C. (1992). Structuring and modularizing algebraic specifications: The PLUSS language, evolutions and perspectives. In *STACS'92. Proceedings*. Springer LNCS 577.

Goldblatt, R. (1982). *Axiomatising the Logic of Computer Programming*. Springer LNCS 130.

Gordon, M. (1988). HOL: A Proof Generating System for Higher-order Logic. In Birtwistle, G. and Subrahmanyam, P., editors, *VLSI Specification and Synthesis*. Kluwer Academic Publishers.

Gordon, M., Milner, R., and Wadsworth, C. (1979). *Edinburgh LCF: A Mechanised Logic of Computation*. Springer LNCS 78.

Graf, P. (1996). *Term Indexing*. Springer LNCS 1053.

Harel, D. (1984). Dynamic logic. In Gabbay, D. and Guenther, F., editors, *Handbook of Philosophical Logic*, volume 2, pages 496–604. Reidel.

Harper, R., MacQueen, D., and Milner, R. (1986). *Standard ML*. ECS-LFCS-86-2. Univ. Edinburgh.

Heisel, M., Reif, W., and Stephan, W. (1989). A Dynamic Logic for Program Verification. In Meyer, A. and Taitslin, M., editors, *Logical Foundations of Computer Science*. Springer LNCS 363.

Heisel, M., Reif, W., and Stephan, W. (1990). Tactical Theorem Proving in Program Verification. In Stickel, M., editor, *10th International Conference on Automated Deduction. Proceedings*. Springer LNCS 449.

Kaplan, S. (1987). A compiler for conditional term rewriting systems. In *2nd Conf. on Rewriting Techniques anf Applications. Proceedings*. Bordeaux, France, Springer LNCS 256.

Kaufmann, M. and Moore, J. S. (1997). An industrial strength theorem prover for a logic based on common lisp. *IEEE Transactions on Software Engineering*, 23(4).

Mosses, P. D. (1996). CoFI : The common framework initiative for algebraic specification. In Ehrig, H., v. Henke, F., Meseguer, J., and Wirsing, M., editors, *Specification and Semantics*. Dagstuhl-Seminar-Report 151. http://www.brics.dk/Projects/CoFI.

Owre, S., Rushby, J. M., and Shankar, N. (1992). PVS: A Prototype Verification System. In Kapur, D., editor, *Automated Deduction - CADE-11. Proceedings*, Springer LNAI 607. Saratoga Springs, NY, USA.

Paulson, L. C. (1994). *Isabelle: A Generic Theorem Prover*. Springer LNCS 828.

Reif, W. (1992). Correctness of Generic Modules. In Nerode and Taitslin, editors, *Symposium on Logical Foundations of Computer Science*. Springer LNCS 620.

Reif, W. (1995). The KIV-approach to Software Verification. In Broy, M. and Jähnichen, S., editors, *KORSO: Methods, Languages, and Tools for the Construction of Correct Software – Final Report*. Springer LNCS 1009.

Reif, W., Schellhorn, G., and Stenzel, K. (1995a). Interactive Correctness Proofs for Software Modules Using KIV. In *Tenth Annual Conference on Computer Assurance*, IEEE press. NIST, Gaithersburg (MD), USA.

Reif, W., Schellhorn, G., and Stenzel, K. (1995b). Tactics in KIV. *Journal on Information Processing and Cybernetics*, 30.

Reif, W. and Stenzel, K. (1993). Reuse of Proofs in Software Verification. In Shyamasundar, R., editor, *Foundation of Software Technology and Theoretical Computer Science. Proceedings*. Springer LNCS 761.

Sannella, D. T. and Wirsing, M. (1983). A kernel language for algebraic specification and implementation. In *Coll. on Foundations of Computation Theory*, Springer LNCS 158. Linköping, Sweden.

CHAPTER 2

PROOF THEORY AT WORK:
PROGRAM DEVELOPMENT IN THE MINLOG SYSTEM

1. INTRODUCTION

The old idea that proofs are in some sense functions, has been made precise by the Curry-Howard-correspondence between proofs in natural deduction and terms in typed λ-calculus. Since the latter can be viewed as an idealized functional programming language, this amounts to an interpretation of proofs as functional programs. This concept and related ones going back to work of Gentzen, Gödel, Kleene and Kreisel are implemented in MINLOG, an interactive proof system designed for generating proof terms and exploring their algorithmic content. Besides tools for interactive proof generation, MINLOG has automatic devices

— to search for purely logical (sub)proofs,
— to check the correctness of a proof,
— to rcmovc dctours in a proof,
— to make a nonconstructive proof constructive,
— to read off witnesses from a constructive proof,
— to adapt an already existing proof to special cases,
— to produce a legible verbalization from a formal proof.

The motivation for the development of MINLOG is to use the proofs-as-programs paradigm to let program development go hand-in-hand with program verification. MINLOG's path to correct programs can be divided into three steps:

1. Specification of the desired properties $A[input, output]$.
2. Interactive proof M of $\forall x \exists y A[x, y]$ (supported by automatic proof search).
3. Extraction $[\![\cdot]\!]$ of computational content of M, i.e. a program $[\![M]\!]$ satisfying $A[x, [\![M]\!](x)]$ for all inputs x.

The advantage of this way of composing programs is that once you have checked your proof-checker correct — and this is easy for a small one like MINLOG's — proof theory guarantees (in principle) the correctness of all extracted programs. Furthermore one can regard a proof as an extremely detailed comment on its extracted program.

W. Bibel, P. H. Schmitt (eds.), Automated Deduction. A basis for applications. Vol. II
© 1998 Kluwer Academic Publishers. Printed in the Netherlands

Since functional programs can be represented directly by terms in the language, MINLOG is also well-suited for program verification.

As MINLOG is based on natural deduction and generates proof terms in a λ-calculus style, it is closely related to implementations of constructive type theory such as Alf, Coq, Lego, NuPrl, or Typelab. In particular it shares many features with the specification and verification environment Typelab described in II.1.3 which – like MINLOG – is tailored for practical use on concrete programming problems. However, whereas Typelab and other systems mainly exploit the rich type theoretic specification language, we have chosen for MINLOG a very simple language (first-order formulas over a simply typed λ-calculus) which needs no 'expert knowledge' to understand. The conceptual simplicity of MINLOG allows the user to quickly access non-trivial proving and programming problems and to use advanced proof theoretic proof manipulation techniques. MINLOG is an open system, where the user may easily add self-defined concepts. Another advantage is that the extracted programs are just simply typed λ-terms with higher type primitive recursion. In systems like NuPrl or Coq proofs and programs are not separated so clearly and in particular programs may have complicated dependent types. A system where proofs and programs are clearly distinguished is PX (Hayashi, 1990) which is based on Feferman's untyped theory of functions and classes. However, in PX arbitrary partial recursive programs may be extracted whereas in MINLOG programs extracted from proofs are always total.

We describe the logical background of MINLOG's kernel in section 2 and its concrete realization in section 3. Further important features will be explained by means of three non-trivial examples:

In our first example coming from practice (section 4) we will use MINLOG to interactively verify a train scheduler. Nondeterminism will lead us to higher order terms. Also we will see the use of an error object, introducing the notion of partial objects.

The next example (section 5) is well-known in computer science: quicksort. Here we see the proofs-as-programs paradigm in a concrete case. The algorithm is *extracted* from a proof of its specification. We will also see how to introduce inductive data types as an extension to MINLOG.

The final example about search in finite trees (section 6) shows two techniques of very-high-level program development:

The fact, that a finite tree fulfilling a property always has a minimal subtree fulfilling it too, has a trivial classical proof. MINLOG is able to automatically turn it into a constructive proof (using a refined A-translation) and to extract a search algorithm.

Finally, if we have some additional information on the data (and hence enrich the specification), we can reuse and shorten the proof, resulting in a different (possibly more efficient) algorithm which in general cannot be obtained by optimizing the original program.

2. LOGIC FOR COMPUTABLE FUNCTIONALS

The theory implemented in MINLOG is Heyting arithmetic in finite types (Troelstra and van Dalen, 1988), arranged around a kernel of minimal logic. Classical arithmetic is contained in it as the fragment without the constructive quantifier $\exists^*$. From the classical point of view this fragment is sufficient since existential statements can be expressed via $\neg \forall \neg$.

2.1. *The term-calculus*

As objects of computational interest we choose the terms of Gödel's System T, coming with types (even if the type information is mostly suppressed in this text). Besides the natural numbers, the booleans are predefined ground types.

Types. $\mathsf{boole} \mid \mathsf{nat} \mid \rho \times \sigma \mid \rho \to \sigma.$

Constants. $\mathsf{true} \mid \mathsf{false} \mid 0 \mid S \mid \mathsf{R}_{\mathsf{boole},\rho} \mid \mathsf{R}_{\mathsf{nat},\rho}.$

The successor S is of type $\mathsf{nat} \to \mathsf{nat}$, the types of the recursors (for each type ρ) will be made precise later. From typed variables x and constants c terms are built according to the standard of simply typed λ-calculus.

Terms. $x \mid c \mid \langle r, s \rangle \mid \pi_0 r \mid \pi_1 r \mid \lambda x r \mid rs.$

General bracketing conventions: products associate to the left, arrows to the right. The scope of a binder is the maximal (minimal) possible one, whenever the bound variable is (not) followed by a dot.

Terms are computed according to the

Conversions.

$$
\begin{aligned}
\pi_0 \langle r, s \rangle &\mapsto r, \\
\pi_1 \langle r, s \rangle &\mapsto s, \\
(\lambda x r)s &\mapsto r_x[s], \\
\mathsf{R}_{\mathsf{boole},\rho}\, rs\,\mathsf{true} &\mapsto r, \\
\mathsf{R}_{\mathsf{boole},\rho}\, rs\,\mathsf{false} &\mapsto s, \\
\mathsf{R}_{\mathsf{nat},\rho}\, rs0 &\mapsto r, \\
\mathsf{R}_{\mathsf{nat},\rho}\, rs(St) &\mapsto st(\mathsf{R}_{\mathsf{nat},\rho}\, rst),
\end{aligned}
$$

where $r_x[s]$ (or $r[s]$ for short) means s substituted for x in r. These rules show that $R_{\text{boole},\rho}rst$ (r, s of type ρ) corresponds to **if** t **then** r **else** s, whereas $R_{\text{nat},\rho}rst$ is the usual recursor with base r (of type ρ) and loop s (of type nat $\to \rho \to \rho$) to be passed through t-times. The first three rules are called β-conversions, the others we call R-conversions.

As is well known (Tait, 1967; Troelstra and van Dalen, 1988), the rewriting procedure stops for each term with a unique *normal form*. Hence the congruence given by the conversions is decidable. In our setting, congruent terms are identified. This *Poincaré-equality* mostly saves us from needing an equality-calculus. (However the user importing self-defined conversions is responsible for their termination together with β and R!)

2.2. *The logical framework*

Now the basis is ready to introduce the objects of interest for theorem proving: formulas. It is crucial but nevertheless easy to distinguish between a term t of type boole and the atomic formula $\text{atom}(t)$ saying 't is true'. In particular we have the atomic formulas

$$
\begin{aligned}
\top &:= \text{atom}(\text{true}), \\
\bot &:= \text{atom}(\text{false}).
\end{aligned}
$$

From the atomic formulas (which are decidable since their characteristic terms are computable) we build the

Formulas. $\text{atom}(t) \mid A \wedge B \mid A \to B \mid \forall x A \mid \exists^* x A,$

where $\exists^*$ is the *strong* or *constructive* existential quantifier — to distinguish from the classical $\exists$ which can be viewed as defined by

$$
\begin{aligned}
\neg A &:= A \to \bot, \\
\exists x A &:= \neg \forall x \neg A.
\end{aligned}
$$

Formulas without $\exists^*$ are called *classical*. For the moment we will concentrate on the classical part. We will meet $\exists^*$ later again.

In the spirit of the Curry-Howard-correspondence we regard proofs in (the $\wedge, \to, \forall$-fragment of) natural deduction as terms in a typed λ-calculus, where the type of a proof term corresponds to (more precisely: *is*) its end formula in natural deduction, and the types of its free variables *are* its free assumptions.

For the sake of clarity we annotate types as superscripts in the definition of proofs. To start with, an assumption A is introduced by an assumption-variable u^A, whereas an axiom A is introduced by a constant c^A. For the moment our only constant is the *truth axiom* $\text{Truth}^\top$.

Proofs. $u^A \mid c^A \mid \langle M^A, N^B \rangle^{A \wedge B} \mid (\pi_0 M^{A \wedge B})^A \mid (\pi_1 M^{A \wedge B})^B \mid$
$(\lambda u^A M^B)^{A \to B} \mid (M^{A \to B} N^A)^B \mid (\lambda x^\rho M^A)^{\forall x^\rho A} \mid (M^{\forall x^\rho A} t^\rho)^{A_x[t]}.$

The usual restriction on the object-variable x applies when building $\lambda x M$.

2.3. *Reasoning in classical and intuitionistic arithmetic*

As we have said nothing special about $\perp$ so far, we have a calculus for minimal logic. To expand it to arithmetic and incorporate constructivity we add

Induction axioms. $\mathsf{Ind}_{\forall x^{\text{boole}} A}$: $A_x[\text{true}] \to A_x[\text{false}] \to \forall x A,$

$\mathsf{Ind}_{\forall x^{\text{nat}} A}$: $A_x[0] \to \forall x(A \to A_x[Sx]) \to \forall x A.$

$\exists^*$ *axioms.* $\exists^{*+}_{x,A}$: $\forall x. A \to \exists^* x A,$

$\exists^{*-}_{x,A,B}$: $\exists^* x A \to \forall x(A \to B) \to B,$

where in the latter x must not occur free in B. The notation $M:A$ is synonymous to M^A.

We formulated induction and existence as axioms since we regard them as extensions of our logical kernel. You might miss negation axioms. But they are already present:

Theorem. (ex-falso-quodlibet.) $\perp \to A$ is provable.

Proof. By induction on A. *Base.* We have to show $\perp \to \text{atom}(t)$. This is an instance of $\forall x. \perp \to \text{atom}(x)$. To prove the latter, we use the boolean induction axiom $\mathsf{Ind}_{\forall x. \perp \to \text{atom}(x)}$. It forces us to prove both, $\perp \to \top$ and $\perp \to \perp$. Using an assumption u of type $\perp$, proofs are simply $\lambda u \text{Truth}$ resp. $\lambda u u$. Summing up, $\mathsf{Ind}_{\forall x. \perp \to \text{atom}(x)} \lambda u \text{Truth} \lambda u u t$ is a proof of $\perp \to \text{atom}(t)$. *Step.* By induction hypothesis we have a proof M of type $\perp \to A$. We get $\perp \to \exists^* x A$ as the type of $\lambda u. \exists^{*+}_{x,A} x(Mu)$. The other cases of the induction step are easy, too.

A similar consideration shows that *stability* $\neg\neg A \to A$ is provable for all classical formulas A. Nevertheless we often use ex-falso-quodlibet and stability as axioms in order to avoid myriads of boolean inductions.

Summing up, with the appropriate choice of a term calculus we get Heyting arithmetic, and restriction to the classical fragment yields Peano arithmetic.

2.4. *Semantics*

The denotational semantics kept in the design of MINLOG is Scott's domain theory (Scott, 1982). A variable is supposed to denote a partial continuous functional of its type, whereas a program constant is interpreted by a computable functional. This is reflected in our formalism where we can work with possibly undefined objects (cf. section 4). Since in some applications no partial objects appear, we in addition allow a second kind of variables ranging only over the subclass of total functionals. For the theory of totality in higher types we refer to (Berger, 1993b). Computability of the constants implies that atomic formulas are decidable and extracted programs are executable (See also section 5).

3. THE MINLOG SYSTEM

The MINLOG system is implemented in SCHEME, a LISP-dialect. Its size is about 1/2 MB, which is rather small compared with other theorem provers. It is an interactive system which however has quite elaborate components for automated proof search and term rewriting.

3.1. *Interactive theorem proving*

For proving a theorem in MINLOG the user first has to specify the framework consisting of a language and axioms. This can be done by declaring new variables and program constants of possibly user-defined types. We allow constants parametric in types and, corresponding to that, axioms parametric in formulas. Canonical examples are the recursion operator and the induction scheme. Further axioms or lemmas to be proven later can be introduced as so-called global assumptions. The user may take anything (hopefully valid) for granted; as long as a global assumption contains no computational information, it will not affect program extraction—as we will see in section 5.

In the sequel we want to give an idea how to work with the MINLOG system. Let us begin with a small example:

$$(\forall x. A[x] \rightarrow B[x]) \rightarrow \forall x A[x] \rightarrow \forall x B[x].$$

We first present a natural deduction proof of this formula in familiar tree notation:

$$\lambda u \frac{\lambda v \frac{\lambda x \frac{\dfrac{u : \forall x. A[x] \to B[x] \qquad x}{A[x] \to B[x]} \qquad \dfrac{v : \forall x A[x] \qquad x}{A[x]}}{\dfrac{B[x]}{\forall x B[x]}}}{\forall x A[x] \to \forall x B[x]}}{(\forall x. A[x] \to B[x]) \to \forall x A[x] \to \forall x B[x]}$$

Below the protocol of the interactive proof is shown. The lines in parentheses are the commands given by the MINLOG user; those starting with a semicolon are the responses of the system, e.g. telling us current goals and actual contexts (some responses are dropped here).

We start with introducing an arbitrary type arb and declare a variable x of type arb as well as two program constants a and b of type arb->boole, displayed with capital letters, representing the formulas $A[x]$ and $B[x]$. The numerical argument 0 in the declaration of the program constants determines their degree of totality. This degree has impact on the execution of possible term rewriting discussed below.

After stating the goal the proof is built up in a backward reasoning fashion. For conciseness, the system displays the atomic formula atom(t) by the boolean term t (since it is clear from the context that a formula is meant). Hence we read (A x) for the internal (atom (a x)).

The symbols ?, ?-KERNEL, ?1 are labels to identify the (sub)goals. In II.1.3 the role of these 'metavariables', which are an official part of the formal language of Typelab, is thoroughly investigated, especially with respect to the problems arising in conjunction with β-reduction. In MINLOG, however, these problems do not occur, since we do not allow normalization for incomplete proofs.

With the command assume introduction rules for implication and universal quantification can be treated (backwards) in one step; the user only has to choose some new labels for the assumptions. The command use-with allows to prove the subgoal ?-KERNEL: $B[x]$ by using the assumption u at x provided one can find a proof for the premise $A[x]$. This is the new goal ?1, which of course can be proved with a further application of an elimination rule.

```
(add-ground-type 'arb 'x)
(add-program-constant 'a (cons-arrow 'arb 'boole) 0 "A")
(add-program-constant 'b (cons-arrow 'arb 'boole) 0 "B")

(set-goal '? (parse-formula "(all x (a x -> b x))  ->
                    all x (a x)  ->  all x (b x)"))
```

```
;?:(all x.A x -> B x) -> all x A x -> all x B x

(assume 'u 'v 'x)
;ok, under these assumptions we have the new goal
;?-KERNEL: B x from
;   u:all x.A x -> B x
;   v:all x A x
;   x

(use-with 'u 'x '?1)
;ok, ?-KERNEL can be obtained from
;?1: A x from
;   u:all x.A x -> B x
;   v:all x A x
;   x

(use-with 'v 'x)
;ok, ?1 is proved. Proof finished.
```

Now let us have a look at the proof term corresponding to that proof. It was generated interactively by editing an incomplete derivation term with holes in it which are the open goals. In our example the (partial) proof term develops as follows:

$$?$$
$$\lambda u, v, x. \text{ ?-KERNEL}$$
$$\lambda u, v, x. \ ux\,?1$$
$$\lambda u, v, x. \ ux(vx).$$

There are also some tools for forward reasoning. For instance, instead of the last two steps in the interactive proof, it is also possible to first instantiate the assumption v with x. Then only the context changes since we have a new assumption $v_i : A(x)$. To finish the proof we can use u at x and v_i. This different proof leads to different (partial) proof terms developing as follows:

$$?$$
$$\lambda u, v, x. \text{ ?-KERNEL}$$
$$\lambda u, v, x. \ (\lambda v_i. \ \text{?-KERNEL-INST})(vx)$$
$$\lambda u, v, x. \ (\lambda v_i. \ uxv_i)(vx).$$

Note that this proof term can be β-normalized to the proof term above (see subsection 3.4).

3.2. *Automated proof-search*

Of course our first example was an easy one that only deals with logic. A useful proof system should be able to prove such a formula without human interaction:

```
(set-goal '? (parse-formula "(all x (a x -> b x))  ->
                             all x (a x)  ->  all x (b x)"))
;?:(all x.A x -> B x) -> all x A x -> all x B x
```

```
(search)
;ok, ? is proved by minimal quantifier logic.
;Proof finished.
```

The automatically generated proof term, $\lambda u_{01}, u_{02}, x. u_{01} x (u_{02} x)$, is equal modulo bound renaming to the former one. It is obtained by breadth-first proof-search in the style of (Miller, 1991). Usually, this device is applied to subgoals provable by pure predicate logic (including the use of global assumptions).

3.3. *Proof presentation*

A proof term can be used to automatically produce a legible presentation, i.e. a TEX-script:

> **Proposition.** $(\forall x. A(x) \to B(x)) \to \forall x A(x) \to \forall x B(x)$.
>
> *Proof.* For $(\forall x. A(x) \to B(x)) \to \forall x A(x) \to \forall x B(x)$ assume $\forall x. A(x) \to B(x)$ [1], $\forall x A(x)$ [2] and x.. Then we have to show $B(x)$. We derive this from $A(x) \to B(x)$ (as an instance of [1]) and have to show $A(x)$. That is an instance of [2].

This purely logical example was chosen to demonstrate the Curry-Howard-correspondence and the interactive use of the logical rules. MINLOG's emphasis, however, lies on concrete mathematical problems; data types and constants should have a fixed denotational semantics expressed operationally by suitable higher type rewriting rules.

3.4. *Normalization-by-evaluation*

The heart of MINLOG is the normalization-by-evaluation mechanism providing an efficient implementation of the operational semantics. Since terms are represented by SCHEME expressions it is quite natural to use SCHEME evaluation instead of explicitly programming a normalization procedure (Berger and Schwichtenberg, 1991; Berger, 1993a; Berger et al., 1998). Terms are normalized with respect to βR-conversion and additional higher order rewrite rules given by the user. Correctness and termination of the additional rules are under control of the user. The problem of termination and confluence of higher type rewrite systems has been analyzed e.g. in (van de Pol and Schwichtenberg, 1995; Nipkow, 1993).

The effect of term rewriting is shown in a second small example. We prove

$$\forall n \, \text{even}(2*n)$$

where odd and even are defined as usual

$$\text{odd}(0) = \text{false} \qquad\qquad \text{even}(0) = \text{true}$$
$$\text{odd}(n+1) = \text{even}(n) \qquad\qquad \text{even}(n+1) = \text{odd}(n)$$

by induction on n. The base case $\text{even}(2*0)$, to be proven first, normalizes to the truth axiom. In the step case we also use term rewriting when applying the command use-with.

```
(set-goal '? (parse-formula "all n even(2*n)"))

(ind)
;ok, ? can be obtained from
;?-STEP: all n.even(2*n) -> even(2*(n+1))
;?-BASE: even(2*0)

(normalize-goal)
;ok, the normalized goal is
;?-BASE-NF: T

(prop)
;ok, ?-BASE-NF is proved by minimal propositional logic.
;The active goal now is
;?-STEP: all n.even(2*n) -> even(2*(n+1))

(assume 'n 'ih)
;ok, under these assumptions we have the new goal
;?-STEP-KERNEL: even(2*(n+1)) from
;    n   ih:even(2*n)

(use-with 'ih)
;ok, ?-STEP-KERNEL is proved. Proof finished.
```

Step by step we get the following (partial) proof terms:

$$?$$
$$\text{nat-ind-at}_{\forall n\,\text{even}(n)}\ \text{?-BASE ?-STEP}$$
$$\text{nat-ind-at}_{\forall n\,\text{even}(n)}\ \text{Truth}^{\top}\ (\lambda n, \text{ih. ih})$$

where nat-ind-at is the constant representing the induction axiom. Note that although the rewrite steps are not shown in the proof term, they are visible ($\searrow^{*}\nearrow$) in the automatically generated TEX-script.

Proposition. $\forall n\,\text{even}(2 * n)$.

Proof. $\forall n\,\text{even}(2 * n)$ is shown by induction on n. The base case follows from the axiom of truth where $\text{even}(2 * 0)\searrow^{*}\nearrow \top$. Step case: Assume n and $\text{even}(2 * n)$ [1] and use $\text{even}(2 * n)$ [1] where $\text{even}(2 * (n+1))\searrow^{*}\nearrow\text{even}(2 * n)$.

Now we have described the basic system. The following sections are devoted to MINLOG's additional features and examples.

4. HIGHER TYPES FOR PROGRAM VERIFICATION: SCHEDULING

As a first example of how to prove something with the MINLOG system we pick a simple but somewhat practical one: a safety property for a distributed system. The example is mathematically quite trivial; however, it hopefully makes clear that in a similar way one can deal with more complex (and more realistic) examples. In particular, we want to demonstrate the following points.

1. Nondeterminism can be described using a choice function (or scheduler) as a parameter (as was done in Boyer and Moore, 1988, p. 87) in a first order context).
2. To make proofs manageable it is essential to use rewrite rules to take care of the computational parts.
3. It is useful to give explicit definitions (as primitive recursive functionals) wherever possible. This turns many of the proof obligations into computational tasks (cf. 2).

4.1. *The problem*

Consider a track divided into segments, with an arbitrary number of trains moving back and forth. For simplicity let us assume that the track is linear and infinite, so the segments can be modelled by the integers. Trains can move in both directions, but they can enter another segment only if it is empty, i.e. not occupied by another train. We want to give a formal proof of a very simple safety property of this system, namely that no train can ever pass another one.

The system we want to describe is nondeterministic: at any time every train can try to move in any direction. We choose an interleaving model to describe the situation, i.e. we assume that the actions take place one after the other. To prove safety properties we do not loose anything by this view.

We model the nondeterministic system using a choice function (or scheduler) as a parameter. This choice function determines at every moment which action is to be taken. Hence the behaviour of the system is uniquely determined once the choice function is given. The description of the system then consists of giving a definition of the state of each agent by primitive recursion over the time (modelled discretely by the natural numbers), with the choice function as a parameter.

It is convenient to split up the choice function into two components: a function giving the active segment, and a function telling the direction in which an activated train can try to move.

4.2. *The language*

An appropriate language has four ground types (sorts), with variables

m, n for natural numbers (used to model discrete time),

p, q for booleans,

i for segments,

j for trains.

The predecessor and successor functions for segments are denoted by prev and next. We also use $=, <, \leq$ for segments.

It will be useful to allow an error object undef (read 'undefined') for every ground type. We use variables with a hat (i.e. $\hat{n}, \hat{p}, \hat{i}, \hat{j}$) to range over arbitrary objects of their type, including undef. Variables without hat range over defined (or total) objects (cf. section 2).

Parameters
The system is determined by its state at time 0, given by the functions initocc: segment $\rightarrow$ train resp. initloc: train $\rightarrow$ segment describing the initial distribution of trains to the track, together with the two parts of the choice function: act (for active-segment) and back (or, if you prefer, forth), a boolean valued function: back(n) telling whether the direction to move at time n is 'back'. For notational convenience we will suppress these four parameters.

Occupancy
Our first perspective is standing at a certain segment of the track, looking back and forth, waiting for a train. The program constant occ (for occupancy) yields for given (parameters,) time n and segment i the train occupying this segment at this time. If there is no train in the segment, the value is to be undefined. We want to define occ in such a way that the following property holds: *A train leaves the actual segment in the actual direction or will reach a segment moving in the actual direction from its neighbour being actual, only if its goal segment is not occupied; and nothing happens besides.* Together with the initialization this results in six clauses, in which $\mathrm{def}(\mathrm{occ}(n,i))$ means that at time n, there is a train that occupies segment i, and $\mathrm{occ}(n+1,i) = \mathrm{undef}^{\mathrm{train}}$ means that at time $n+1$, no train occupies segment i.

$$\mathrm{occ}(0,i) = \mathrm{initocc}(i),$$
$$\mathrm{back}(n) \rightarrow \mathrm{act}(n) = i \rightarrow \neg\mathrm{def}(\mathrm{occ}(n,\mathrm{prev}(i))) \rightarrow$$
$$\mathrm{occ}(n+1,i) = \mathrm{undef}^{\mathrm{train}},$$
$$\mathrm{back}(n) \rightarrow \mathrm{act}(n) = \mathrm{next}(i) \rightarrow \neg\mathrm{def}(\mathrm{occ}(n,i)) \rightarrow$$
$$\mathrm{occ}(n+1,i) = \mathrm{occ}(n,\mathrm{next}(i)),$$
$$\mathrm{back}(n) \rightarrow \mathrm{act}(n) = i \rightarrow \mathrm{def}(\mathrm{occ}(n,\mathrm{prev}(i))) \rightarrow$$
$$\mathrm{occ}(n+1,i) = \mathrm{occ}(n,i),$$
$$\mathrm{back}(n) \rightarrow \mathrm{act}(n) = \mathrm{next}(i) \rightarrow \mathrm{def}(\mathrm{occ}(n,i)) \rightarrow$$
$$\mathrm{occ}(n+1,i) = \mathrm{occ}(n,i),$$
$$\mathrm{back}(n) \rightarrow \mathrm{act}(n) \neq \mathrm{next}(i) \rightarrow \mathrm{act}(n) \neq i \rightarrow$$
$$\mathrm{occ}(n+1,i) = \mathrm{occ}(n,i).$$

The clauses for the other direction are similar. So we define occ accordingly, by primitive recursion using the above case distinctions. The recursion equations can then be used as rewrite rules.

Location
Now let us assume, we are sitting in a train, looking where we move. The program constant loc (for location) yields for given (parameters,) time n and

train j the segment where the train is located at this time. For loc we require the following property: *The train being in the active segment moves one
segment in the actual direction only if this segment is not* occupied(!)*, and
nothing happens besides.* In clauses:

$$\mathsf{loc}(0, j) = \mathsf{initloc}(j),$$
$$\mathsf{act}(n) = \mathsf{loc}(n, j) \to \mathsf{back}(n) \to \neg\mathsf{def}(\mathsf{occ}(n, \mathsf{prev}(\mathsf{loc}(n, j)))) \to$$
$$\quad \mathsf{loc}(n+1, j) = \mathsf{prev}(\mathsf{loc}(n, j)),$$
$$\mathsf{act}(n) = \mathsf{loc}(n, j) \to \mathsf{back}(n) \to \mathsf{def}(\mathsf{occ}(n, \mathsf{prev}(\mathsf{loc}(n, j)))) \to$$
$$\quad \mathsf{loc}(n+1, j) = \mathsf{loc}(n, j),$$
$$\mathsf{act}(n) \neq \mathsf{loc}(n, j) \to \mathsf{loc}(n+1, j) = \mathsf{loc}(n, j).$$

The clauses for the other direction are similar; we define loc accordingly, by
primitive recursion.

There is one important point to be made here. The occ-function and the
whole first perspective could have been avoided if we had used an existential
quantifier in the equations above, as follows:

$$\mathsf{act}(n) = \mathsf{loc}(n, j) \to \mathsf{back}(n) \to \neg \exists j_1 \mathsf{loc}(n, j_1) = \mathsf{prev}(\mathsf{loc}(n, j)) \to$$
$$\quad \mathsf{loc}(n+1, j) = \mathsf{prev}(\mathsf{loc}(n, j)),$$
$$\mathsf{act}(n) = \mathsf{loc}(n, j) \to \mathsf{back}(n) \to \exists j_1 \mathsf{loc}(n, j_1) = \mathsf{prev}(\mathsf{loc}(n, j)) \to$$
$$\quad \mathsf{loc}(n+1, j) = \mathsf{loc}(n, j).$$

However, since we want to shift as much as possible of the burden of proving
something to our rewrite apparatus, we make this existential quantifier explicit by introducing an appropriate function. This is a general pattern: whenever in a given problem there is a functional dependency of some data from
others, introduce a function to describe this dependency in an explicit form.
This function can then be used in a rewriting mechanism, which generally
makes proving much simpler.

4.3. *The specification and its proof*

Now we can formulate the safety property we want to prove, namely that no
train can ever pass another one:

$$\mathsf{initocc}(\mathsf{initloc}(j_1)) = j_1 \to \mathsf{initocc}(\mathsf{initloc}(j_2)) = j_2 \to$$
$$\mathsf{initloc}(j_1) < \mathsf{initloc}(j_2) \to \forall n \, \mathsf{loc}(n, j_1) < \mathsf{loc}(n, j_2).$$

Rather than describing the interactive construction of the proof in the MIN-
LOG system, we restrict ourselves to some hints: The proof is by induction

on n. In order to keep it comprehensible, it is crucial to think of appropriate lemmas first. In the present example, beside some basic arithmetical facts, the following lemmas are proved separately and then used:

$$\text{prev}(\text{loc}(n, j)) \leq \text{loc}(n+1, j),$$
$$\text{loc}(n+1, j) \leq \text{next}(\text{loc}(n, j)),$$
$$\text{back}(n) \rightarrow \text{loc}(n+1, j) \leq \text{loc}(n, j),$$
$$\neg\text{back}(n) \rightarrow \text{loc}(n, j) \leq \text{loc}(n+1, j),$$
$$\text{initocc}(\text{initloc}(j)) = j \rightarrow \text{occ}(n, \text{loc}(n, j)) = j.$$

The size of the proof term finally obtained is about two pages, and the effort to generate it (without using any automated search) was 42 interactions. The amount for each lemma was approximately the same.

4.4. *Final remarks on the scheduling example*

1. The constants occ and loc are defined by primitive recursion with function parameters, i.e. within Gödel's system T. Since we only have used the recursion equations as rewrite rules, termination as well as preservation of the values is guaranteed.
2. The possibility to use e.g. i as well as $\hat{i}$ provides additional expressive power, which can be quite useful: cf. the use of $\text{def}(\text{occ}(n, i))$ for 'is occupied by a (real) train'.
3. The ground type segment was described axiomatically, by means of properties of prev and next like $\text{prev}(\text{next}(i)) = i$. If instead we have a fixed ground type like the integers for segments and constants for particular segments, we obtain an executable specification. It can be executed for completely concrete data, but also for partially given ones (partial evaluation).

5. PROGRAM EXTRACTION QUICKSORT

In the previous section, we showed how MINLOG can be used to verify that a given program (loc) meets its specification. Now we will explain how to obtain correct programs in MINLOG using program extraction. From a constructive proof one can read off a witness, i.e. a term that realizes the proven formula. First, we will explain what it means exactly that a term (a program) realizes a formula (we use the modified realizability interpretation (Kreisel, 1959)) and how a realizing term can be obtained. Then we will use this technique to get the well-known quicksort algorithm. This example also shows another extension of the basic MINLOG system at work: the Simultaneous Free Algebras. The SFA implementation allows the user to add inductively defined data types (in this example lists of natural numbers) as new ground types to the system and automatically provides the corresponding induction schemes and recursors, as well as an interface for the (otherwise fairly complicated) creation of functionals defined by recursion over this type. Similar considerations concerning program extraction and its application to sorting problems can already be found in (Turner, 1991). In the present section we want to show that such a task can be carried out rather elegantly in the MINLOG system.

5.1. *Program extraction*

A *Harrop formula* is a formula not containing the constructive existential quantifier $\exists^*$ strictly positive. In the following let A, B stand for Harrop formulas and C, D for non–Harrop formulas. First of all we define the type $\tau(C)$ of potential realizers of C:

$$
\begin{aligned}
\tau(\exists^* x^\rho A) &:= \rho, \\
\tau(\exists^* x^\rho C) &:= \rho \times \tau(C), \\
\tau(\forall x^\rho C) &:= \rho \to \tau(C), \\
\tau(A \to C) &:= \tau(C), \\
\tau(D \to C) &:= \tau(D) \to \tau(C), \\
\tau(A \wedge C) &:= \tau(C), \\
\tau(C \wedge A) &:= \tau(C), \\
\tau(C \wedge D) &:= \tau(C) \times \tau(D).
\end{aligned}
$$

Now we can define the *modified realizability interpretation* $t^{\tau(C)}$ **mr** C ('t realizes C') and ε **mr** A:

$$
\begin{aligned}
t \text{ mr } \exists^* x A &:= \varepsilon \text{ mr } A_x[t], \\
t \text{ mr } \exists^* x C &:= \pi_1 t \text{ mr } C_x[\pi_0 t], \\
\varepsilon \text{ mr } \forall x A &:= \forall x. \varepsilon \text{ mr } A, \\
t \text{ mr } \forall x C &:= \forall x. t x \text{ mr } C, \\
\varepsilon \text{ mr } (A \to B) &:= \varepsilon \text{ mr } A \to \varepsilon \text{ mr } B, \\
\varepsilon \text{ mr } (C \to A) &:= \forall x. x \text{ mr } C \to \varepsilon \text{ mr } A, \\
t \text{ mr } (A \to C) &:= \varepsilon \text{ mr } A \to t \text{ mr } C, \\
t \text{ mr } (D \to C) &:= \forall x. x \text{ mr } D \to t x \text{ mr } C, \\
\varepsilon \text{ mr } (A \wedge B) &:= \varepsilon \text{ mr } A \wedge \varepsilon \text{ mr } B,
\end{aligned}
$$

$$t \text{ mr } (A \wedge C) \quad := \quad \varepsilon \text{ mr } A \wedge t \text{ mr } C,$$
$$t \text{ mr } (C \wedge A) \quad := \quad t \text{ mr } C \wedge \varepsilon \text{ mr } A,$$
$$t \text{ mr } (C \wedge D) \quad := \quad \pi_0 t \text{ mr } C \wedge \pi_1 t \text{ mr } D.$$

Finally, the most important part: how to get a realizing term from a constructive proof M^C. Informally, the structure of the extracted term $[\![M]\!]$ is *exactly* the same as that of the constructive part of the proof term M. In other words, to get the extracted term, we simply 'cut out' those parts of the proof term which prove Harrop formulas and replace the constants (representing axioms in our logic system) in the remaining part with corresponding extracted terms. For example induction axioms are replaced by recursors. Then it is easy to show $[\![M]\!]$ mr C.

To define the extracted term more formally, assume we have assigned to any assumption variable u^C a new variable $x_u^{\tau(C)}$.

$$[\![u^C]\!] \quad := \quad x_u^{\tau(C)},$$
$$[\![\lambda x^\rho M^C]\!] \quad := \quad \lambda x [\![M]\!],$$
$$[\![\lambda u^A M^C]\!] \quad := \quad [\![M]\!],$$
$$[\![\lambda u^C M^D]\!] \quad := \quad \lambda x_u^{\tau(C)} [\![M]\!],$$
$$[\![\langle M^C, N^A \rangle]\!] \quad := \quad [\![M]\!],$$
$$[\![\langle N^A, M^C \rangle]\!] \quad := \quad [\![M]\!],$$
$$[\![\langle M^C, N^D \rangle]\!] \quad := \quad \langle [\![M]\!], [\![N]\!] \rangle,$$
$$[\![M^{\forall x^\rho C} t^\rho]\!] \quad := \quad [\![M]\!] t,$$
$$[\![M^{A \to C} N^A]\!] \quad := \quad [\![M]\!],$$
$$[\![M^{D \to C} N^D]\!] \quad := \quad [\![M]\!] [\![N]\!],$$
$$[\![\pi_1(M^{A \wedge C})]\!] \quad := \quad [\![M]\!],$$
$$[\![\pi_0(M^{C \wedge A})]\!] \quad := \quad [\![M]\!],$$
$$[\![\pi_i(M^{C \wedge D})]\!] \quad := \quad \pi_i([\![M]\!]).$$

Now all that is left is to find the extracted terms for the constructive axioms:

$$[\![(\exists^{*+}_{x^\rho, A})]\!] \quad := \quad \lambda x x,$$
$$[\![(\exists^{*+}_{x^\rho, C})]\!] \quad := \quad \lambda x, y^{\tau(C)} . \langle x, y \rangle,$$
$$[\![(\exists^{*-}_{x^\rho, A, C})]\!] \quad := \quad \lambda x, y^{\rho \to \tau(C)} . y x,$$
$$[\![(\exists^{*-}_{x^\rho, D, C})]\!] \quad := \quad \lambda y^{\rho \times \tau(D)}, z^{\rho \to \tau(D) \to \tau(C)} . z (\pi_0 y) (\pi_1 y),$$
$$[\![\text{Ind}_{\forall p^{\text{boole}} C}]\!] \quad := \quad R_{\text{boole}, \tau(C)},$$
$$[\![\text{Ind}_{\forall n^{\text{nat}} C}]\!] \quad := \quad R_{\text{nat}, \tau(C)}.$$

5.2. *Simultaneous free algebras*

Using SFAs, we can add new ground types defined by simultaneous induction. Formally, an SFA is defined from a list of new types $\iota_1, \ldots, \iota_n$ and for each ι_i a list of constructors $c_{i1}, \ldots, c_{in_i}$ together with their types. Each constructor c_{ij} has a type $\rho_{ij1} \to \ldots \to \rho_{ijn_{ij}} \to \iota_i$ with each ρ_k being constructed using $\to$ only and with the ι_k occurring only strictly positive in them. The objects of the types $\iota_1, \ldots, \iota_n$ are freely generated from these constructors, which is expressed by the induction schemes that are automatically added to the system together with these types. Here are some examples:

–   The type seq of lists of natural numbers is generated from the constructors ε: seq and $\frown$: nat $\to$ seq $\to$ seq and is introduced by

```
(introduce-algebras
 '((seq ((empty)
         (add (n nat) (s seq)))))))
```

The induction axiom for a formula $A[s^{\text{seq}}]$ reads

$$A[\varepsilon] \to \forall n, s\, (A[s] \to A[n \frown s]) \to \forall s A[s].$$

–   To define the type tree of finitely branching trees, we have to simultaneously define the type treelist. The constructors are $\{\cdot\}$: treelist $\to$ tree , ε: treelist and $\frown$: tree $\to$ treelist $\to$ treelist. It is introduced by

```
(introduce-algebras
 '((treelist ((empty)
              (add (t tree) (l treelist))))
   (tree ((tr (l treelist))))))
```

and the induction principle for formulas $A[t^{\text{tree}}]$ and $B[l^{\text{treelist}}]$ is

$$\forall l\, (B[l] \to A[\{l\}]) \to B[\varepsilon] \to \forall t, l\, (A[t] \to B[l] \to B[t \frown l]) \to$$
$$\forall t A[t] \wedge \forall l B[l].$$

–   Finally the type inftree of trees branching over the natural numbers uses the constructors ε: inftree and lim: (nat $\to$ inftree) $\to$ inftree.

```
(introduce-algebras
 '((inftree ((empty)
             (lim (f (c-arrow 'nat 'inftree)))))))
```

The induction axiom for $A[t^{\text{inftree}}]$ is

$$A[\varepsilon] \to \forall f^{\text{nat} \to \text{inftree}}\, (\forall n A[f n] \to A[\text{lim } f]) \to \forall t A[t].$$

Functionals defined by structural recursion over an inductive type are usually given as a set of equations, one equation for each constructor. For example, the length of an element of the type seq would be defined:

$$\mathsf{len}(\varepsilon) \quad := \quad 0,$$
$$\mathsf{len}(n^{\frown}s) \quad := \quad 1 + \mathsf{len}(s).$$

We could enter a term that does exactly the computation given above using the appropriate recursor. The term would look like this:

```
(((alg-rec-at 'seq '(seq) '(nat)) 0)
(lambda (n^) (lambda (s^) (lambda (m^)
((plus-nat m^) 1)))))
```

This certainly is not a convenient way of entering the equations above. Therefore, the SFA extension also provides a simple interface for entering recursion equations. The command

```
(def-rec 'len (c-arrow 'seq 'nat) 0)
```

tells MINLOG that we want to define a constant len of type seq → nat.

```
; Please feed in (using define-vars) new variables
; for the following types:
; nat
```

The interface asks for one or more variable names that will be used to represent step values in the recursion equations.

```
(define-vars 'll)
; len empty =? C[]
```

Now we are asked to enter the right-hand side of the first equation.

```
(define-rhs (parse-term "0"))
; len(add n^ s^) =? C[n^,s^,ll^]
; where ll^ = len s^
```

Finally, we have to give the right-hand side of the second equation. Here we can use the free variables $\hat{n},\hat{s}$ (for the arguments of the constructor) and $\hat{ll}$ (for the step value).

```
(define-rhs (parse-term "ll^ + 1"))
; OK, COMPLETED; You have added the following
; recursion equations:
; len empty = 0
; len(add n^ s^) = (len s^)+1
```

From now on, the constant len will be used in all in- and outputs to represent the rather lengthy term given above.

5.3. *The quicksort example*

We will now use seq to get the quicksort algorithm from a proof of its specification. To this end, we introduce some more recursively defined functionals that we will need:

- $s_1 * s_2$ concatenates the two lists s_1 and s_2.
- $\#_m(s)$ counts the occurrences of m in s.
- $s \geq m$ checks whether m is a lower bound for the elements in s:

$$\varepsilon \geq m \quad := \quad \text{true},$$
$$(n \widehat{\ } s) \geq m \quad := \quad \textbf{if} \quad m \leq n \quad \textbf{then} \quad s \geq m \quad \textbf{else} \quad \text{false} \quad \textbf{fi}.$$

- Likewise, $s < m$ checks whether m is a proper upper bound.
- $s_{\geq m}$ computes the sublist of s of elements $\geq m$:

$$\varepsilon_{\geq m} \quad := \quad \varepsilon,$$
$$(n \widehat{\ } s)_{\geq m} \quad := \quad \textbf{if} \quad m \leq n \quad \textbf{then} \quad n \widehat{\ } (s_{\geq m}) \quad \textbf{else} \quad s_{\geq m} \quad \textbf{fi}.$$

- $s_{<m}$ is defined correspondingly.
- sorted(s) checks whether the list s is sorted:

$$\text{sorted}(\varepsilon) \quad := \quad \text{true},$$
$$\text{sorted}(n \widehat{\ } s) \quad := \quad \textbf{if} \quad s \geq n \quad \textbf{then} \quad \text{sorted}(s) \quad \textbf{else} \quad \text{false} \quad \textbf{fi}.$$

Using the structural induction scheme, we can easily prove the following fundamental properties of these functionals:

1. $\text{len}(s) \leq 0 \rightarrow s = \varepsilon$
2. $\text{len}(s_{<n}) \leq \text{len}(s)$
3. $\text{len}(s_{\geq n}) \leq \text{len}(s)$
4. $(\forall m \#_m(s_1) = \#_m(s_2)) \rightarrow s_1 \geq n \rightarrow s_2 \geq n$
5. $(\forall m \#_m(s_1) = \#_m(s_2)) \rightarrow s_1 < n \rightarrow s_2 < n$
6. $s_{\geq n} \geq n$
7. $s_{<n} < n$
8. $\text{sorted}(s_1) \rightarrow s_1 \geq n \rightarrow \text{sorted}(s_2) \rightarrow s_2 < n \rightarrow \text{sorted}(s_2 * s_1)$
9. $s \geq n \rightarrow \text{sorted}(s) \rightarrow \text{sorted}(n \widehat{\ } s)$
10. $\#_n(s) = \#_n(s_{<m}) + \#_n(s_{\geq m})$
11. $\#_n(s_1 * s_2) = \#_n(s_1) + \#_n(s_2)$

Now we formulate the specification of the desired algorithm: we want to show that every list of natural numbers can be sorted, i.e. for every list s there is a list t such that t is sorted and every number n occurs in t as often as in s. Formally:

$$\forall s \exists^* t. \, \text{sorted}(t) \wedge \forall n \#_n(s) = \#_n(t).$$

The idea is to take, given a list $s = n' \frown s'$, the first element n' and split s' up into $s'_{<n'}$ and $s'_{\geq n'}$, sort these shorter lists and put the results together in the right order. In other words, we prove the goal by induction over the *length* of the given list, so we first prove

$$\forall m, s. \, \text{len}(s) \leq m \rightarrow \exists^* t. \, \text{sorted}(t) \wedge \forall n \#_n(s) = \#_n(t)$$

by induction on m and then get the original goal by instantiating m with $\text{len}(s)$. The proof is simple; we show it anyway to demonstrate that this example can be carried out in MINLOG without running into technical difficulties.

Base $m = 0$. Let s with $\text{len}(s) \leq 0$ be given. Then $s = \varepsilon$ by lemma 1. Let $t := \varepsilon$, then $\text{sorted}(t) \searrow^*_{\swarrow} \top$, we have to show $\forall m \#_m(s) = \#_m(t)$. This follows from $s = \varepsilon = t$.

Step $m \rightarrow m + 1$. Let s with $\text{len}(s) \leq m + 1$ be given.

Note: we make a case distinction here ($s = \varepsilon$ or $s = n' \frown s'$). Formally, this is achieved by using the structural induction scheme for seq, without making use of the induction hypothesis in the step case.

Case $s = \varepsilon$. See $m = 0$.

Case $s = n' \frown s'$. Now $\text{len}(s') \leq m \searrow^*_{\swarrow} \text{len}(s) \leq m + 1$, which we have just assumed. Using lemma 2 and 3 we conclude $\text{len}(s'_{<n'}) \leq m$ and $\text{len}(s'_{\geq n'}) \leq m$. Using the induction hypothesis for $s'_{<n'}$ and $s'_{\geq n'}$ we get t_1 and t_2 with (1) $\text{sorted}(t_1)$, (2) $\forall n \#_n(s'_{<n'}) = \#_n(t_1)$, (3) $\text{sorted}(t_2)$ and (4) $\forall n \#_n(s'_{\geq n'}) = \#_n(t_2)$. Let $t := t_1 * (n' \frown t_2)$. We have to show $\text{sorted}(t)$ and $\forall n \#_n(s) = \#_n(t)$.

$\text{sorted}(t)$: lemma 7 yields $s'_{<n'} < n'$ and hence, using lemma 5 and (2), we get (5) $t_1 < n'$. In the same way, using lemma 6 and 4 instead of 7 and 5, we get $t_2 \geq n'$ and conclude (6) $n' \frown t_2 \geq n'$ using normalization again, as well as (7) $\text{sorted}(n' \frown t_2)$ by lemma 9 and (3). Finally with lemma 8, (7), (6), (1) and (5) we get $\text{sorted}(t_1 * (n' \frown t_2))$.

To show $\forall n \#_n(s) = \#_n(t)$, let n be given.

Case $n = n'$: $\#_n(n' \frown s') \searrow^*_{\swarrow} \#_n(s') + 1 \overset{\text{L.10}}{=} \#_n(s'_{<n'}) + \#_n(s'_{\geq n'}) + 1$
$\overset{(2),(4)}{=} \#_n(t_1) + \#_n(t_2) + 1 \searrow^*_{\swarrow} \#_n(t_1) + \#_n(n' \frown t_2) \overset{\text{L.11}}{=} \#_n(t_1 * n' \frown t_2)$.

Case $n \neq n'$: $\#_n(n' \frown s') \searrow^*_{\swarrow} \#_n(s') \overset{\text{L.10}}{=} \#_n(s'_{<n'}) + \#_n(s'_{\geq n'})$
$\overset{(2),(4)}{=} \#_n(t_1) + \#_n(t_2) \searrow^*_{\swarrow} \#_n(t_1) + \#_n(n' \frown t_2) \overset{\text{L.11}}{=} \#_n(t_1 * n' \frown t_2)$.

Here is the program extracted from this proof M. We show an edited version of the original SCHEME expression (we have split the expression into three parts and renamed the bound variables):

$$
\begin{aligned}
[\![M]\!] &= \lambda s.\, g\, \mathrm{len}(s)\, s, \qquad \text{where} \\
g &= \mathsf{R}_{\mathrm{nat,seq}\to\mathrm{seq}}(\lambda s\,\varepsilon)(\lambda n\, h) \qquad \text{and} \\
h &= \lambda f^{\mathrm{seq}\to\mathrm{seq}}.\, \mathsf{R}_{\mathrm{seq,seq}}\, \varepsilon\,(\lambda n\, \lambda s_1\, \lambda s_2.\,(f\,(s_{1\,<n})) * (n\widehat{\ }(f\,(s_{1\,\geq n})))).
\end{aligned}
$$

$(g\,n)\colon \mathrm{seq} \to \mathrm{seq}$ is a quicksort function able to deal with lists of length up to n and is defined by recursion over n:

$$
\begin{aligned}
g\,0\,s &= \varepsilon \\
g\,(n+1)\,s &= h\,(g\,n)\,s.
\end{aligned}
$$

If f is a function that can sort lists of length up to n, then $(h\,f)$ is a function that sorts lists of length up to $n+1$:

$$
\begin{aligned}
h\,f\,\varepsilon &= \varepsilon \\
h\,f\,(n\widehat{\ }s) &= (f\,(s_{<n})) * (n\widehat{\ }(f\,(s_{\geq n}))).
\end{aligned}
$$

6. ADVANCED FEATURES: A-TRANSLATION AND PRUNING

There are a number of proofs of $\forall\exists$–statements which — although very short and elegant — do not immediately yield a program, since they contain non-constructive arguments. However, it is well-known that such classical proofs under certain circumstances can be translated into constructive proofs, and hence yield algorithms via program extraction (cf. section 5). There is a substantial literature on that subject, and the MINLOG system supports a variant, known as 'A-translation', which goes back to work of Friedman (1978) and Leivant (1985). We will explain this translation by means of an example concerning minimization on finite binary trees. We will also use this example to discuss a second technique for program development: the 'pruning' operation going back to Goad (1980).

6.1. *Search through binary trees*

In our example we consider finite binary trees with leaves labelled by integers:

$$
\mathrm{leaf}(n) \mid \langle s_1, s_2 \rangle.
$$

We let P be a decidable property of trees. We wish to prove constructively that for every tree t with property $P[t]$ there is a subtree s of t satisfying $P[s]$

which is minimal, i.e. for no proper subtree s' of s, $P[s']$ holds. Of course we are interested in the program extracted from such a proof.

Let $s \subseteq t$ mean that s is a subtree of t and define

$$Q[s] \quad :\equiv \quad \forall s' \subseteq s . s' \neq s \to \neg P[s'],$$
$$A_0[t,s] \quad :\equiv \quad s \subseteq t \land P[s] \land Q[s].$$

Then our goal can be stated formally as

$$\forall t . P[t] \to \exists^* s A_0[t,s].$$

6.2. *From classical proofs to programs*

We will first prove the goal classically, i.e. deduce from $P[t]$ the classical existence statement

$$\neg \forall s \neg A_0[t,s]$$

and then apply Friedman's A-translation which gives us a constructive proof containing a quite clever algorithm.

The classical proof
Here and in 6.2.2 and 6.2.3 we fix a tree t and suppress it notationally, for sake of better readability. Hence

$$A_0[s] :\equiv A_0[t,s].$$

Roughly, the classical proof goes as follows: assume $P[t]$ and $\forall s \neg A_0[s]$. We have to derive a contradiction. Using the second assumption we can easily prove $\forall s . s \subseteq t \to Q[s]$ by induction on s. Now, setting $s := t$, we obtain $Q[t]$ and hence $A_0[t]$ ($\equiv A_0[t,t]$) using the assumption $P[t]$. But this contradicts our assumption $\forall s \neg A_0[s]$.

A-translation
In order to apply the A-translation to this proof we first have to look at the shape of the proof in some detail. Note that we only used induction on trees and the following facts about P, Q and $\subseteq$:

ax$_1$: $\forall n . Q[\text{leaf}(n)]$

ax$_2$: $\forall s_1, s_2 . \neg P[s_1] \to \neg P[s_2] \to Q[s_1] \to Q[s_2] \to Q[\langle s_1, s_2 \rangle]$

ax$_3$: $\forall s . s \subseteq s$

ax$_4$: $\forall s_1, s_2, s . \langle s_1, s_2 \rangle \subseteq s \to s_1 \subseteq s \land s_2 \subseteq s.$

If we view Q as a primitive predicate then these are Π-formulas, i.e. universal formulas with a quantifier-free kernel. Moreover, the kernel of our goal, $A_0[s]$, is quantifier-free. Now, the A-translation works for a situation like this. Let us briefly explain its crucial idea at our classical proof.

Recall that above we have proved $\bot$ from the axioms $ax_1, \ldots, ax_4$, the assumptions $P[t]$ and $\forall s . A_0[s] \to \bot$. Observe also that the symbol $\bot$, for falsity, didn't play a special role in the proof (e.g. we used neither ex-falso-quodlibet nor stability); in particular the *proof* was entirely constructive (although the *formula* proven was, of course, a non-constructive existence statement). Therefore we may replace everywhere in the proof the formula $\bot$ by our constructive goal-formula

$$A :\equiv \exists^* s A_0[s]$$

and obtain a correct and constructive derivation of the formula A from the assumptions $P[t]$ and $\forall s . A_0[s] \to A$. But: the latter formula is an instance of an $\exists^{*+}$ axiom. Hence we get A constructively from the axioms and $P[t]$ alone.

Unfortunately there are some complications: (1) Of course, we want to be allowed to use ex-falso-quodlibet ($\bot \to A$) and stability ($\neg\neg A \to A$) in our proofs and (2) the translated proof also uses the translated axioms; for instance

$$ax_2' : \forall s_1, s_2 . (P[s_1] \to A) \to (P[s_2] \to A) \to Q[s_1] \to Q[s_2] \to Q[\langle s_1, s_2 \rangle]$$

which is not provable, and, in general, will even be false. A way out of these problems is to first replace in the classical proof every atomic formula by its double negation (Gödel's negative translation) and afterwards replace $\bot$ by A. Then (1) ex-falso-quodlibet and stability are translated into formulas provable in minimal logic and (2) each assumption ax_i is translated into a formula which follows constructively from ax_i. Of course, this modified translation (i.e. Gödel's translation followed by the replacement $\bot \mapsto A$) also affects the formula A_0, but still transforms the formula $\forall s . A_0[s] \to \bot$ into a provable formula.

Remarks: 1. Friedman's original translation (Friedman, 1978) replaces every atomic formula R by $R \vee A$ and not, as we did, by $(R \to A) \to A$. But clearly the formulas $R \vee A$ and $(R \to A) \to A$ are constructively equivalent assuming decidability of R. We have chosen the latter variant, since in MINLOG we prefer reasoning with implications rather with disjunctions.

2. Another way to see the A-translation, is to say that in constructive logic we may pass from a proof of $\neg\forall x \neg D[x]$ (D quantifier-free), or equivalently $\neg\neg\exists^* D[x]$, to a proof of $\exists^* x D[x]$. This is known as Markov's rule.

The translated proof

In MINLOG we have implemented a refinement of the A-translation which does not replace all atomic formulas R by $(R \to A) \to A$. In our example this is only necessary for formulas $Q[\cdot]$; in general, it can be decided easily whether an atomic formula has to be replaced or not. For more information on this refinement we refer to (Berger and Schwichtenberg, 1995).

Now, the MINLOG system transforms our classical proof into a constructive one. We show this automatically generated proof in tree form below. Due to lack of space we graphically contracted consecutive applications of elimination rules to one rule. Similarly for consecutive introduction rules. A double line means that the conclusion follows from the premises by some elimination rules. The subproofs named M, M' etc. will play a role in section 6.2.4 only.

To make the proof tree easier to understand we give also an informal description: assume $P[t]$ and let

$$B[s] :\equiv s \subseteq t \to (Q[s] \to A) \to A$$

(again we suppressed t). We first prove $\forall s B[s]$ by induction on s. The base, $B[\mathsf{leaf}(n)]$, is easy. To prove the step we assume $B[s_1]$ and $B[s_2]$. In order to show $B[\langle s_1, s_2 \rangle]$ we further assume $\langle s_1, s_2 \rangle \subseteq t$ and $Q[\langle s_1, s_2 \rangle] \to A$. We have to show A. *Case $P[s_1]$*: then by i.h., $B[s_1]$, we have $(Q[s_1] \to A) \to A$. Hence it suffices to show $Q[s_1] \to A$. So assume $Q[s_1]$. We have $s_1 \subseteq t$, $P[s_1]$ and $Q[s_1]$, i.e. $A_0[s_1]$. Hence A, by existence introduction. *Case $\neg P[s_1]$*: *Sub-case $P[s_2]$*: similar to case $P[s_1]$, but using i.h., $B[s_2]$. *Sub-case $\neg P[s_2]$*: we use again the i.h., $B[s_1]$. Hence it suffices to prove $Q[s_1] \to A$. So assume $Q[s_1]$ and show A. Using the i.h., $B[s_2]$, we see that in fact we may also assume $Q[s_2]$. Now we have $\neg P[s_1]$, $\neg P[s_2]$, $Q[s_1]$ and $Q[s_2]$. Hence $Q[\langle s_1, s_2 \rangle]$, by ax_2. Now we remember that $Q[\langle s_1, s_2 \rangle] \to A$ holds. Hence A. This completes the inductive proof of $\forall s B[s]$. Setting $s := t$ we get $(Q[t] \to A) \to A$. So it finally suffices to prove $Q[t] \to A$. So assume $Q[t]$. Since we assumed $P[t]$ and also $t \subseteq t$ holds, we have $A_0[t]$ ($\equiv A_0[t,t]$). Hence A, by existence introduction. This completes the proof.

Note that in the proof above $\exists$-introduction and case analysis on P occur which both were not used in the classical proof. The explanation is that $\exists$-introduction is used for proving the translation of the wrong assumption, $\forall s \neg A_0[s]$, and case analysis is needed for proving ax_2 from its translation.

$$\cfrac{\cfrac{\text{ax}_1 \quad n}{u_2 : Q[\text{leaf}(n)]}}{\cfrac{\forall n. B[\text{leaf}(n)]}{}\; n, u_1, u_2}$$

$$\cfrac{\cfrac{\text{ax}_4, u_3}{s_1 \sqsubseteq t}\quad u_5 \quad u_6}{\cfrac{\text{ih}_1\quad \cfrac{s_1 \sqsubseteq t}{P[s_1] \to A}\quad \cfrac{M'' : A}{A}\; u_5}{A_0[s_1]}}\; u_6$$

$$\cfrac{A_0[s]}{A}\quad \equiv\quad s \sqsubseteq t \wedge P[s] \wedge Q[s]$$
$$A \quad\equiv\quad \exists^* s\, A_0[s]$$
$$B[s] \quad\equiv\quad s \sqsubseteq t \to (Q[s] \to A) \to A$$

$$\cfrac{\forall s. B[s] \qquad t}{\forall s_1, s_2. B[s_1] \to B[s_2] \to B[\langle s_1, s_2 \rangle]}\;\text{ind}$$
$$\cfrac{M : A}{(Q[t] \to A) \to A}$$
$$A$$

$$\text{ax}_0:\quad P[t]$$
$$\text{ax}_1:\quad \forall n. Q[\text{leaf}(n)]$$
$$\text{ax}_2:\quad \forall s_1, s_2. \neg P[s_1] \to \neg P[s_2] \to Q[s_1] \to Q[s_2] \to Q[\langle s_1, s_2 \rangle]$$
$$\text{ax}_3:\quad \forall s.\, s \sqsubseteq s$$
$$\text{ax}_4:\quad \forall s_1, s_2, s.\, \langle s_1, s_2 \rangle \sqsubseteq s \to s_1 \sqsubseteq s \wedge s_2 \sqsubseteq s$$

$$\cfrac{M' : A}{\neg P[s_1] \to A}\; u_9 \quad \text{cases}_{P[s_1]}$$

$$\cfrac{\text{ax}_4, u_3}{s_2 \sqsubseteq t}\quad u_7 \quad u_8$$
$$\cfrac{\text{ih}_2 \quad \cfrac{s_2 \sqsubseteq t}{Q[s_2] \to A}}{A}\; u_8$$
$$\cfrac{A_0[s_2]}{A}\; u_7$$
$$\cfrac{P[s_2] \to A}{A}$$

$$\cfrac{\text{ih}_2\quad \cfrac{s_2 \sqsubseteq t}{Q[s_2] \to A}}{\cfrac{A}{\neg P[s_2] \to A}\; u_{10}}\;\text{cases}_{P[s_2]}$$
$$\cfrac{N : A}{Q[s_1] \to A}\; u_{11}$$
$$\cfrac{N' : A}{Q[s_2] \to A}\; u_{12}$$

$$\cfrac{\text{ax}_2\quad s_1\; s_2\quad u_9\; u_{10}\; u_{11}\; u_{12}}{u_4\quad Q[\langle s_1, s_2 \rangle]}$$

$$\cfrac{\text{ax}_3\quad t}{\cfrac{t \sqsubseteq t}{A_0[t]}\quad \text{ax}_0\quad u_0}\; u_0$$
$$\cfrac{Q[t] \to A}{A}$$

$$u_0 : Q[t] \qquad\qquad u_3 : \langle s_1, s_2 \rangle \sqsubseteq t \qquad\qquad u_5 : P[s_1] \qquad\qquad u_7 : P[s_2] \qquad\qquad u_9 : \neg P[s_1] \qquad\qquad u_{11} : Q[s_1]$$
$$u_1 : \text{leaf}(n) \qquad\qquad u_4 : Q[\langle s_1, s_2 \rangle] \to A \qquad\qquad u_6 : Q[s_1] \qquad\qquad u_8 : Q[s_2] \qquad\qquad u_{10} : \neg P[s_2] \qquad\qquad u_{12} : Q[s_2]$$
$$u_2 : Q[\text{leaf}(n)] \qquad\qquad u_4 : Q[\langle s_1, s_2 \rangle] \to A$$

The extracted program
From the translated proof we obtain the following program:

$$
\begin{aligned}
f(t) &= g(t,t) \\
g(\text{leaf}(n),t) &= t \\
g(\langle s_1, s_2 \rangle, t) &= \textbf{if} \quad P[s_1] \\
&\qquad \textbf{then} \quad g(s_1, s_1) \\
&\qquad \textbf{else} \quad \textbf{if} \quad P[s_2] \\
&\qquad\qquad\qquad \textbf{then} \quad g(s_2, s_2) \\
&\qquad\qquad\qquad \textbf{else} \quad g(s_1, g(s_2, t)) \quad \textbf{fifi}.
\end{aligned}
$$

6.3. *Pruning*

The idea of Goad's pruning operation is the following: consider the constructive proof above obtained from the A-translation. At many points of this proof we have derived the formula A. Suppose M is such a subproof deriving A, and assume that M contains again a subproof M' deriving A, too. Now it is tempting to simplify the whole proof by replacing M by the smaller proof M'. However, in general this will be impossible since M' may depend on an assumption u^B which is bound in M by an $\rightarrow$-introduction. In λ-term notation

$$
M \rightarrow M[\lambda u^B M'[u^B]].
$$

Now assume that we are interested only in trees t satisfying $P[t]$ and some additional condition $C[t]$, i.e. we look for a proof of

$$
\forall t . P[t] \wedge C[t] \rightarrow \exists^* s A_0[t,s]
$$

and expect from this hopefully simpler proof a hopefully optimized program adapted to the restriction $C[t]$. Such a simplified proof may be obtained as follows: in the subproof $M'[u^B]$ considered above we replace the assumption u^B by a proof K^B which uses the additional assumption $C[t]$ (and possibly other assumptions valid at the particular occurrence of u), and then replace the modified proof $M[\lambda u^B M'[K^B]]$ by $M'[K^B]$. Of course, there is no guarantee that $M'[K^B]$ is indeed simpler than $M[\lambda u^B M'[u^B]]$, but in most cases it will be.

In the following we will consider three pruning conditions $C_1[t]$, $C_2[t]$, $C_3[t]$, and study their effect on the proof and the extracted program. It will turn out that not only are the new programs simpler than the one in 6.2.4, but they also yield different results. This shows that it is essential that the pruning operation is done on *proofs*, since optimized *programs* will always compute the same result on the restricted inputs.

First example

Assume we enrich our specification by the additional information

$$C_1[t] :\equiv \forall s_1, s_2 . \langle s_1, s_2 \rangle \subseteq t \to P[s_1] \to P[s_2],$$

i.e. if P holds for a left subtree, than also for the right one. (The choice of the somewhat arbitrary formula C_1 is motivated by the effect it later will have.) Remember that in the constructive proof described informally in 6.2.3 we had a *case* $\neg P[s_1]$ with the goal A. Then there was a *sub-case* $\neg P[s_2]$ again with goal A. In this sub-case we used the assumption $\neg P[s_1]$. This assumption can now be proved using the pruning condition $C_1[t]$ and the assumptions $\neg P[s_1]$ and $\langle s_1, s_2 \rangle \subseteq t$ which are both valid at that point. This means that the case analysis according to whether $P[s_1]$ holds can be removed from the proof, since we have shown that in fact $\neg P[s_1]$ holds. In the proof tree the simplified proof is obtained by replacing the subproof M by M', where the assumption u_9 is replaced by a proof using $C_1[t]$, u_3 and u_{10}.

The program extracted from the simplified proof is the following:

$$
\begin{aligned}
f(t) &= g(t,t) \\
g(\text{leaf}(n), t) &= t \\
g(\langle s_1, s_2 \rangle, t) &= \textbf{if} \quad P[s_1] \\
&\qquad \textbf{then} \quad g(s_1, s_1) \\
&\qquad \textbf{else} \quad g(s_1, g(s_2, t)) \quad \textbf{fi}.
\end{aligned}
$$

Second example

Next consider

$$C_2[t] :\equiv \forall s . s \subseteq t \to P[s].$$

This has an extreme effect on the proof, since already the outer case analysis on $P[s_1]$ is decided positively. Hence we get a proof without case analysis. This means that in the proof tree we replace the subproof M by M'', where the assumption u_5 is replaced by a proof using $C_2[t]$, u_3 and ax_4. The extracted program is simply

$$
\begin{aligned}
f(t) &= g(t,t) \\
g(\text{leaf}(n), t) &= t \\
g(\langle s_1, s_2 \rangle, t) &= g(s_1, s_1).
\end{aligned}
$$

Third example
Our last pruning condition is

$$C_3 :\equiv \forall s.\, \neg P[s] \to Q[s]$$

which is in fact an extra condition on P. We look again at the *case* $\neg P[s_1]$ and therein at the *sub-case* $\neg P[s_2]$. To prove A we used both induction hypotheses, $B[s_1]$ and $B[s_2]$. We ended up in a situation where we had to prove A under the extra assumptions $Q[s_1]$ (from ih$_1$) and $Q[s_2]$ (from ih$_2$). Now from $\neg P[s_1]$ and $\neg P[s_2]$ and the pruning condition C_3 we can *prove* $Q[s_1]$ and $Q[s_2]$. Hence we do not need to use the induction hypotheses at that point. For the proof tree this means that the subproof N is replaced by N', where the assumptions u_{11} and u_{12} are replaced by a proof using C_3, u_9 and u_{10}. The effect on the extracted program is that the nested recursive call disappears:

$$
\begin{aligned}
f(t) &= g(t,t)\\
g(\mathsf{leaf}(n),t) &= t\\
g(\langle s_1,s_2\rangle,t) &= \textbf{if}\quad P[s_1]\\
&\qquad \textbf{then}\quad g(s_1,s_1)\\
&\qquad \textbf{else}\quad \textbf{if}\quad P[s_2]\\
&\qquad\qquad\quad \textbf{then}\quad g(s_2,s_2)\\
&\qquad\qquad\quad \textbf{else}\quad t\quad \text{fifi}.
\end{aligned}
$$

ACKNOWLEDGEMENTS

We are grateful to Felix Joachimski, Karl-Heinz Niggl and Klaus Weich for their contributions to the MINLOG system which were used in many places of this presentation.

REFERENCES

Berger, U.: 1993a, 'Program extraction from normalization proofs'. In: M. Bezem and J. Groote (eds.): *Typed Lambda Calculi and Applications*, Vol. 664 of *Lecture Notes in Computer Science*. pp. 91–106, Springer Verlag, Berlin, Heidelberg, New York.
Berger, U.: 1993b, 'Total Sets and Objects in Domain Theory'. *Annals of Pure and Applied Logic* **60**, 91–117.

Berger, U., M. Eberl, and H. Schwichtenberg: 1998, 'Normalization by evaluation'. Submitted to: B. Möller and J.V. Tucker (eds.): *Prospects for hardware foundations*. NADA volume, *Lecture Notes in Computer Science*.

Berger, U. and H. Schwichtenberg: 1991, 'An inverse of the evaluation functional for typed λ–calculus'. In: R. Vemuri (ed.): *Proceedings of the Sixth Annual IEEE Symposium on Logic in Computer Science*. pp. 203–211, IEEE Computer Society Press, Los Alamitos.

Berger, U. and H. Schwichtenberg: 1995, 'Program Extraction from Classical Proofs'. In: D. Leivant (ed.): *Logic and Computational Complexity, LCC '94*, Vol. 960 of *Lecture Notes in Computer Science*. pp. 77–97, Springer Verlag, Berlin, Heidelberg, New York.

Boyer, R. S. and J. S. Moore: 1988, *A Computational Logic Handbook*, Vol. 23 of *Perspectives in Computing*. Academic Press, Inc.

Friedman, H.: 1978, 'Classically and intuitionistically provably recursive functions'. In: D. Scott and G. Müller (eds.): *Higher Set Theory*, Vol. 669 of *Lecture Notes in Mathematics*. pp. 21–28, Springer Verlag, Berlin, Heidelberg, New York.

Goad, C. A.: 1980, 'Computational uses of the manipulation of formal proofs'. Ph.D. thesis, Stanford University. Stanford Department of Computer Science Report No. STAN–CS–80–819.

Hayashi, S.: 1990, 'An introduction to PX'. In: G. Huet (ed.): *Logical Foundations of Functional Programming*. Addison–Wesley, pp. 432–486.

Kreisel, G.: 1959, 'Interpretation of analysis by means of constructive functionals of finite types'. In: A. Heyting (ed.): *Constructivity in Mathematics*. North–Holland, Amsterdam, pp. 101–128.

Leivant, D.: 1985, 'Syntactic Translations and Provably Recursive Functions'. *The Journal of Symbolic Logic* **50**(3), 682–688.

Miller, D.: 1991, 'A logic programming language with lambda–abstraction, function variables and simple unification'. *Journal of Logic and Computation* **1**(4), 497–536.

Nipkow, T.: 1993, 'Orthogonal Higher–Order Rewrite Systems are Confluent'. In: M. Bezem and J. Groote (eds.): *Typed Lambda Calculi and Applications*, Vol. 664 of *Lecture Notes in Computer Science*. pp. 306–317, Springer Verlag, Berlin, Heidelberg, New York.

Scott, D.: 1982, 'Domains for denotational semantics'. In: E. Nielsen and E. Schmidt (eds.): *Automata, Languages and Programming*, Vol. 140 of *Lecture Notes in Computer Science*. pp. 577–613, Springer Verlag, Berlin, Heidelberg, New York.

Tait, W. W.: 1967, 'Intensional Interpretations of Functionals of Finite Type I'. *The Journal of Symbolic Logic* **32**(2), 198–212.

Troelstra, A. S. and D. van Dalen: 1988, *Constructivism in Mathematics. An Introduction*, Vol. 121, 123 of *Studies in Logic and the Foundations of Mathematics*. North–Holland, Amsterdam.

Turner, R.: 1991, *Constructive Foundations for functional languages*. McGraw–Hill.

van de Pol, J. and H. Schwichtenberg: 1995, 'Strict functionals for termination proofs'. In: M. Dezani-Ciancaglini and G. Plotkin (eds.): *Typed Lambda Calculi and Applications*, Vol. 902 of *Lecture Notes in Computer Science*. pp. 350–364, Springer Verlag, Berlin, Heidelberg, New York.

M. STRECKER, M. LUTHER, F. VON HENKE

CHAPTER 3

INTERACTIVE AND AUTOMATED PROOF CONSTRUCTION IN TYPE THEORY

1. INTRODUCTION

This chapter gives a survey of TYPELAB, a specification and verification environment that integrates interactive proof development and automated proof search. TYPELAB is based on a constructive type theory, the Calculus of Constructions, which can be understood as a combination of a typed λ-calculus and an expressive higher-order logic. Distinctive features of the type system are dependent function types (Π types) for modeling polymorphism and dependent record types (Σ types) for encoding specifications and mathematical theories.

Type theory provides a homogeneous theoretical framework in which the construction of a function and the construction of a proof can be considered to be essentially the same activity. There is, however, a practical difference in that the development of a function requires more insight and therefore usually has to be performed under human guidance, whereas proof search can, to a large extent, be automated. Internally, TYPELAB exploits the homogeneity provided by type theory, while externally offering an interface to the human user which conceals most of the complexities of type theory. Interactive construction of proof objects is possible whenever desired; metavariables serve as placeholders which can be refined incrementally until the desired object is complete. For procedures which can reasonably be automated, high-level tactics are available. In this respect, TYPELAB can be understood as a proof assistant which, in addition to the manipulations of formulae traditionally performed by theorem provers, permits to carry out operations on entities such as functions and types.

From a different perspective, TYPELAB can be viewed as a programming environment in which, apart from the execution of programs in the style of functional language interpreters, properties of programs can be specified and verified and in which complex developments can be carried out. Even though these features are currently not advanced very much beyond the stage of a research prototype, it is possible to enter expressions at the top level and evaluate them by reduction to normal form. Function definition is so far limited to (higher-order) primitive recursive functions and no efficient compilation is

W. Bibel, P. H. Schmitt (eds.), Automated Deduction. A basis for applications. Vol. II
© 1998 *Kluwer Academic Publishers. Printed in the Netherlands*

currently available, but a wide range of practically relevant functions can be coded in a natural style.

TYPELAB supports program development by stepwise refinement: Declarations of types, functions and axioms can be bundled up to specifications or mathematical theories. Specifications, as internal objects of the logic, can be handled in complete analogy to other entities of the logic. In particular, they can be parameterized, possibly by other specifications, and can be the domain or range of functions, which in this case can be interpreted as refinement mappings.

The main focus of this chapter is the theory on which the proof development component of TYPELAB is based. For an illustration of program development in TYPELAB, the report (Luther and Strecker, 1998) should be consulted. Metavariables play an important role in the interactive construction of proof objects. A closer analysis of term operations such as reduction in a language with metavariables reveals that substitutions in metavariables have to be delayed until an instantiation for the metavariable has been found. This suggests a calculus with a notion of explicit substitutions, which is described in greater detail in Section 2. In Section 3, it will be shown how this machinery, developed primarily for interactive proof construction, provides a foundation for automation of proof search. A comparison with related systems and final remarks (Section 4) conclude this chapter.

2. A CALCULUS WITH METAVARIABLES

Metavariables are useful as placeholders for incomplete objects that have to be constructed in the course of a proof. The concept of such a proof object has to be understood in a broad sense: A proof object can be a function that is part of a realization of a specification, it can be a witness of an existentially quantified variable, or a term which encodes logical reasoning.

Take, for example, the proposition $\exists f : T \to T.f(x) = x$. According to the propositions-as-types principle[1], a proof of this proposition requires the construction of a term t which can be understood as an encoding of the individual proof steps and whose type is the proposition to be proved. The proposition and the term t depend on a list of assumptions and variable declarations, a *context* Γ, which states, among others, that T is a type and that x is of type T. We write $\Gamma \vdash ?t_0 : \exists f : T \to T.f(x) = x$ to express that we want to construct a term of the given type. The metavariable $?t_0$ will be further refined

---

[1] Some of the following motivation makes use of terminology introduced in later sections.

as the proof progresses. For example, a reasonable next step is to introduce a metavariable $?f$ as placeholder for the function f, so that the above problem is reduced to proving $\Gamma \vdash ?f : T \to T$ and $\Gamma \vdash ?t_1 : ?f(x) = x$. A partial solution for $?t_0$ is a term of the form $t[?f, ?t_1]$. The proof can be completed by instantiating $?f$ with the term $\lambda z : T.\ z$ and using reflexivity of equality to provide a solution for $?t_1$.

For the seemingly disparate notions of proof problem sketched above, type theory provides a unifying framework. For this convenience, a price has to be paid: in a dependently typed logic such as the Calculus of Constructions that TYPELAB is based on, metavariables can depend on one another. The solution for one metavariable, such as $?f$ in the above example, can determine the type of another metavariable, such as $?t_1$, and thus influence the set of valid solutions for $?t_1$. A priori, it is not clear which dependencies among metavariables should be admitted, and whether a partial solution provided for a metavariable can safely be accepted. The sections below try to clarify these questions by developing a calculus with metavariables.

An additional difficulty arises from the fact that metavariables not only depend on a type, but also on a context that determines which variables can legally occur in the solution of the metavariable. For example, the metavariable $?f$ depends on the context Γ in which the variable x is declared. Instead of providing the closed term $\lambda z : T.\ z$ as solution for $?f$, we could as well have chosen the term $\lambda z : T.\ x$ in which x occurs free. However, a solution $\lambda z : T.\ y$, with y not declared in Γ, would be ill-typed and therefore not admissible as solution for $?f$, apart from being useless in this particular case.

In the following, we will examine more closely the behaviour of instantiations. The reader should be aware of one peculiarity: When instantiating a metavariable, all occurrences of the metavariable are replaced by the solution term *without renaming bound variables* (see Definition 2.11 for the details). Thus, variables which are free in a solution term are possibly captured after instantiation. The reason for chosing this non-standard definition of instantiation is that, intuitively speaking, the decision of what is a "free" or a "bound" variable in usual presentations of type theory is a matter of perspective, but not an absolute notion. Variables in a well-typed term t that are not locally bound (for example by λ-abstraction) are still bound in those contexts in which t is typeable. Thus, whenever a metavariable depends on a context, as sketched above, and an instantiation can reference variables of this context, there has to be a possibility of propagating this instantiation through bindings without renaming bound variables. It should be noted that the traditional approach to dealing with this problem is based on a functional encoding of scopes, which, in a type-theoretic framework, has some disadvantages (see Section 2.4 for a discussion).

There are mainly two problems when dealing with context-dependent meta-variables, illustrated by the following examples:

Example 2.1

Commutativity of instantiation and reduction: Assume that metavariable $?n_1$ is defined to be of type T in a context containing $x : T$, that is, $x : T \vdash ?n_1 : T$. In a naive approach, first reducing the term $trm_1 := (\lambda x : T. \ ?n_1) \ t$ to $?n_1$ and then instantiating the result with term x yields the result x. First instantiating $?n_1$ to x and then reducing yields t (the variable x bound by λ-abstraction is the same object as the variable x bound in the context of the metavariable).

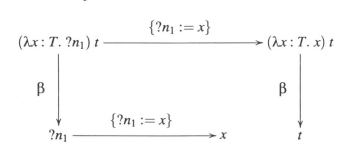

Note that this problem is not caused by a particular type system, but arises in any calculus in which there is a notion of β-reduction and in which metavariables depend on a context.

Example 2.2

Keeping track of type information: Consider a metavariable $?n_2$ defined with the following context and type:

$$A : Type, T : Type, x : T \vdash ?n_2 : T$$

Consider the term $trm_2 := (\lambda T : Type. \ \lambda x : T. \ ?n_2) \ A$ in context A:Type. When first instantiating $?n_2$ with x and then reducing, the resulting $\lambda x : A. \ x$ is easily seen to have type $A \to A$. When first reducing trm_2, however, the question arises what the type of the resulting term $\lambda x : A. \ ?n_2$ should be. $A \to T$ is certainly not correct, as T does not even occur in context A:Type. Claiming that $?n_2$ has type A is also problematic, since then, the term $?n_2$ would have different types (A resp. T) in different contexts. As opposed to the first problem, this difficulty is directly related to the type system and arises in a similar form in any calculus with dependent types.

To overcome these difficulties, we will keep track of substitutions that have been carried out in metavariables. This leads to a notion of *explicit substitution*. After reduction, trm_1 becomes $?n_1^\frown [x := t]$, whereas trm_2 becomes

$\lambda x : A. \ ?n_2^\frown [T := A]$. The calculus developed in the following solves the above problems, and it will be shown that it has desirable meta-theoretical properties such as confluence and strong normalization.

A calculus with metavariables and explicit substitutions for Martin-Löf's monomorphic type theory is presented by Magnusson (1995), and some algorithms such as unification are defined and shown to be correct. However, the properties of the calculus are not examined. Ensuring confluence and termination is not just a routine matter; some straightforward definitions of simply typed λ-calculus with explicit substitutions lack these properties, see for example (Lescanne, 1994; Melliès, 1995). More recently, Muñoz (1997) has presented a calculus with explicit substitutions for the Calculus of Constructions which avoids these problems by labeling terms before reduction. As opposed to our calculus, substitutions are marked with types to make reconstruction of typing derivations possible.

Section 2.1 is concerned with the behaviour of – not necessarily well-typed – terms, Section 2.2 defines typing rules for metavariables. Alternative approaches are conceivable, for example using a functional encoding of scopes which avoids dependence of metavariables on contexts. Consequently, no explicit substitutions have to be taken into account. In Section 2.4, it will be shown that both representations have the same strength in that there is a one-to-one correspondence between them. From a practical perspective, however, a functional encoding of scopes would usually become very clumsy for all those metavariables nested deeply inside terms, for example proof obligations generated for theorems within specifications

2.1. *Term calculus*

The term language of the logic is built up from a set V of variables, from type constants *Prop* and *Type$_i$* ($i \geq 0$), denoting universes of propositions and types, of dependent function types (Π types) which generalize universal quantification and dependent record types (Σ types) which are used to model specifications. On the level of elements of types, we have λ-abstraction, function application, pairing and projection. In addition to these term constructors, already present in the core logic as defined in (Luo, 1994), a set M of *metavariables* is introduced. Metavariables $?n, ?m, \ldots \in M$ are placeholders for terms to be constructed. For the sake of readability, we use a presentation of terms with named variables, subject to the usual naming conventions. From this, a more precise notation such as de Bruijn indices can easily be derived.

The syntax of the calculus with metavariables is then defined by the grammar of Figure 1. One of the productions for terms T permits to build terms of the form $?n^\frown \sigma$, which expresses that the application of substitution σ to

$$
\begin{aligned}
T \quad ::= \quad & V \\
| \quad & Prop \mid Type_i \\
| \quad & \Pi V : T.T \mid \lambda V : T.T \mid (T\ T) \\
| \quad & \Sigma V : T.T \mid pair_T(T,\ T) \mid \pi_1(T) \mid \pi_2(T) \\
| \quad & M^\frown S \\
\\
S \quad ::= \quad & [\,] \mid [V := T] :: S
\end{aligned}
$$

Figure 1. Grammar of language with metavariables and explicit substitutions

metavariables $?n$ has been delayed. Substitutions are generated by production S. By this construction, a substitution effectively becomes a part of a term and can be reasoned about in the calculus. This notion of *explicit substitutions* has to be distinguished from the traditional notion of substitutions, which are defined as meta-operations on terms.

It should be remarked that substitutions can only be attached to metavariables and not to arbitrary terms. This distinguishes our calculus from (to our best knowledge) all calculi of explicit substitutions presented in the literature. Note that our purpose for using explicit substitutions is, quite modestly, to compensate for problems arising from substitutions into metavariables and not to provide an abstract machine model for reduction in λ-calculi.

To improve readability, we sometimes write $\forall x : A.B$ instead of $\Pi x : A.\ B$, or $A \rightarrow B$ if x does not occur in B. Metavariables with empty substitution, $?n^\frown[\,]$, will often be abbreviated as $?n$.

For proving properties of the term calculus, such as confluence, it will be convenient to assume that there is a function *svars* which associates to each $?n \in M$ a list of variables that may be substituted into $?n$, the substitutions for other variables being discarded (cf. Definition 2.4). When considering well-typed terms in Section 2.2, it will turn out that this list of variables is naturally given by the context on which $?n$ depends.

We can now define explicit (or *internal*) substitutions more formally as follows:

Definition 2.3 (Internal Substitution)
An internal substitution is a list of the form $[x_1 := t_1, \ldots x_n := t_n]$ associating terms to variables. A substitution $\sigma \equiv [x_1 := t_1, \ldots x_n := t_n]$ is *valid* for a metavariable $?n$ if:

1. All the variables x_i are distinct.
2. All the variables x_i are contained in $svars(?n)$.

3. If x_j occurs before x_j in σ, then x_i occurs before x_j in $svars(?n)$.

The *domain* of a substitution is defined to be the set $dom(\sigma) := \{x_1, \ldots x_n\}$.

Since substitutions in our calculus are only internalized as attachment to metavariables to express that a substitution is delayed, we cannot completely dispense with an external notion of substitution.

Definition 2.4 (External Substitutions)
A substitution $\sigma := \{x := s\}$ homomorphically maps over terms as usual. On metavariables, it is defined as follows (we assume that $x \notin \{y_1 \ldots y_n\}$ and will from now on enforce this requirement when applying substitutions):

- If $x \notin svars(?n)$:
  $$(?n^\frown[y_1 := t_1, \ldots y_n := t_n])\sigma := ?n^\frown[y_1 := t_1\sigma, \ldots y_n := t_n\sigma]$$
- If $x \in svars(?n)$ and x is declared between y_i and y_{i+1} in $svars(?n)$:
  $$(?n^\frown[y_1 := t_1, \ldots y_n := t_n])\sigma :=$$
  $$?n^\frown[y_1 := t_1\sigma, \ldots y_i := t_i\sigma, x := s, y_{i+1} := t_{i+1}\sigma, \ldots y_n := t_n\sigma]$$

In order to distinguish external substitutions from the internalized explicit substitutions, we write the former in braces like $\{\ldots\}$ and the latter in brackets like $[\ldots]$. However, the letters σ and τ will be used indiscriminately for both kinds of substitutions.

As in standard λ-calculus, β-reduction is now defined by $(\lambda x : T. M) N \rightarrow_\beta M\{x := N\}$. For example, $(\lambda x : T. x) N \rightarrow_\beta N$ and $(\lambda x : T. ?n^\frown[]) N \rightarrow_\beta ?n^\frown[x := N]$, if $x \in svars(?n)$. Similarly, we define π-reduction, which simplifies projections applied to pairs, by $\pi_i(pair_T(M_1, M_2)) \rightarrow_\pi M_i$ for $i = 1, 2$. From these base relations, more complex relations such as one-step reduction $\rightarrow_1$ (the closure of $\rightarrow_\beta \cup \rightarrow_\pi$ compatible with term structure), parallel one-step reduction $\Rightarrow_1$, reduction $\twoheadrightarrow$ (the reflexive-transitive closure of $\rightarrow_1$) and convertibility $\simeq$ can be defined (see e.g. (Barendregt, 1984) for details).

The Church-Rosser property ensures that two diverging computation paths can always be joined again. If reductions of the strict part of $\twoheadrightarrow$ always terminate (which indeed they do, see Proposition 2.18), then normal forms are unique. This way, convertibility of two terms can be decided, by reducing them to normal form and comparing the normal forms syntactically. The following result can be established by an adaptation of a method originally developed by Martin-Löf and Tait:

Proposition 2.5 (Reduction is Church-Rosser)
The reduction relation $\twoheadrightarrow$ satisfies the Church-Rosser property.

2.2. *Typing*

Typing judgements of the form $\Gamma \vdash t : T$ are used to express that term t is of type T in context Γ, where Γ is a list of declarations of the form $x_1 : T_1, \ldots, x_n : T_n$. Typing judgements are inductively generated by typing rules such as the following (λ) rule, which expresses how elements of dependent function types are introduced.

$$\frac{\Gamma, x : A \vdash M : B}{\Gamma \vdash \lambda x : A.M : \Pi x : A.B} \; (\lambda)$$

By the propositions-as-types principle, which interprets propositions as types and elements of propositions as proofs, and by equating Π-abstraction and $\forall$-abstraction, the rule can also be understood as saying that a λ-abstraction provides a proof trace for the introduction of a universal quantifier.

As opposed to the situation encountered in less complex calculi (such as the simply-typed λ-calculus), there can be intricate dependencies among metavariables in calculi with dependent types. In particular, the type of one metavariable can depend on the value assigned to another one, and the well-typedness of a context can depend on the value assigned to a metavariable.

Before stating typing rules and examining their properties, some restrictions on dependencies among metavariables have to be imposed which are strong enough to make verification of the correctness of solutions for metavariables possible. The restrictions should be sufficiently liberal so that dependencies among metavariables can be exploited for proof search.

Example 2.6
Consider the ($\exists R$) rule (cf. Section 3.3):

$$\frac{\Gamma \vdash ?n_1 : A \qquad \Gamma \vdash ?n_2 : P(?n_1)}{\Gamma \vdash ?n_0 : \exists x : A.P(x)} \; (\exists R)$$

Application of this rule introduces two metavariables $?n_1$ and $?n_2$, where $?n_2$ depends on $?n_1$ since $?n_1 \in MV(P(?n_1))$ (here, $MV(.)$ is the set of metavariables occurring in a term or context). If there is a declaration of the form $h : P(a)$ in Γ, then $?n_2$ can be solved with h, leading to an assignment $?n_1 := a$ as a side-effect (see the description of unification in Section 3.1).

Example 2.7
Starting from the (academic) formula $\exists P : Prop, Q : P \rightarrow Prop, x : P.(P \rightarrow Q(x))$, eliminating existential quantifiers (as in the example above) and introducing assumptions, one obtains the following set of metavariables:

$\vdash ?P : Prop$

$\vdash ?Q : ?P \to Prop$

$\vdash ?x : ?P$

$h : ?P \vdash ?n : ?Q(?x).$

One step that suggests itself now is to equate $?n$ with h and consequently $?P$ with $?Q(?x)$, leading to a new proof problem with $\vdash ?Q : ?Q(?x) \to Prop$ and $\vdash ?x : ?Q(?x)$, in which the dependency of metavariables is cyclic. There is no intuitive interpretation of such a proof problem, nor does it seem clear that tentative solutions of such a proof problem can effectively be verified.

In the sequel, we will permit proof problems with metavariable dependencies of the first kind, but will exclude circularities of the second kind.

Let us emphasize again that a metavariable depends on a context Γ and has a type T, as expressed by the more suggestive notation $\Gamma \vdash ?n : T$. In the course of a proof, metavariables occurring in Γ or T can be instantiated. Therefore, a context and a type are not invariantly assigned to a metavariable $?n$ by functions depending only on $?n$, but they are determined by the proof problem under consideration, as reflected by the following definition.

Definition 2.8 ((Valid) Proof Problem)

A proof problem P is a triple $(M_P, ctxt_P, type_P)$ consisting of:

- A finite set of metavariables M_P
- A function $ctxt_P$ assigning a context to each $?n \in M_P$, such that $dom(ctxt_P(?n)) = svars(?n)$.
- A function $type_P$ assigning a term to each $?n \in M_P$

The subscripts in $M_P, ctxt_P, type_P$ will be omitted whenever P is clear from the context.

For a proof problem P and $?n_1, ?n_2 \in M_P$, the relation $<_P$ is defined as:

$?n_1 <_P ?n_2$ iff $?n_1 \in MV(ctxt(?n_2))$ or $?n_1 \in MV(type(?n_2))$.

Let $\ll_P$ be defined as the transitive closure of $<_P$.

A proof problem P is called *valid* if $\ll_P$ is an irreflexive partial order.

The proof problem P_1 with the set of metavariables $\{?n_1, ?n_2\}$ as given in Example 2.6 is valid, with $?n_1 \ll_{P_1} ?n_2$ since $?n_1 \in MV(P(?n_1))$. Regarding Example 2.7, the proof problem $P_2 = (M_2, ctxt_2, type_2)$ has $M_2 = \{?x, ?Q\}$, $ctxt_2(?x), ctxt_2(?Q)$ the empty context and $type_2(?x) = (?Q\ ?x)$ and $type_2(?Q)$ $= ?Q(?x) \to Prop$. P_2 is not valid, since $?x \ll_{P_2} ?x$.

Unless stated otherwise, we will only consider valid proof problems in the following. Since, for every valid proof problem P, the set M_P is finite and the order $\ll_P$ is transitive and irreflexive, it is also well-founded on M_P. This fact can be used to show termination of certain functions involving metavariables.

$$\frac{ctxt(?n) \vdash type(?n) : Type_j}{ctxt(?n) \vdash ?n^\frown[\,] : type(?n)} \text{ (MV-base)}$$

$$\frac{z \notin dom(\Gamma) \cup dom(\Delta) \cup dom(\sigma) \quad \Gamma \vdash T : Type_j \quad \Gamma, \Delta \vdash ?n^\frown\sigma : N}{\Gamma, z : T, \Delta \vdash ?n^\frown\sigma : N} \text{ (MV-weak)}$$

$$\frac{\Gamma \vdash t : T \quad \Gamma, x : T, \Delta \vdash ?n^\frown\sigma : N}{\Gamma, \Delta\{x := t\} \vdash (?n^\frown\sigma)\{x := t\} : N\{x := t\}} \text{ (MV-β-Red)}$$

Figure 2. Typing rules for Metavariables

In particular, the definitions made in the following can be shown to be well-defined.

The typing rules for metavariables with explicit substitutions are shown in Figure 2. Typing rules for metavariables are added to the typing rules of the base logic (Luo, 1994) in a "modular" fashion, that is, without making a modification of the base rules necessary. The typing rules for metavariables can be motivated as follows:

MV-base A metavariable $?n$ with empty substitution is type-correct in case its defining type $type(?n)$ and, consequently, its defining context $ctxt(?n)$ are well-typed.

MV-weak The weakening rule, which is admissible for the base logic, is explicitly added for metavariables.

MV-β-Red This rule simulates the behaviour of β-reduction. To illustrate its effect, we resume Example 2.2 which leads to a type-incorrect term when treated naively.

Assume, then, that the term $((\lambda T : Type. \lambda x : T. ?n) A)$ has to be reduced to normal form, where $A : Type, T : Type, x : T \vdash ?n : T$. Note that the type of this term is $A \to A$. β-reduction yields the term $\lambda x : A. ?n^\frown[T := A]$. The derivation of its type reflects the procedure of β-reduction – with the sole difference that $T : Type$ and $x : T$ are not bound locally by λ-abstraction, but globally in the context:

$$\frac{\dfrac{A : Type \vdash A : Type \quad A : Type, T : Type, x : T \vdash ?n^\frown[\,] : T}{A : Type, x : A \vdash ?n^\frown[T := A] : A} \text{ (MV-β-Red)}}{A : Type \vdash \lambda x : A. ?n^\frown[T := A] : A \to A} \text{ (λ)}$$

The observation suggested by this example – the type of terms is invariant under reduction – is confirmed by Proposition 2.10 below.

At first glance, the rules do not appear to have a form appropriate for type-checking. In particular, the rule (MV-β-Red) seems to require guessing a substitution $\{x := t\}$. It can however be shown that typechecking of judgements $\Gamma \vdash ?n^\frown \sigma : T$ is indeed possible by starting with the judgement $ctxt(?n) \vdash ?n^\frown[] : type(?n)$ and incrementally building up the expected substitution σ by applying rules in the forward direction.

Altogether, the extended calculus preserves the pleasant properties of the base calculus, such as the following:

Proposition 2.9 (Decidability of Type Inference and Type Checking)
- Given a term t and a context Γ, there is an algorithm that determines whether t is well-typed in Γ or not and, in the first case, computes the principal type of t in Γ.
- Given a term t, a type T and a context Γ, there is an algorithm that determines whether $\Gamma \vdash t : T$ holds or not.

Proposition 2.10 (Subject Reduction)
If $\Gamma \vdash M : A$ and $M \twoheadrightarrow N$, then $\Gamma \vdash N : A$.

2.3. *Solutions of metavariables*

Definition 2.11 (Instantiation)
An instantiation[2] ι_P for a valid proof problem $P = (M_P, ctxt_P, type_P)$ is a function mapping the metavariables in M_P to terms, subject to the requirement that for every metavariable $?n \in M_P$:

- if $\iota_P(?n) = t$ and $?m \in MV(t)$, then $\iota_P(?m) = ?m$

The notation for instantiations is comparable to the notation for substitutions: If $M_P = \{?n_1, \ldots ?n_k\}$, then ι_P is written as $\{?n_1 := t_1, \ldots ?n_k := t_k\}$. Whenever P is understood from the context, the subscript is omitted from ι_P. Instantiations are inductively extended to terms as follows:

- $\iota(x) = x$ for variables x.
- $\iota(Prop) = Prop$, $\iota(Type_i) = Type_i$
- $\iota(Qx : T.M) = Qx : \iota(T).\iota(M)$ for $Q \in \{\lambda, \Pi, \Sigma\}$
- $\iota(f\, a) = (\iota(f)\, \iota(a))$

---

[2] The term *instantiation* has been chosen to distinguish instantiation of metavariables from substitution of variables

- $\iota(pair_T(t_1, t_2)) = pair_{\iota(T)}(\iota(t_1), \iota(t_2))$
- $\iota(\pi_i(t)) = \pi_i(\iota(t))$ for $i = 1, 2$
- $\iota(?n^\frown[x_1 := t_1 \ldots x_k := t_k]) = \iota(?n)\{x_1 := \iota(t_1) \ldots x_k := \iota(t_k)\}$

This definition can be extended to contexts in an obvious manner.

Note in particular that bound variables are not renamed: As opposed to substitutions $\{x := s\}$, which must not be applied to terms in which x is bound (see Definition 2.4), variables occurring in the term t of an instantiation $\{?n := t\}$ have to be interpreted relative to the defining context of $?n$. When an instantiation maps over a metavariable, the substitution saved up so far is carried out on the solution of the metavariable.

Definition 2.12 (Instantiation of proof problems)
Assume that $P = (M_P, ctxt_P, type_P)$ is a valid proof problem and ι an instantiation for P. The instantiation $\iota(P) = (M'_P, ctxt'_P, type'_P)$ is defined to be the proof problem consisting of:

- $M'_P := \{?n \in M_P | \iota(?n) = ?n\}$
- $ctxt'_P$ is defined as the function which, for $?n \in M'_P$, yields $\iota(ctxt_P(?n))$
- $type'_P$ is defined as the function which, for $?n \in M'_P$, yields $\iota(type_P(?n))$

Intuitively, if P is a proof problem, then $\iota(P)$ is the proof problem that remains after providing a partial solution ι.

Example 2.13
Assume the proof problem P is given by:

$$A : Type, P : A \to Prop, a : A, h : (P\, a) \vdash ?n_1 : A$$

$$A : Type, P : A \to Prop, a : A, h : (P\, a) \vdash ?n_2 : (P\, ?n_1)$$

and the instantiation ι_P by $\{?n_1 := a\}$. Then $P' := \iota_P(P)$ is the proof problem consisting of:

$$A : Type, P : A \to Prop, a : A, h : (P\, a) \vdash ?n_2 : (P\, a)$$

It can be solved by another instantiation $\iota_{P'} := \{?n_2 := h\}$.

The notion of instantiation is not related to type correctness. This is remedied by the following definition:

Definition 2.14 (Valid / Well-typed Instantiation)
Let $P = (M_P, ctxt_P, type_P)$ bc a valid proof problem.

- An instantiation ι is called *valid* if $\iota(P)$ is a valid proof problem.

- An instantiation ι is called *well-typed* if, for every $?n \in M_P$,

$$\iota(ctxt_P(?n)) \vdash \iota(?n) : \iota(type_P(?n))$$

As a consequence of the following proposition, it is sufficient to verify instantiations "locally", i.e. for metavariables only, to ensure type-correctness "globally" for all terms to which instantiations are applied.

Proposition 2.15 (Instantiation preserves typing)
Let P be a valid proof problem, Γ, t and T be a context resp. terms in which at most metavariables from P occur, and ι a well-typed instantiation for P.

- If $\Gamma \vdash t : T$ holds, then also $\iota(\Gamma) \vdash \iota(t) : \iota(T)$.
- $\iota(P)$ is a well-typed proof problem.

These definitions and propositions provide the foundations of incremental proof construction. They ensure that, whenever metavariables are solved with valid, well-typed instantiations, no type-incorrect terms can result.

2.4. *Functional encoding of scopes*

In the following, a functional representation of scopes will be examined. The general idea is to replace a metavariable $?n$ of type T which depends on assumptions $x_1 : T_1, \ldots, x_k : T_k$ by a metavariable $?F$ which is of functional type $\Pi x_1 : T_1 \ldots \Pi x_k : T_k.T$ and which does not depend on assumptions.

Example 2.16
This procedure can best be illustrated by an example. Consider the following proof problem:

$$A : Type, P : A \rightarrow Prop \vdash ?n_0 : \forall a : A. \exists x : A.(P\,a) \rightarrow (P\,x)$$

This problem can be decomposed into the subproblems:

$$A : Type, P : A \rightarrow Prop, a : A \vdash ?n_1 : A$$

$$A : Type, P : A \rightarrow Prop, a : A \vdash ?n_2 : (P\,a) \rightarrow (P\,?n_1)$$

It can easily be verified that $?n_1 := a$ and $?n_2 := \lambda y : (P\,a).\,y$ are well-typed instantiations. The problem can also be stated with metavariables $?F_1$ and $?F_2$ which are the functional analogues of $?n_1$ and $?n_2$ and which are defined by:

$$\vdash ?F_1 : \Pi A : Type, P : A \rightarrow Prop, a : A.A$$

$$\vdash ?F_2 : \Pi A : Type, P : A \rightarrow Prop, a : A.(P\,a) \rightarrow (P\,(?F_1\,A\,P\,a))$$

The solutions of this proof problem are $?F_1 := \lambda A : Type, P : A \to Prop, a : A.a$ and $?F_2 := \lambda A : Type, P : A \to Prop, a : A.\lambda y : (P\ a).\ y$.

This functional translation can be defined more formally as a mapping taking terms t to terms $\bar{t}$ and proof problems P to proof problems $\bar{P}$. Furthermore, it can be shown that this mapping is type-preserving and that there is a one-to-one correspondence between solutions of P and solutions of $\bar{P}$.

A benefit of a functional representation of scopes is that substitutions cannot take effect in metavariables: $?F\sigma$ will always be the same as $?F$, since all the variables in the domain of σ cannot occur free in a solution of $?F$. Thus, under a functional representation, there is no need to explicitly record substitutions applied to metavariables – altogether, the calculus essentially behaves like a calculus without metavariables. These advantages have to be traded against the difficulties incurred when dealing with metavariables that depend on many local assumptions, such as the metavariables created for theorems in specifications or metavariables created as type-correctness conditions in refinement mappings.

A functional encoding of metavariables is used in many algorithms and systems dealing with proof search in higher-order logic. Only recently have there been attempts (Dowek et al., 1995) to restate unification algorithms for the simply typed λ-calculus in terms of calculi of explicit substitutions. The transformation of metavariables into a functional encoding is closely related to "lifting" as presented by Paulson (1989). Miller (1992) examines methods of exchanging existential and universal quantifiers and develops a similar technique called "raising".

By showing that the reduction relations are preserved under the functional translation, it can be shown that, since the calculus without metavariables is strongly normalizing, so is the calculus with metavariables and explicit substitutions. The following lemma shows how to simulate reductions in the calculus with metavariables by reductions in the calculus with metavariables in functional representation, which we will call "essentially metavariable-free".

Lemma 2.17
If $M \Rightarrow_1 M'$ and $M \neq M'$, then $\overline{M} \Rightarrow_1 \overline{M'}$ and $\overline{M} \neq \overline{M'}$.

Proof. By induction on the generation of $\Rightarrow_1$.

Proposition 2.18 (Strong Normalization)
The calculus with metavariables and explicit substitutions is strongly normalizing.

Proof. Assume, to the contrary, that there is a well-typed term M which permits an infinite sequence of reduction steps: $M \equiv M_0 \Rightarrow_1 M_1 \Rightarrow_1 \ldots$, with $M_i \neq M_{i+1}$. It can be shown that the term $\overline{M}$ is well-typed in the essentially metavariable-free calculus, and by Lemma 2.17, there is an infinite sequence of well-typed terms $\overline{M} \equiv \overline{M_0} \Rightarrow_1 \overline{M_1} \Rightarrow_1 \ldots$, with $\overline{M_i} \neq \overline{M_{i+1}}$. This contradicts the strong normalization property of the essentially metavariable-free calculus.

3. Automated Proof Search

In this section, automation of proof search will be examined. The kind of automation we are aiming at should integrate well with the interactive proof construction presented in Section 2, and it should lead to practically useful procedures, even if completeness has to be sacrificed.

Previous work includes investigations carried out by Pym and Wallen (1991) on proof search in the $\lambda\Pi$ calculus. The proof procedure does not make appeal to typing rules as presented in Section 2.2 to ensure that solutions of metavariables are well-typed in the metavariable-free fragment, but instead uses a complex consistency criterion which is based on permutability of rules. Dowek (1993) develops a complete search procedure for all systems of the λ-cube. For the Calculus of Constructions, this proof procedure has a possibly infinite branching factor which can be avoided when giving up completeness. Even though both approaches (Pym and Wallen, 1991) and (Dowek, 1993) use context-dependent metavariables as we do, the problems related to substituting into metavariables are not addressed by a calculus of explicit substitutions.

Contrary to the procedures defined by Pym, Wallen and Dowek, we have deliberately chosen a formulation which includes the usual logical connectives and is not restricted to the elementary term constructors of the logic, viz. essentially Π-abstraction and its non-dependent version, implication. This makes this calculus amenable to traditional proof search techniques and optimizations.

In Section 3.1, we will briefly review unification. In Section 3.2, a sequent (or tableau) calculus will be developed. Usually, in sequent calculi an eigenvariable condition has to be satisfied. The discussion in Section 3.3 will show that this proviso need not be enforced explicitly, since all derivations with type-correct instantiations in the sense of Section 2.3 respect it. The main emphasis in the following is on conveying an idea of how the apparatus developed so far can be used for automated proof search, and not on providing details of efficient search procedures.

3.1. *Unification*

Let us briefly point out how unification can be defined in our framework so as to produce well-typed instantiations in the sense of Definition 2.14.

Unification of terms t_1, t_2 consists in finding an instantiation for the meta-variables occurring in t_1 or t_2 such that t_1 and t_2 are equal modulo convertibility. In order to be able to develop a correctness criterion, we suppose that the metavariables of t_1 and t_2 stem from a well-typed proof problem P_0. It is furthermore assumed that t_1 and t_2 are well-typed in a context Γ, even though their types need not agree at the outset.

Definition 3.1 (Unification Problem)
A *unification problem* $\langle P_0, \Gamma \vdash t_1 \overset{?}{=} t_2 \rangle$ is a pair consisting of a well-typed proof problem P_0 and a unification equation $\Gamma \vdash t_1 \overset{?}{=} t_2$, where Γ is a valid context, t_1 and t_2 are terms well-typed in Γ and all the metavariables of Γ, t_1, t_2 are among the metavariables of P_0.

The rules defining the unification algorithm will use a judgement of the form $\langle P_0, \Gamma \vdash t_1 \overset{?}{=} t_2 \rangle \Rightarrow P_1; \iota_1$, which expresses that the unification problem $\langle P_0, \Gamma \vdash t_1 \overset{?}{=} t_2 \rangle$ can be solved by instantiation ι_1, leaving open the metavariables of the proof problem P_1. As an example, consider the unification rule for λ-abstraction in Figure 3.

$$\frac{\langle P_0, \Gamma \vdash A_1 \overset{?}{=} A_2 \rangle \Rightarrow P_1; \iota_1 \qquad \langle P_1, \iota_1(\Gamma, x_1 : A_1 \vdash B_1 \overset{?}{=} B_2\{x_2 := x_1\}) \rangle \Rightarrow P_2; \iota_2}{\langle P_0, \Gamma \vdash \lambda x_1 : A_1. B_1 \overset{?}{=} \lambda x_2 : A_2. B_2 \rangle \Rightarrow P_2; \iota_2} \ (\lambda\text{-}\lambda)$$

$$\frac{\begin{array}{c} ctxt(?n) \vdash t : T \\ \langle P_0, \Gamma \vdash type(?n) \overset{?}{=} T \rangle \Rightarrow P_1; \iota_1 \\ valid(\iota_1 \cup \{?n := t\}, P_1) \\ \iota_2 := \iota_1 \cup \{?n := t\} \qquad P_2 := \iota_2(P_1) \end{array}}{\langle P_0, \Gamma \vdash ?n \overset{?}{=} t \rangle \Rightarrow P_2; \iota_2} \ (\text{MV-term})$$

Figure 3.   First-order unification

In order to unify a metavariable $?n$ and an arbitrary term t (rule (MV-term)), the type of $?n$ and the type T of t are first unified, yielding an instantiation ι_1. It still has to be verified that the instantiation $\iota_1 \cup \{?n := t\}$ is valid, in particular that it passes the occurs check implicit in Definition 2.11 and that the resulting proof problem P_2 satisfies the acyclicity condition of Definition 2.8. It can then be concluded that the instantiation $\iota_2 := \iota_1 \cup \{?n := t\}$ is valid and well-typed in the sense of Definition 2.14.

Altogether, the rules satisfy the following invariants, which can be understood as stating the correctness of the unification algorithm.

Proposition 3.2 (Invariants of Unification)
Assume $\langle P_0, \Gamma \vdash t_1 \overset{?}{=} t_2 \rangle$ is a unification problem. If $\langle P_0, \Gamma \vdash t_1 \overset{?}{=} t_2 \rangle \Rightarrow P_1; \iota_1$, then:

 - ι_1 is a valid, well-typed instantiation for P_0.
 - $\iota_1(t_1) \simeq \iota_1(t_2)$
 - P_1 is a valid, well-typed proof problem.

To some extent, the rules can be adapted to incorporate higher-order aspects, for example unification in the simply-typed λ-calculus (see (Huet, 1975; Nipkow, 1993) and also (Pfenning, 1991)). This also entails some modifications of the invariants of Proposition 3.2. Thus, completeness of unification can be achieved for a larger part of the language than the essentially first-order fragment considered above.

An invariant of our unification algorithm is that two terms are only unified after a unifier for their types has been established. In some cases, this prevents simplifications such as leaving equations as constraints that can be solved later on in the course of the proof. This requirement is relaxed in the procedures for the calculus LF of Elliott (1989) and Pym (1990). This, however, leads to considerably more complex algorithms.

3.2. *Tableau-style proof search*

This section develops a tableau calculus appropriate for proof search. For a better understanding, the tableau calculus is first given in a familiar presentation as a set of sequent rules. A formulation as proof transformation system establishes the connection with the terminology of the preceding sections. This section only considers the quantifier-free fragment, whereas Section 3.3 discusses problems related to quantifiers. For lack of space, only some of the rules are presented.

Although the core language of TYPELAB does not include the usual logical connectives and the existential quantifier as term constructors, these can easily be defined by an encoding which dates back to (Prawitz, 1965). For example, $A \wedge B$ can be defined as $\Pi R : Prop.(A \to B \to R) \to R$. It can be shown that this encoding corresponds to the usual definition of the connectives by means of natural deduction rules. Thus, for the above definition of conjunction, the following rules are derivable (with appropriately defined $andI$, $andEl$ and $andEr$):

$$\frac{\Gamma \vdash a : A \quad \Gamma \vdash b : B}{\Gamma \vdash (andI \; a \; b) : A \wedge B} \; (\wedge I)$$

$$\frac{\Gamma \vdash p : A \wedge B}{\Gamma \vdash (andEl \; p) : A} \; (\wedge El) \qquad \frac{\Gamma \vdash p : A \wedge B}{\Gamma \vdash (andEr \; p) : B} \; (\wedge Er)$$

For the other connectives, similar rules are derivable.

Natural deduction rules can be divided into introduction rules (e.g. $(\wedge I)$) which say how a connective can be constructed, and elimination rules (e.g. $(\wedge El)$ and $(\wedge Er)$) which say how a connective can be decomposed. A well-known disadvantage of natural deduction is the lack of a subformula property in its elimination rules. Thus, for performing "backwards" proof search, applying rules from the conclusion towards the premises, formulae have to be guessed, such as formula B in $(\wedge El)$ and formula A in $(\wedge Er)$.

Sequent calculi as defined by Gentzen (1934) replace the schema of introduction-elimination rules by a schema of Left-Right rules in which connectives are decomposed on the left resp. right side of a sequent. In our case, the interpretation of $\Gamma \vdash ?n : T$ as a sequent is obvious. The Right rules are identical to the introduction rules of the natural deduction calculus. An application of a Left rule to a formula F, or more precisely, to a context entry $x : F$ in Γ does not lead to a removal of this context entry from Γ, since x might appear elsewhere in Γ. Rather, new context entries are added at the end of Γ. This twist of representation makes it more appropriate to call our calculus a tableau calculus.

The rules for conjunction are displayed in Figure 4. Associated with each rule is a term (cf. Figure 5) which spells out how a solution of the metavariable $?n_0$ in the conclusion of the rule can be obtained from solutions of the metavariables ($?n_1$ and $?n_2$ resp. $?n_1$) of the premises. The correctness of $(\wedge R)$ and $(\wedge L)$, when interpreted as derived rules, can immediately be verified by typechecking their solution terms.

More formally, tableau-style proof search can be related to the notions developed in Section 2 by presenting the rules in the form of a proof transformation system. A transformation rule $P_0; \iota_0 \Longrightarrow P_1; \iota_1$ (possibly containing side conditions) expresses that proof problem P_0 and instantiation ι_0 can be transformed to proof problem P_1 and instantiation ι_1 when applying a proof

$$\frac{\Gamma \vdash ?n_1 : A \quad \Gamma \vdash ?n_2 : B}{\Gamma \vdash ?n_0 : A \wedge B} \quad (\wedge R)$$

$$\frac{\Gamma, p : A \wedge B, \Gamma', p_A : A, p_B : B \vdash ?n_1 : G}{\Gamma, p : A \wedge B, \Gamma' \vdash ?n_0 : G} \quad (\wedge L)$$

Figure 4. Tableau rules for conjunction

$(\wedge R)$ $?n_0 := (andI \ ?n_1 \ ?n_2)$
$(\wedge L)$ $?n_0 := (\lambda p_A : A, p_B : B . ?n_1) \ (andEl \ p) \ (andEr \ p)$

Figure 5. Solutions associated with the proof rules

rule backwards. The transformation rules corresponding to the proof rules of Figure 4 are displayed in Figure 6[3].

$P \cup \{\Gamma \vdash ?n_0 : A \wedge B\}; \iota_0$
$\iota_1 := \iota_0 \uplus \{?n_0 := (andI \ ?n_1 \ ?n_2)\}$
$\implies \iota_1(P \cup \{\Gamma \vdash ?n_1 : A, \ \Gamma \vdash ?n_2 : B\}); \iota_1$

$P \cup \{\Gamma, p : A \wedge B, \Gamma' \vdash ?n_0 : G\}; \iota_0$
$\iota_1 := \iota_0 \uplus \{?n_0 := (\lambda p_A : A, p_B : B . ?n_1) \ (andEl \ p) \ (andEr \ p)\}$
$\implies \iota_1(P \cup \{\Gamma, p : A \wedge B, \Gamma', p_A : A, p_B : B \vdash ?n_1 : G\}); \iota_1$

Figure 6. Rules of Transformation System

We can state invariants of the proof transformation system, similar to the invariants for unification (Proposition 3.2) but technically more involved. As a special case, the following proposition expresses that an automated proof search procedure which is started on a goal $\Gamma \vdash ?n_0 : G$ and succeeds in solving it completely, constructs a well-typed proof term t for $?n_0$.

Proposition 3.3 (Correctness of Proof Search)

If $\{\Gamma \vdash ?n_0 : G\}; \{\} \implies \{\}; \{?n_0 := t\}$, then $\Gamma \vdash t : G$.

---

[3] The combination $\iota \uplus \kappa$ of two instantiations ι, κ performs the necessary instantiations of κ within ι which ensure the idempotency of the resulting instantiation.

$$\frac{\Gamma \vdash ?n_1 : A \qquad \Gamma \vdash ?n_2 : P(?n1)}{\Gamma \vdash ?n_0 : \exists x : A.P(x)} \ (\exists R) \qquad \frac{\Gamma, x : A \vdash ?n_1 : B(x)}{\Gamma \vdash ?n_0 : \Pi x : A.\ B(x)} \ (\Pi R)$$

$$\frac{\Gamma, p : \exists x : T.P(x), \Gamma', y : T, p' : P(y) \vdash ?n_1 : G}{\Gamma, p : \exists x : T.P(x), \Gamma' \vdash ?n_0 : G} \ (\exists L)$$

$$\frac{\Gamma, p : \Pi x : A.\ B(x), \Gamma' \vdash ?n_1 : A}{\Gamma, p : \Pi x : A.\ B(x), \Gamma', p' : B(?n_1) \vdash ?n_2 : G}{\Gamma, p : \Pi x : A.\ B(x), \Gamma' \vdash ?n_0 : G} \ (\Pi L)$$

Figure 7. Tableau rules for quantifiers

So far, we have neglected questions regarding completeness. The introductory remarks of Section 3 have given some plausibility to the claim that procedures which are complete for the whole logic become too unwieldy to be practically useful. The identification of suitable fragments which lend themselves to complete proof methods is a topic of current research.

3.3. *Quantifiers in sequent rules*

In traditional presentations of sequent calculi, the rules $(\forall R)$ and $(\exists L)$ are usually stated with an "eigenvariable condition". Typically, the rule $(\forall R)$ is then given by:

$$\frac{\Gamma \vdash B(z)}{\Gamma \vdash \forall x.\ B(x)} \ (\forall R)$$

under the proviso that the fresh variable z does not occur in Γ. In this section, we will examine how the eigenvariable condition is enforced in our calculus. The quantifier rules are displayed in Figure 7.

Since the argument leading to Proposition 3.3 can be used to show the correctness of the associated proof transformation system, this section does not provide an additional proof, but rather an illustration for one aspect of correctness. Since most of the considerations are independent of typing, we will omit type information whenever convenient.

Example 3.4

As an illustration, we will use the following formulas throughout this section, namely $\forall x.\exists y.x = y$, which is valid, and $\exists x.\forall y.x = y$, which is not.

In traditional tableau calculi, Skolemization is used to eliminate a universal quantifier by keeping track of existential quantifiers on which it depends.

Skolemization[4] expresses that the formula $\exists x.\forall y.P(x,y)$ is valid iff $\exists x.P(x, f(x))$ is valid, where f is a fresh function constant. There are two impediments to using Skolemization in our framework. The first is that it is hard to justify the introduction of a new function constant f proof-theoretically. When considering, in a typed calculus, the transition from $\exists x : A.\forall y : B. P(x,y)$ to $\exists x : A.P(x,f(x))$, we make a claim as to the existence of a function $f : A \to B$, for which we have no direct evidence. The second reason for not using Skolemization, of a more practical nature, is that it can blow up formulae and make them difficult to understand.

The method that is implicit in our approach is to describe the dependence of existential variables on universal variables. A proof obligation $x_1 : T_1, \dots, x_k : T_k \vdash ?n : T$ expresses that the existential variable $?n$ occurs in the scope of the universal variables $x_1, \dots, x_k$ and can only be solved by terms in which at most $x_1, \dots, x_k$ occur free. The examples demonstrate the procedure: In the first example, the proof succeeds because $?y$ can be unified with x.

$$\frac{\dfrac{x : T \vdash ?y : T \qquad x : T \vdash ?n_2 : x = ?y}{x : T \vdash ?n_1 : \exists y : T.x = y} \ (\exists R)}{\vdash ?n_0 : \forall x : T.\exists y : T.x = y} \ (\forall R)$$

In the second example, however, $?x$ does not unify with y because y does not occur in the context of $?x$.

$$\frac{\vdash ?x : T \qquad \dfrac{y : T \vdash ?n_2 : ?x = y}{\vdash ?n_1 : \forall y : T.?x = y} \ (\forall R)}{\vdash ?n_0 : \exists x : T.\forall y : T.x = y} \ (\exists R)$$

Again, this dependence could be made explicit by a functional encoding of scopes, as for example in the ISABELLE system (Paulson, 1994). The observations of the above example can be generalized to the following proposition, which expresses that no well-typed instantiation of metavariables can violate the eigenvariable condition:

Proposition 3.5 (Eigenvariable Condition)
Assume $\Gamma, x : A$ is a valid context with occurrences of metavariables $?m_1, \dots, ?m_k$ in Γ. Then there is no well-typed instantiation ι of $?m_1, \dots, ?m_k$ such that x occurs free in $\iota(\Gamma)$

---

[4] The dual variant of Skolemization in which existential quantifiers are eliminated and which preserves satisfiability is commonly found in refutational theorem proving.

Proof. Since $\Gamma, x : A$ is a valid context and ι a valid instantiation, by Proposition 2.15, $\iota(\Gamma)$ is a valid context and thus cannot contain a free occurrence of x.

Surprisingly enough, our approach to avoid Skolemization in proof search is reminiscent of a method developed by Bibel (1982) for first order logic, even though the path taken to arrive at this solution is quite different. It is worth noting that our approach gives a rather direct criterion to verify that eigenvariable conditions are respected, as opposed to indirect criteria implicit in the methods of Pym and Wallen (1991) and Shankar (1992) which encode permutabilities of rule applications in their proof search procedures. Even though rule permutabilities are not used to ensure correctness, they can be exploited to optimize proof search. In particular, they can help to recognize when goals cannot be satisfied even by application of alternative proof rules, thus avoiding useless backtracking. Details are a topic of current research.

4. CONCLUSIONS

This chapter has given a survey of the theory underlying the prover component of the TYPELAB system. Metavariables play a central role in that they permit to incrementally construct entities which are relevant in the software development and verification process, such as specifications, functions and proofs. Metavariables can occur nested deeply inside terms, depending on a great number of local assumptions. Therefore, common techniques such as a functional encoding of scopes are not appropriate here. Some problems arising from a naive use of metavariables have been identified, and a calculus with explicit substitutions has been presented to solve them (Section 2). This calculus has desirable properties such as confluence and strong normalization, and it provides a foundation for an automation of proof search (Section 3) in that it directly ensures some conditions such as the eigenvariable proviso for quantifier rules.

Other systems based on a type theory are NUPRL (Constable et al., 1986), COQ (Barras et al., 1997), LEGO (Pollack, 1994) and ALF (Magnusson and Nordström, 1994). NUPRL and COQ provide powerful automation for fragments of the logic and permit to extract programs from proofs (Paulin-Mohring and Werner, 1993), but do not allow for a direct construction of objects with the aid of metavariables. In LEGO and ALF, proof construction essentially consists in finding appropriate instantiations for metavariables, further automation is not available. TYPELAB aims at a synthesis of these approaches.

REFERENCES

Barendregt, H.: 1984, *The Lambda Calculus.* Elsevier Science Publishers.

Barras et al., B.: 1997, 'The Coq Proof Assistant Reference Manual, Version 6.1'. INRIA Rocquencourt – CNRS - ENS Lyon.

Bibel, W.: 1982, 'Computationally Improved Versions of Herbrand's Theorem'. In: J. Stern (ed.): *Proc. Herbrand Colloquium.* pp. 11–28, North Holland.

Constable et al., R.: 1986, *Implementing Mathematics with the Nuprl Proof Development System.* Prentice-Hall.

Dowek, G.: 1993, 'A complete proof synthesis method for the cube of type systems'. *Journal of Logic and Computation* **3**(3), 287–315.

Dowek, G., T. Hardin, and C. Kirchner: 1995, 'Higher-order unification via explicit substitutions'. In: D. Kozen (ed.): *Proceedings LICS'95.* pp. 366–374. extended abstract.

Elliott, C. M.: 1989, 'Higher-order unification with dependent function types'. In: N. Dershowitz (ed.): *Proc. 3rd Intl. Conf. on Rewriting Techniques and Applications.* pp. 121–136, Springer LNCS 355.

Gentzen, G.: 1934, 'Untersuchungen über das logische Schließen'. *Mathematische Zeitschrift* **39**, 176–210 and 405–431.

Huet, G.: 1975, 'A unification algorithm for typed Lambda-calculus'. *Theoretical Computer Science* pp. 27–57.

Lescanne, P.: 1994, 'From $\lambda\sigma$ to $\lambda\nu$ a journey through calculi of explicit substitutions'. In: *Proc. POPL'94.* pp. 60–69.

Luo, Z.: 1994, *Computation and Reasoning.* Oxford University Press.

Luther, M. and M. Strecker: 1998, 'A guided tour through TYPELAB'. Technical Report 98-03, Universität Ulm.

Magnusson, L.: 1995, 'The Implementation of ALF - a Proof Editor based on Martin-Löf's Monomorphic Type Theory with Explicit Substitution'. Ph.D. thesis, Chalmers University of Technology.

Magnusson, L. and B. Nordström: 1994, 'The ALF proof editor and its proof engine'. In: H. Barendregt and T. Nipkow (eds.): *Types for Proofs and Programs.* pp. 213–237.

Melliès, P.-A.: 1995, 'Typed λ-calculi with explicit substitutions may not terminate'. In: *Typed Lambda Calculi and Applications.* Springer LNCS 902.

Miller, D.: 1992, 'Unification under a mixed prefix'. *Journal of Symbolic Computation* **14**, 321–358.

Muñoz, C.: 1997, 'Un calcul de substitutions pour la représentation de preuves partielles en théorie de types'. Thèse de doctorat, Université Paris 7.

Nipkow, T.: 1993, 'Functional Unification of Higher-Order Patterns'. In: *Proc. 8th IEEE Symp. Logic in Computer Science.* pp. 64–74.

Paulin-Mohring, C. and B. Werner: 1993, 'Synthesis of ML programs in the system Coq'. *J. Symbolic Computation* **11**, 1–34.

Paulson, L.: 1989, 'The Foundation of a Generic Theorem Prover'. *Journal of Automated Reasoning* **5**, 363–397.

Paulson, L.: 1994, *Isabelle - a generic theorem prover*. Springer LNCS 828.

Pfenning, F.: 1991, 'Unification and anti-unification in the Calculus of Constructions'. In: *Sixth Annual Symposium on Logic in Computer Science*. pp. 74–85,
IEEE Computer Society Press.

Pollack, R.: 1994, 'The Theory of LEGO – A proof checker for the Extended Calculus of Constructions'. Ph.D. thesis, University of Edinburgh.

Prawitz, D.: 1965, *Natural Deduction – A proof-theoretic study*. Almqvist & Wik-
sells.

Pym, D.: 1990, 'Proofs, Search and Computation in General Logic'. Ph.D. thesis,
University of Edinburgh.

Pym, D. and L. Wallen: 1991, 'Proof-search in the $\lambda\Pi$-calculus'. In: G. Huet and G.
Plotkin (eds.): *Logical Frameworks*. Cambridge University Press, pp. 311–340.

Shankar, N.: 1992, 'Proof Search in the Intuitionistic Sequent Calculus'. In: D. Kapur
(ed.): *Proc. CADE-11*. Springer LNCS 607.

CHAPTER 4

INTEGRATING AUTOMATED AND INTERACTIVE
THEOREM PROVING

1. INTRODUCTION

Automated and *interactive* theorem proving are the two main directions in the field of deduction. Most chapters of this book belong to either the one or the other, whether focusing on theory, on methods or on systems. This reflects the fact that, for a long time, research in computer-aided reasoning was divided into these two directions, driven forward by different communities. Both groups offer powerful tools for different kinds of tasks, with different solutions, leading to different performance and application profiles. Some important examples are: ACL2 (Kaufmann and Moore, 1988), HOL (Gordon, 1988), IMPS (Farmer et al., 1996), Isabelle (Paulson, 1994), KIV (Reif et al., 1997) (see also Chapter II.1.1), NQTHM (Boyer and Moore, 1979), and PVS (Owre et al., 1992) for the interactive (or tactical) theorem proving community; and KoMeT (Bibel et al., 1994), Otter (Wos et al., 1992), Protein (Baumgartner and Furbach, 1994), Setheo (Goller et al., 1994), Spass (Weidenbach et al., 1996), and $_3TAP$ (Beckert et al., 1996) for the automated theorem proving community.

In this chapter we present a project to *integrate interactive and automated theorem proving*. Its aim is to combine the advantages of the two paradigms. We focus on one particular application domain, which is deduction for the purpose of software verification. Some of the reported facts may not be valid in other domains. We report on the integration concepts and on the experimental results with a prototype implementation.

Automatic provers are very fast for the majority of the problems they can solve at all. With increasing complexity, response time increases dramatically. Beyond a certain problem size, automated theorem provers produce reasonable results only in very exceptional cases. Interactive theorem provers on the other hand can be used even in very large case studies. For small problems, they do, however, require many user interactions, particularly when combinatorial exhaustive search has to be performed.

Concerning software verification and the typical proof tasks arising there, the gap between both methods (if applied naively) is even more dramatic. There are essentially two reasons for that phenomenon. First, the theories

W. Bibel, P. H. Schmitt (eds.), Automated Deduction. A basis for applications. Vol. II
© 1998 *Kluwer Academic Publishers. Printed in the Netherlands*

occurring in verification projects are very large (hundreds of axioms). Second, the majority of these axioms use equality. Such theories are not well handled by automatic provers. We present techniques to relieve these problems.

We investigate a conceptual integration of interactive and automated theorem proving for software verification that goes beyond a loose coupling of two proof systems. Our concrete application domain turned out to have an enormous influence on the integration concepts. We have implemented a prototype system combining the advantages of both paradigms. In large applications, the integrated system incorporates the proof engineering capabilities of an interactive system and, at the same time, eliminates user interactions for those goals that can be solved by the efficient combinatorial proof search embodied in an automated prover. We report on the integration concept, on the encountered problems, and on experimental results with the prototype implementation. Furthermore, the current directions of our ongoing research are described.

The technical basis for the integration are the systems KIV (Reif, 1995) and $_3TAP$ (Beckert et al., 1996), both of which were developed in the research groups of the authors at Ulm and Karlsruhe. KIV (Karlsruhe Interactive Verifier) is an advanced verification system which has been applied in large realistic case studies in academia and industry for many years now. $_3TAP$ (Three-valued Tableau-based Automated Theorem Prover) is an automated tableau prover for full first-order logic with equality. It does not require normal forms, and it is easily extensible. Although we experimented with these particular systems, the conceptual results carry over to other provers.

Based on statistics from case studies in KIV, we estimate that in our application domain up to 30% of all user interactions needed by an interactive prover could be saved in principle by a first-order theorem prover. Current provers, however, are far from this goal, because they are in general not prepared for deduction in large software specifications (i.e., very large search spaces) or for typical domain specific reasoning. In Section 2 we describe these and other problems, and in the Sections 3 to 6 we present the solutions we came up with so far.

Many of our decisions are based on experimental evidence. Therefore, we put a lot of effort in a sophisticated verification case study: Correct compilation of Prolog programs into Warren Abstract Machine code ((Schellhorn and Ahrendt, 1997) and Chapter III.2.7). We use it as a reference or benchmark. Parts of it are repeated every now and then to evaluate the success of our integration concepts, see Section 7.

In realistic applications in software verification, proof attempts are more likely to fail than to go through. This is because specifications, programs, or user-defined lemmas typically are erroneous. Correct versions usually are

only obtained after a number of corrections and failed proof attempts. Therefore, the question is not only how to produce powerful theorem provers but
also how to integrate proving and error correction. Current research on this
and related topics is discussed in Section 8.

There are different approaches of combining interactive methods with automated ones. Their relation to our approach is the subject of Section 9. Finally, in Section 10 we draw conclusions.

2. IDENTIFYING THE PROBLEMS OF INTEGRATION

Theorem proving with an interactive system typically proceeds by simplifying goals using backward reasoning with proof rules (tactics). Many proof
rules may be applied automatically, but usually the tactics corresponding to
the main line of argument in the proof must be supplied interactively. To allow the proof engineer to keep track of the details of a proof, system response
time to the application of tactics should be short.

In the case of software verification, the initial goals contain programs. The
tactics to reduce these goals make use of first-order lemmas and ultimately
reduce the goal to a first-order formula. Usually, these first-order goals require
interaction as well as the program goals. Using an (ideal) automated theorem
prover would relieve the proof engineer from a lot of interaction. Therefore,
the scenario we considered was to use $_3TAP$ as a *tactic* to prove first-order
goals, thus exploiting its capability of fast combinatorial search. A suitable
interface was implemented such that $_3TAP$ can be called either interactively
or by KIV's heuristics. Termination of this tactic was guaranteed simply by
imposing a time limit (usually between 15 seconds and 1 minute).

Based on this first, loosely integrated version, we started to experiment
with using the automatic theorem prover to solve first-order theorems encountered in software verification. As expected, the automatic theorem prover
initially could not meet the requirements found in software verification. Virtually no relevant theorem could be proved. Analysis of the proof attempts
identified a number of reasons, some expected and some unexpected. The
most important ones are:

Interface Automated theorem provers usually do not support separate input
of axioms and goal. Instead, one is forced to prove the combined goal
"axioms imply theorem" with universally quantified axioms and theorem. Most theorem provers do some preprocessing on formulas to speed
up proof search. In $_3TAP$, links between potentially unifiable terms are
computed, whereas in many other theorem provers formulas are converted to clauses. We found that preprocessing 200 axioms with $_3TAP$

takes about 30 sec. (the same holds for other provers we tested, see Chapter III.2.9). Preprocessing the axioms at every proof attempt is obviously unacceptable for interactive theorem proving.

Correctness Management Automated provers do not record which assumptions were actually needed in a proof. But such information is necessary if the interactive theorem prover does not rely on strict bottom-up proving. In KIV, for example, the correctness management prevents cycles in the dependencies of lemmas and invalidates only a minimal set of previous work when goals or specifications are changed.

Different Logics Most automated theorem provers only support formulas of first-order logic without sorts (e.g. Otter) or even only clauses (e.g. Setheo) while interactive theorem provers often support more expressive logics like higher-order logic (e.g. HOL, Isabelle or PVS) or type theory (e.g. NuPRL (Constable et al., 1986) or Typelab, see Chapter II.1.3).

Inductive Theorems Many theorems can only be proved inductively from the axioms, but most automated theorem provers (including $_3T^AP$) are not capable of finding inductive proofs.

Large Theories Automated provers are tuned to prove theorems over small theories and a small signature. Moreover, the given axioms are always relevant to prove the goal. In contrast, theories used in software verification usually contain hundreds of axioms of which only few are relevant for finding a particular proof. Still, all axioms contribute to the search space.

Domain Characteristics In our application domain, which is software verification, specifications are well structured and have specific properties (e.g. sorted theories, equality reasoning is important). Also, theorems used as lemmas often have an operational meaning in the interactive theorem prover (e.g. equations oriented as rewrite rules). Automated theorem provers do not exploit these properties.

The first two items above are technical problems requiring mere changes to the interface. Now, preprocessing of axioms is done by $_3T^AP$ when the user of the integrated system selects a specification to work on. A separate command for initiating proof attempts refers to the preprocessed axioms by naming the specification in which they occur. Embedding the automated prover in the correctness management is done by converting proofs found by $_3T^AP$ to proofs in the sequent calculus used in KIV.

To handle the problem of different logics, a suitable subset of the formulas treated by the interactive system must be selected and, in general, translated to first-order logic. For those automatic provers which only support clauses, even the first-order formulas have to be transformed by standard encoding techniques like the ones described in (Plaisted and Greenbaum, 1986).

In our case, the solution is simpler, since the logic used in KIV is actually an extension of many-sorted first-order logic by program formulas (Dynamic Logic), and $_3T\!AP$ supports full many-sorted first-order logic. For a discussion of some of the issues that have to be solved for higher-order logic (polymorphism, currying, problems arising from the identification of boolean terms with formulas) see (Archer et al., 1993).

The presence of inductive theorems is a fundamental problem. It can be mitigated by adding previously derived (inductive) theorems as lemmas. On the one hand, this reduces the number of theorems which require inductive proofs in our experiments to about 10% of all theorems (see Chapter III.2.9). On the other hand, there are roughly as many potentially useful or necessary lemmas as there are axioms, which adds considerably to the problem of large theories.

This leaves us with two problems, namely handling large theories and exploiting domain characteristics (of software verification). Both will be tackled in the following sections. The number of potentially relevant axioms in proving a goal is minimized by exploiting a specification's structure (Section 3). Measures we have taken to exploit the characteristics of software verification are presented next. Finally, Section 6 deals with the conversion of proofs from $_3T\!AP$ to KIV.

3. REDUCTION OF THE AXIOM SET

Specifications in KIV are built up from elementary first-order theories with the usual operations of algebraic specification: union, enrichment, parameterization, and actualization. Their semantics is the whole class of models (loose semantics). Reachability constraints like "nat generated by 0, succ" are used to define induction principles. Typical specifications used to formally describe software systems contain several hundred axioms.

Structuring operations are not used arbitrarily in formal specifications of software systems. Almost all enrichments "**enrich** SPEC **by** Δ" are *hierarchy persistent*. This property means that every model of SPEC can be extended to a model of the whole (enriched) specification. Hierarchy persistency cannot be checked syntactically, but is usually guaranteed by a modular implementation of the specification (Reif, 1995).

Hierarchy persistency can be exploited to define simple, syntactic criteria for eliminating many irrelevant axioms, which then no longer must be
passed to the automated theorem prover. This enables theorem proving in
large theories, and is described in seperately in Chapter (III.2.9). There, the
reduction technique used to identify the axioms in question is described in
greater detail. It is sufficient, for example, to work merely with the axioms of
the minimal specification whose signature contains that of the theorem.

4. EQUALITY HANDLING

4.1. *Incremental Equality Reasoning*

KIV specifications—like most real word problems—make heavy use of equality. It is, therefore, essential for an automated deduction system that is integrated with an interactive prover to employ efficient equality reasoning techniques.

Part of $_3TAP$ is a special equality background reasoner that uses a completion-based method (Beckert, 1994) for solving E-unification problems extracted from tableau branches. This equality reasoner is much more efficient
than just including the equality axioms. In addition to the mere efficiency
of the tableau-based foreground reasoner and that of the equality reasoner,
the interaction between them plays a critical role for their combined efficiency. It is a difficult problem to decide when it is useful to call the equality
reasoner and how much time to allow for its computations. Even with good
heuristics at hand, one cannot avoid calling the equality reasoner either too
early or too late. This problem is aggravated in the framework of integration
by the fact that most equalities present on a branch are actually not needed
to close it, such that computing a completion of all available equalities not
only is expensive, but quite useless. These difficulties can (at least partially)
be avoided by using incremental methods for equality reasoning (Beckert
and Pape, 1996). These are algorithms that—after a futile try to solve an
E-unification problem—allow to store the results of the equality reasoner's
computations and to reuse them for a later call (with additional equalities).
Then, in case of doubt, the equality reasoner can be called early without running the risk of doing useless computations. In addition, an incremental reasoner can reuse data for different extensions of a set of equalities.

Fortunately, due to the inherently incremental nature of $_3TAP$'s algorithm
for solving E-unification problems, it was possible to design and implement
an incremental version of it: rewrite rules and unification problems that are
extracted from new literals on a branch can simply be added to the data of

the background reasoner. Previously, information computed by the equality reasoner of $_3TAP$ could not be reused, and the background reasoner was either called (a) on exhausted tableau branches (i.e., no expansion rule is applicable; this meant that because of redundant equalities even simple theorems could not be proved) or (b) it was called each time before a branching rule was applied (which usually lead to early calls and repeated computation of the same information). Now $_3TAP$ avoids both problems. The incremental equality reasoner may be called each time before a disjunctive rule is applied without risking useless computations.

4.2. *Generating Precedence Orders from Simplifier Rules*

As in most interactive theorem provers, in KIV simplification is a key issue. *Simplifier rules* are theorems over a data structure, which have been marked by the proof engineer for use in the simplifier. The way of use is determined by their syntactic shape. The most common simplifier rules are conditional rewrite rules of the form $\phi \rightarrow \sigma = \tau$, intended for rewriting instances of σ to instances of τ if ϕ is provable.

The fact that simplifier rules have an operational meaning within an interactive prover can be used in several ways to guide proof search in automated systems. One example is the automatic generation of useful precedence orders of function symbols (otherwise, an order must be provided manually or an arbitrary default order is used). Such orders are used in many refinements of calculi in theorem proving, hence our results are of general interest.

In $_3TAP$, the precedence order is used to orient equations with a lexicographic path order based on it. The more equations occurring during a proof can be ordered, the smaller the search space becomes.

A first attempt to generate an order from the rewrite rules of KIV is to define $f > g$ for top-most function symbols f and g that occur on the left and on the right side of a rewrite rule, respectively, provided $f \neq g$. The attempt to generate a total order from this information (by a topological sort), however, typically fails due to a number of conflicts such as the following (*cons, car, cdr* and *append* are the usual list operations):

append(cons(a,x),y) = cons(a,append(x,y))
x $\neq$ nil $\rightarrow$ cons(car(x),append(cdr(x),y)) = append(x,y)

The first rule suggests, in accordance with intuition, that *append* > *cons*, while the second one suggests the contrary. To avoid such conflicts, one excludes rewrite rules of the form $\phi \rightarrow \sigma = \tau$, where τ can be embedded into σ (as is the case in the second rule above) from the order generation. We tested the resulting algorithm with five specifications from existing case studies in

KIV, each having approximately 100 rewrite rules. The result was that the function symbols could *always* be topologically sorted, except one conflicting pair of rewrite rules. Additional restrictions on the order could be extracted by considering the first *differing* function symbols in rewrite rules instead of the top-most ones (no additional conflicts arose). For maximal flexibility KIV passes on only a partial order (if a conflict is found, no information on this function symbol is generated). The partial order is made total and used in the equality handling module of $_3TAP$. For yet another use, see the following section.

Our considerations show that rewrite rules, which are used in interactive theorem proving to "simplify" terms (relative to the user's intuition), can be translated rather directly into information used in automated theorem proving. Experiments showed that with suitable simplifier rules and a similar algorithm as the one above, one obtains an analogous result for predicate symbols (provided that the equality symbol is considered to be special).

5. RESTRICTING THE SEARCH SPACE

5.1. *Problem-Specific Orders*

Calculi which incorporate search space restrictions based on atom and literal orders are relatively well investigated in the domain of resolution theory (Fermüller et al., 1993). In order to employ such restrictions in $_3TAP$, we could build on recent work on order-based refinements for tableau calculi (Hähnle and Klingenbeck, 1996; Hähnle and Pape, 1997). In fact this latter research was partially motivated by the integration of paradigms discussed in the present article.

Ordered tableaux constitute a refinement of free variable tableaux (Fitting, 1990). They have a number of advantages that become particularly important in the context of software verification: They are defined for *unrestricted first-order formulas* (Hähnle and Klingenbeck, 1996) (in contrast to mere clausal normal form) and they are compatible with another important refinement called *regularity* (Letz et al., 1992). It is possible to extend ordering restrictions to theory reasoning (Hähnle and Pape, 1997). Moreover, ordered tableaux are *proof confluent*: every partial tableau proof can be extended to a complete proof provided the theorem to be proven is valid. This property is an essential prerequisite for automated search for *counter examples*, see Section 8 below. Finally, *problem-specific knowledge* can be used to choose an order, which not only restricts the search space but, more importantly, rearranges it favorably.

The last point is difficult to exploit in general, but in the KIV-$_3T^AP$ integration one can take advantage of the same information computed already to reduce the number of axioms (Section 3) and to provide meaningful simplification orders for equality handling (Section 4.2). The hierarchy of specifications and the implicit hierarchy of function symbols within each specification defines a partial order $<$ of the function and predicate symbols occurring in a problem. Such an order $<$ can be naturally extended to an *A-order* $\prec$: a binary, irreflexive, transitive relation on atoms which is stable under substitutions. Ordered tableaux can be characterized by restricting branch extension as follows: a formula ϕ can be used to extend a tableau branch B iff either (i) ϕ has an $\prec$-maximal connection into B (i.e., the connection literal of ϕ occurs $\prec$-maximally in ϕ) or (ii) ϕ has an $\prec$-maximal connection into another input formula ψ (i.e., the connection literals of both ϕ and ψ occur $\prec$-maximally).

One particular *A-order*, based on $<$, is $\stackrel{<}{\sim}$. For two terms/atoms $s = f(s_1, \ldots, s_n)$, $t = g(t_1, \ldots, t_m)$ is $s \stackrel{<}{\sim} t$ iff (i) either $f < g$ or $f = g$, $n = m$, and $t_i \stackrel{<}{\sim} s_i$ for $1 \leq i \leq n$ *and* (ii) the variables of s are a subset of the variables of t. The *automatically* computed $\stackrel{<}{\sim}$ often prevents literals from unrelated hierarchies to be maximal and, as a consequence, a formula ϕ has no maximal connection into a branch with only literals from a hierarchy unreachable from the maximal literals of ϕ. Even when $\stackrel{<}{\sim}$ does not perfectly reflect the hierarchy within a problem, completeness of the calculus still guarantees that a proof can be found, although proof search might not be influenced as favorably.

5.2. *Pragmatic Connectives*

As explained above, the automated part of the integrated system is used to prove those sub-tasks that are of first-order nature. In our context of software verification, first-order formulas appear in very different situations: as axioms of a specified theory, as rewrite lemmas, or as subgoals in a proof of a theorem involving programs. Each of these contexts carries its own pragmatics concerning the way formulas are used in proofs. So it is a basic task to enable the automated prover to make use of as much pragmatic information as possible. One solution is to add new logical connectives to the logic of the automated (tableau) prover.

Expansion of a disjunctive formula causes a splitting of the current branch, after which one of the two resulting branches has to be chosen as the new branch in focus. This choice heavily affects the search space. Consider, for example, the formula $p(x) \lor q(x)$: its expansion generates two (sub-)branches, say B_1 and B_2, with leaves $p(x)$ and $q(x)$, respectively. Suppose there are many different instantiations of x that allow to close B_1, but only few instantiations that allow to close B_2. In this case, the search for an instantiation that closes *both* branches should be done by first searching for an instantiation that allows to close B_2 and then checking whether it allows to close B_1 as well; obviously, a much smaller search space is spanned this way.

In general, one has no information that allows to decide which branch should be closed first, so the disjunctive connectives ($\vee$, $\rightarrow$ and $\leftrightarrow$) are treated in a standard way; by default, the left argument is handled first (some theorem provers take the relative size of p and q into account, which may or may not be beneficial). On the other hand, specific knowledge on the role of p or q in the proof would allow to rearrange and optimize the search space.

Typical candidates for this are implications that are intended to *exclude exceptions*. Consider the formula $n \neq 0 \rightarrow property(n)$. It states a property holding for all natural numbers *except* 0. Assume this formula occurs on the branch in focus and $property(n)$ is a complex formula. Expanding the implication, standard treatment puts the new branch B_1 with leaf $n = 0$ in focus. As each of the substitutions $\{n \leftarrow 1\}, \{n \leftarrow 2\}, \ldots$ closes B_1, the natural numbers are enumerated by backtracking. It is much better to examine the branch B_2 containing $property(n)$ first, and then check that the instantiation of n used to close B_2 is not equal to 0.

For this we added a version of implication to the $_3T\!AP$ logic, called **if_then**, whose declarative semantics is the same as that of usual implication, but that carries the pragmatic information that the branch associated with the then-part should be closed first. In the integrated system, the control specific distinction between logically equivalent connectives is made by KIV, used by $_3T\!AP$, but hidden to the user, protecting him or her from being confused by operational semantics.

6. CONVERTING PROOFS

Usually, automated and interactive theorem provers use quite different calculi, supporting machine-oriented proof search and, respectively, the intuition of a human user. To deal with this gap, we chose a dual approach, switching when necessary between the interactive part of the system (based on a sequent calculus) and the automated part of the system (based on free variable tableaux). Doing this we exploit the full power of both methods.

The resulting question is: which kind of information do both system parts have to exchange when they co-operate to construct a proof? First of all, there is *static* information depending on the signature, axioms, lemmas, and the specification structure, but not depending on the formula to prove. The *dynamically* exchanged information, depending on a current proof task, was at first a request to prove a formula, responded by *yes* or *no* (if not timed out). Even this short response is extremely valuable, because it can save the user time and effort otherwise spent in a boring and potentially long proof task.

But there are good reasons for a more informative response, providing the whole proof (if one was found) or at least the assumptions used in it. One important reason is the correctness management used in KIV, which automatically guarantees the absence of cycles in lemma dependencies and automatically invalidates proofs affected by the modification of a lemma. To integrate the proofs found by $_3TAP$ into this management, KIV needs more than just a *yes*. The second reason (which requires the complete proof as a response) is the replay mechanism of KIV, which is able to rebuild the branches remaining valid of an invalidated proof tree. That is why all proofs found by $_3TAP$ should be fully transformable into KIV proof trees. Finally, embedding $_3TAP$ proofs into the KIV calculus makes it possible to visualize the overall proof in a single framework, enabling the user to understand what is going on without requiring him or her to deal with two different calculi.

Therefore, $_3TAP$ proofs must be translated into KIV proofs. Since the first uses a tableau calculus and the second a sequent calculus, the main task seems to be the translation of tableaux into sequent proofs. But this is not difficult, because there is a one-to-one correspondence between tableaux and sequent proofs, provided we consider appropriate instances of the calculi consisting of corresponding sets of rules. In our case, they do *not* correspond one-to-one because they are designed for different purposes: automated search and interaction. These demand different features. For an interactive proof construction, the rules have a great degree of non-determinism (e.g. an arbitrary term has to be guessed) and, at the same time, they have to be simple. On the other hand, in the context of automated search, nothing is more important as the reduction of search space, which requires a very small degree of freedom, postponing decisions and removing non-determinism as much as possible.

A typical example for that difference is the cut-rule, which is used in an interactive calculus to enable case distinctions over arbitrary formulas. Such a rule is not feasible in an automatic proof procedure. Therefore, $_3TAP$'s tableau calculus offers only a kind of restricted cut, called lemma generation, which handles disjunctions by adding the negation of one extension to the other extension. Here, the transformation is easy because lemma generation can be very simply simulated by cuts.

Much more interesting is the different handling of instantiations. KIV uses a "ground" calculus, where the system or the user has to guess a term as an instance of an $\forall$-quantified variable when applying a γ-rule. Given this, the handling of $\exists$-quantified variables by the δ-rule is very easy since Skolem *constants* suffice. On the other hand, $_3TAP$ uses a *free variable* calculus. Here, the γ-rules introduce free variables, representing terms not yet guessed. As a consequence, the δ-rules have to introduce new Skolem terms containing the free variables of the formula. Older versions of these rules (Fitting, 1990) correspond one-to-one to rules of a simple "ground" calculus. But the research in the field of tableaux yielded several improvements. One is a modified δ-rule (δ^{++}), which is proven to allow exponential shorter proofs in extreme cases (Beckert et al., 1993). Here, we decided to provide a "ground" δ-rule that corresponds to δ^{++}. The rule itself has a simple description, but its soundness proof turned out to be difficult.

Another improvement of the free variable tableaux is the usage of *universal variables*. These are special free variables to which arbitrarily many instances can be assigned (see Chapter I.1.1). They have no counterpart in the simple "ground" calculus. To bridge this gap, proofs using universal variables are transformed into ground proofs by collecting all instances assigned to a universal variable and building several ground copies of the affected subproof, one for each instance. The crucial point here is the analysis and handling of dependencies between free variable instances (Stenz, 1997).

7. EVALUATION: THE WAM CASE STUDY

To evaluate our project results and to demonstrate improvements in the integrated system, we chose to verify selected compilation steps from Prolog to the Warren Abstract Machine (WAM) as our major case study. The algorithms and a mathematical analysis of the compiler were already provided (Börger and Rosenzweig, 1995), allowing us to concentrate on the development of a formal specification and on theorem proving. A first analysis (Schmitt, 1994) of the associated correctness problems showed that verification of the first compilation step poses tasks challenging for both KIV and $_3\mathcal{T\!A\!P}$, which should lead to synergy effects. One important challenge for the interactive prover was the fact that, due to its complexity, the correctness proof could be developed only in an incremental process of failed proof attempts, error correction and re-proof (evolutionary verification). For the use of the automated prover an important aspect was, that a large number of standard data types (lists, sets, tuples, etc.) is required, thus a large number of first-order theorems had to be proved. These theorems were used as benchmarks to evaluate improvement of the integrated system.

The content of the case study is described in Chapter III.2.7, together with our solution for the problem of finding large invariants. Here, we just mention that verification revealed some formal gaps in the analysis of (Börger and Rosenzweig, 1995). Moreover, the operational semantics of WAM code, as formalized in that paper, was found to be non-deterministic, allowing non-termination of programs that should terminate according to Prolog semantics. Without giving an explanation here, we just show one example. The Prolog program consisting of the clauses "p :- fail." and "p.", when asked the query "?- p.", has more than one behavior: it *may* terminate with "yes", but it may as well run forever.

It took three man months to verify the two correctness theorems including all required lemmas. The final proofs of the two main theorems in Dynamic Logic required 846 interactions with the user. The algebraic specifications contained 207 axioms; 184 first-order lemmas were used. These were relatively easy to prove compared with the complexity of the main theorems, and usually the first attempt to state these lemmas was correct. Nevertheless, additional 350 interactions were required. About 90% of these interactions could potentially be saved by the use of an (ideal) automated theorem prover (the remaining interactions involve induction). In reality, first tests with the 58 theorems of the top-level specification showed that $_3TAP$ was only able to prove 5 of these, giving a ratio of only 8%. With the improvements of the integrated system we have implemented so far, the integrated system is now able to prove 21 (36%) theorems fully automatic. The most significant gain in the productivity of $_3TAP$ (from 9 to 21) was achieved by the reduction of the set of axioms, which often left only 10–20 relevant axioms for each proof.

8. CURRENT DIRECTIONS OF RESEARCH

Beyond the work described so far, current research concentrates on deepening the integration of the automated and the interactive proving paradigm by exploiting further domain specifics found in software verification. Four aspects of this work are described in the following.

Tableau Rewriting Using Simplifiers Simplifiers of the form $(\forall x)(p(x) \to \psi(x))$ are frequently used in KIV specifications. Typical examples are *definitions*[1], where the formula ψ is used to specify the meaning of predicate symbol p, such as in $(\forall x)(even(x) \to (\exists y)(x = 2 * y))$.

Simplifiers carry pragmatic information. In proofs, a simplifier is used solely to deduce an instance $\psi(t)$ of the conclusion from a given instance $p(t)$ of the premise, thus it is known (and part of the pragmatics of the simplifier)

---

[1] Definitions are usually equivalences of the form $(\forall x)(p(x) \leftrightarrow \psi(x))$; they replace the two simplifiers $(\forall x)(p(x) \to \psi(x))$ and $(\forall x)(\neg p(x) \to \neg\psi(x))$, but *not* the implication $(\forall x)(\psi(x) \to p(x))$, which is (in general) not a simplifier.

that the contrapositive $(\forall x)(\neg\psi(x) \rightarrow \neg p(x))$ is not needed in proofs. Note, that although both simplifiers and if-then formulas are formal implications, the concept of a simplifier is complementary to that of the if-then operator and their pragmatics is completely different. Consequently, the search space should be organized differently.

As there is a strong relationship between simplifiers and equality reduction rules, the improvement one can hope to gain is similar to that of using reduction rules for rewriting terms (as compared to using equalities in an unrestricted way).

Extending the Application Domain   Until now we have considered the use of automated theorem provers for goals occurring as lemmas or explicit subgoals during interactive correctness proofs of software systems. There are, however, a lot of other first-order theorems hidden in goals containing programs, which can be proved using automated systems. Some typical subgoals of this kind are:

- to test the applicability of a conditional rewrite rule;
- to check disjunctive goals containing programs for first-order validity;
- to check a program conditional (or its logical complement) for validity in the current context, which allows to omit the else- (then-) branch;
- to determine the reachability of a recursive call (which can be expressed in a first-order formula).

For all these proof tasks, KIV uses its general simplifier (possibly involving user interaction). But all these tasks are suitable for trying $_3\mathcal{TAP}$ as they do not require induction. Compared to the first-order theorems we proved until now, they have a very different characteristic: most of these goals are not theorems. Another new characteristic is that not only a lot of axioms (and usable lemmas) are irrelevant to the proof, but also the goals themselves contain a lot of subformulas which are irrelevant. For routine application of $_3\mathcal{TAP}$ these new problems have to be solved first.

Proof Engineering   Beside a high degree of automation, an important criterion for the efficient verification of large software systems is the existence of powerful *proof engineering* tools. These allow incremental development of proofs and thus to cope with failures and resulting corrections and changes. Major components are:

- methods for the analysis of proofs,
- methods to construct counter examples,
- methods to reuse proofs and proof attempts.

There exists a method in KIV to reuse proofs on *program* changes (Reif and Stenzel, 1993). But often it is also necessary to correct *first-order formulas*, e.g. for the incremental development of the *coupling invariant* in the WAM case study. Therefore, one of our major efforts in the future will be to improve the possibilities to analyze and to reuse proofs with changed first-order formulas.

Computing Counter Examples for Non-Valid Problems First-order problems deriving from program verification often do not constitute provable formulas. The reasons range from bugs in the object program or the specification to erroneous tactical decisions supplied earlier by the user (e.g., too weak induction hypothesis). Also, extending the application of an automated prover, as discussed above, involves non-theorems. It is, therefore, an extremely desirable feature of an automated theorem prover to provide counter examples for non-valid formulas.

In general, of course, falsifying interpretations for non-valid first-order formulas are not computable, but in the limited context of program verification additional observations (which have so far been rarely named explicitly) simplify the situation considerably: if a program contains bugs it does not meet its specification for certain inputs. These critical inputs can be obtained from finite counter examples. In practice, such counter examples tend to be small, so there is no need to search for large instantiations.

We can exploit several features specifically present in problems derived from program verification. Counter examples correspond to initial models finitely generated by a known set of function symbols and constants. Moreover, variables are sorted, which further restricts the number of models. On the other hand, one needs to generate models relative to an equality theory.

9. OTHER APPROACHES TO INTEGRATION

There are several approaches described in the literature, which have in common the aim to strengthen interactive proving methods (and tools) by adding automatic methods (and tools) which are able to handle a relevant amount of subtasks.

Apart from first-order logic theorem proving, the most important reasoning methods being integrated in interactive frameworks are: computer algebra (Harrison and Théry, 1993; Ballarin et al., 1995), model checking (Joyce and Seger, 1994; Owre et al., 1996) and decision procedures (Boyer and Moore, 1988). Each of them is typically applied to some particular domain(s). *Computer algebra* is incorporated into proof systems solving problems of mathematical or physical origin. These applications are very different from the problems occurring in software verification, on which we focused in this chapter. *Model checking* is often used in the context of hardware verification. This method is also applied to verify software, but then restricted to (sub-)systems that can be adequately represented by (not too many) finite states. Although the amount of potential applications is growing due to new techniques of abstraction, most problems arising in software verification are infinite-state. Finally, for our domain the *decision procedures* are the most relevant of the listed methods. They are not meant to be flexible but tailor-made for fixed theories that are used very often. Here, the crucial question is: how many datastructures we use often are decidable and, apart from that, unchanging? In the experience of our different case studies, most specifications had often to be modified. But, of course, there are a few decidable theories we use regularly (e.g. fragments of natural numbers, integers), where decision procedures would help.

Computer algebra, model checking, and decision procedures differ from our approach in that the incorporated algorithms handle *decidable* problems. Moreover, the correctness management problem (keeping everything consistent, while the underlying theories are changed and corrected continually) is not relevant for these approaches, because they handle only *fixed* standard theories (e.g. linear arithmetic, calculus and boolean algebra).

In contrast, the problem of impairing correctness of the interactive prover by "believing" the results of connected systems has been an issue in the approaches mentioned above (in particular for computer algebra). In comparison, this issue is less significant for the joint work of two provers.

Relevant work in the area of automated theorem proving are attempts to aggregate several automated provers under a common homogeneous user interface, e.g. ILF (Dahn et al., 1995) (for conceptual issues see also Chapter II.4.14), and experiments that extend existing automated provers by an interactive component, e.g. INKA (Hutter and Sengler, 1996).

We are also aware of several attempts to integrate automated theorem proving techniques with interactive systems, that have been made so far:

The homogeneous approach, which implements an automated theorem prover in the implementation language of the interactive system was used in the HOL system (Schneider et al., 1992). The automated theorem prover FAUST uses a variant of the sequent calculus. The sequent calculus proofs resulting from its application are converted back to HOL proofs. The approach resulted in speed-ups and reduced the necessary interactions. Nevertheless, a prototypical implementation like that of FAUST cannot compete with the results of efficient implementations of automated theorem provers.

Our heterogeneous approach, which combines two different existing provers, has previously been used to combine the HOL system with a resolution prover. The work described in (Archer et al., 1993) concentrates on the selection of a sub-language of higher-order logic suitable for the resolution prover and on proof visualization.

A direct comparison of a heterogeneous and a homogeneous approach can be found in (Busch, 1994). The first is a combination of the higher-order logic proof checker LAMBDA with the automatic first-order logic prover SEDUCT. The second is an extension of LAMBDA by rules suitable for first-order theorem proving, together with heuristics for quantifier instantiation.

Finally we should mention the recent integration of the interactive type theory prover Typelab (see Chapter II.1.3) with the equality prover Discount (Strecker and Sorea, 1997) and current work in progress on combining KIV and INKA in the VSE system (Hutter et al., 1995) and NuPRL with KoMeT.

All these experiments (except the VSE system) are based on higher-order logic languages, in which algorithms are denoted in a functional style. In contrast, our logical framework, the Dynamic Logic, deals with imperative programs combined with first-order formulas. Another difference is the main purpose of our system. It is not designed for formalizing mathematics, but specialized to verify modular programs w.r.t. structured algebraic specifications, i.e. structured first-order theories. Our work differs from others in heavily making use of this application domain, especially for reducing the search space of the automatic prover (see reduction of the axiom set, generating precedence orders and other issues above).

10. CONCLUSION

Ever since we began the project of integrating interactive and automatic theorem provers, we had the strong feeling to be working on a strategic and promising topic. Already the expected benefit from linking the two theorem proving programs provided sufficient motivation to set to work with extra effort. But this was only the beginning.

We anticipated to identify problems that are not particular to our two theorem proving systems, but would arise in any attempt to combine the two theorem proving paradigms. And in fact we can now name typical trouble spots: the difficulties of automated provers to cope with large sets of axioms, the mismatch between first-order automated theorem proving and higher-order interactive provers, the use of pragmatic information to guide proof search, and the internal communication between the interactive and the automated prover. We have made substantial progress towards finding solutions, which again have significance beyond the special situation we are dealing with.

From this experience and from observing the literature and the focus of conferences we see a new research area emerging that might be called "integrated theorem proving".

REFERENCES

Archer, M., G. Fink, and L. Yang: 1993, 'Linking Other Theorem Provers to HOL using PM: Proof Manager'. In: L. Claesen and M. Gordon (eds.): *Higher Order Logic Theorem Proving and its Applications: 4th International Workshop, HUG'92.*

Ballarin, C., K. Homann, and J. Calmet: 1995, 'Theorems and Algorithms: An Interface between Isabelle and Maple'. In: A. H. M. Levelt (ed.): *International Symposium on Symbolic and Algebraic Computation.* pp. 150–157.

Baumgartner, P. and U. Furbach: 1994, 'PROTEIN: A Prover with a Theory Extension Interface'. In: A. Bundy (ed.): *Proc. 12th CADE, Nancy, France*, Vol. 814 of *LNCS.* pp. 769–773.

Beckert, B.: 1994, 'A Completion-Based Method for Mixed Universal and Rigid E-Unification'. In: A. Bundy (ed.): *Proc. 12th CADE, Nancy, France*, Vol. 814 of *LNCS.* pp. 678–692.

Beckert, B., R. Hähnle, P. Oel, and M. Sulzmann: 1996, 'The Tableau-Based Theorem Prover $_3T^AP$, Version 4.0'. In: M. McRobbie and J. Slanley (eds.): *Proc. 13th CADE, New Brunswick/NJ, USA*, Vol. 1104 of *LNCS.* pp. 303–307.

Beckert, B., R. Hähnle, and P. H. Schmitt: 1993, 'The Even More Liberalized δ-Rule in Free Variable Semantic Tableaux'. In: G. Gottlob, A. Leitsch, and D. Mundici (eds.): *Proc. 3rd Kurt Gödel Colloquium (KGC), Brno, Czech Republic.* pp. 108–119.

Beckert, B. and C. Pape: 1996, 'Incremental Theory Reasoning Methods for Semantic Tableaux'. In: P. Miglioli, U. Moscato, D. Mundici, and M. Ornaghi (eds.): *Proc. 5th TABLEAUX, Terrasini/Palermo, Italy*, Vol. 1071 of *LNCS.* pp. 93–109.

Bibel, W., S. Brüning, U. Egly, and T. Rath: 1994, 'KoMeT'. In: A. Bundy (ed.): *Proc. 12th CADE, Nancy, France*, Vol. 814 of *LNCS.* pp. 783–787.

Börger, E. and D. Rosenzweig: 1995, 'The WAM—Definition and Compiler Correctness'. In: C. Beierle and L. Plümer (eds.): *Logic Programming: Formal Methods and Practical Applications.* North-Holland, pp. 21–90.

Boyer, R. and J. Moore: 1988, 'Integrating decision procedures into heuristic theorem provers: A case study with linear arithmetic.'. *Machine Intelligence* **11**. Oxford University Press.

Boyer, R. S. and J. S. Moore: 1979, *A Computational Logic.* Academic Press.

Busch, H.: 1994, 'First-Order Automation for Higher-Order-Logic Theorem Proving'. In: T. F. Melham and J. Camilleri (eds.): *Higher Order Logic Theorem Proving and its Applications: 7th International Workshop*, Vol. 859 of *LNCS.* pp. 97–112.

Constable, R. L., S. F. Allen, H. M. Bromley, W. R. Cleaveland, J. F. Cremer, R. W. Harper, D. J. Howe, T. B. Knoblock, N. P. Mendler, P. Panagaden, J. T. Sasaki, and

S. F. Smith: 1986, *Implementing Mathematics with the Nuprl Proof Development System*. Prentice Hall.

Dahn, B., J. Gehne, T. Honigmann, L. Walther, and A. Wolf: 1995, 'Integrating Logical Function with ILF'. System description, Humboldt-Universität zu Berlin.

Farmer, W. M., J. D. Guttman, and F. J. T. Fábrega: 1996, 'IMPS: an update description'. In: M. McRobbie and J. Slanley (eds.): *Proc. 13th CADE, New Brunswick/NJ, USA*. pp. 298–302.

Fermüller, C., A. Leitsch, T. Tammet, and N. Zamov: 1993, *Resolution Methods for the Decision Problem*, Vol. 679 of *LNCS*. Springer-Verlag.

Fitting, M.: 1990, *First-Order Logic and Automated Theorem Proving*. Springer, New York, first edition. Second edition appeared in 1996.

Goller, C., R. Letz, K. Mayr, and J. Schumann: 1994, 'SETHEO V3.2: Recent Developments – System Abstract'. In: A. Bundy (ed.): *Proc. 12th CADE, Nancy, France*, Vol. 814 of *LNCS*. pp. 778–782.

Gordon, M.: 1988, 'HOL: A Proof Generating System for Higher-order Logic'. In: G. Birtwistle and P. Subrahmanyam (eds.): *VLSI Specification and Synthesis*. Kluwer Academic Publishers.

Hähnle, R. and S. Klingenbeck: 1996, 'A-Ordered Tableaux'. *J. of Logic and Computation* **6**(6), 819–834.

Hähnle, R. and C. Pape: 1997, 'Ordered Tableaux: Extensions and Applications'. In: D. Galmiche (ed.): *Proc. International Conference on Automated Reasoning with Analytic Tableaux and Related Methods, Pont-à-Mousson, France*, Vol. 1227 of *LNCS*. pp. 173–187.

Harrison, J. and L. Théry: 1993, 'Extending the HOL Theorem Prover with a Computer Algebra System to Reason about the Reals'. In: J. J. Joyce and C.-J. H. Seger (eds.): *Higher Order Logic Theorem Proving and its Applications: 6th International Workshop, HUG'93, Vancouver, Canada*, Vol. 780 of *LNCS*. pp. 174–184.

Hutter, D., B. Langenstein, F. Koob, W. Reif, C. Sengler, W. Stephan, M. Ullmann, M. Wittmann, and A. Wolpers: 1995, 'The VSE Development Method - A Way to Engineer High-Assurance Software Systems'. In: B. Gotzheim (ed.): *GI/ITG Tagung Formale Beschreibungstechniken für verteilte Systeme*.

Hutter, D. and C. Sengler: 1996, 'INKA: The Next Generation'. In: M. McRobbie (ed.): *Proc. 13th CADE, New Brunswick/NJ, USA*, Vol. 1104 of *LNCS*. pp. 288–292.

Joyce, J. and C. Seger: 1994, 'The HOL-Voss System: Model-Checking inside a General-Purpose Theorem-Prover'. In: J. J. Joyce and C.-J. H. Seger (eds.): *Higher Order Logic Theorem Proving and its Applications: 6th International Workshop, HUG'93*. pp. 185 – 198.

Kaufmann, M. and J. Moore: 1988, 'Design Goals of ACL2'. Technical report 101, Computational Logic Inc.

Letz, R., J. Schumann, S. Bayerl, and W. Bibel: 1992, 'SETHEO: A High-Perfomance Theorem Prover'. *J. of Automated Reasoning* **8**(2), 183–212.

Owre, S., S. Rajan, J. Rushby, N. Shankar, and M. Srivas: 1996, 'PVS: Combining Specification, Proof Checking, and Model Checking'. In: R. Alur and T. A.

Henzinger (eds.): *Computer-Aided Verification, CAV '96*, Vol. 1102 of *LNCS*. New Brunswick, NJ, pp. 411–414.

Owre, S., J. M. Rushby, and N. Shankar: 1992, 'PVS: A Prototype Verification System'. In: D. Kapur (ed.): *Proc. 11th CADE, Saratoga Springs/NY, USA*, Vol. 607 of *LNCS*. Saratoga, NY, pp. 748–752.

Paulson, L. C.: 1994, *Isabelle: A Generic Theorem Prover*. Springer LNCS 828.

Plaisted, D. A. and S. Greenbaum: 1986, 'A Structure-Preserving Clause Form Translation'. *Journal of Symbolic Computation* 2, 293–304.

Reif, W.: 1995, 'The KIV-approach to Software Verification'. In: M. Broy and S. Jähnichen (eds.): *KORSO: Methods, Languages, and Tools for the Construction of Correct Software—Final Report*, LNCS 1009. Springer-Verlag.

Reif, W., G. Schellhorn, and K. Stenzel: 1997, 'Proving System Correctness with KIV 3.0'. In: *Proc. 14th CADE, Townsville, Australia*, Vol. 1249 of *LNCS*. pp. 69–72.

Reif, W. and K. Stenzel: 1993, 'Reuse of Proofs in Software Verification'. In: R. Shyamasundar (ed.): *Proc. FST TCS, Bombay, India*, Vol. 761 of *LNCS*. pp. 284–293.

Schellhorn, G. and W. Ahrendt: 1997, 'Reasoning about Abstract State Machines: The WAM Case Study'. *Journal of Universal Computer Science (JUCS)* 3(4), 377–380. Available at the URL: http://hyperg.iicm.tu-graz.ac.at/jucs/.

Schmitt, P. H.: 1994, 'Proving WAM Compiler Correctness'. Interner Bericht 33/94, Universität Karlsruhe, Fakultät für Informatik.

Schneider, K., R. Kumar, and T. Kropf: 1992, 'Integrating a first-order automatic prover in the HOL environment'. In: *Higher Order Logic Theorem Proving and its Applications: 4th International Workshop, HUG'91*.

Stenz, G.: 1997, 'Beweistransformation in Gentzenkalkülen'. Diplomarbeit, Fakultät für Informatik, Universität Karlsruhe.

Strecker, M. and M. Sorea: 1997, 'Integrating an Equality Prover into a Software Development System based on Type Theory'. In: G. Brewka, C. Habel, and B. Nebel (eds.): *Proc. KI'97*, Vol. 1303 of *LNCS*. pp. 147–158.

Weidenbach, C., B. Gaede, and G. Rock: 1996, 'SPASS & FLOTTER, Version 0.42'. In: M. McRobbie and J. Slanley (eds.): *Proc. 13th CADE, New Brunswick/NJ, USA*, Vol. 1104 of *LNCS*. pp. 141–145.

Wos, L., R. Overbeek, E. Lusk, and J. Boyle: 1992, *Automated Reasoning, Introduction and Applications (2nd ed.)*. McGraw Hill.

Part 2

Representation and Optimization Techniques

Editors: Detlef Fehrer, Jörg H. Siekmann

INTRODUCTION

People who write papers on theorem proving have unfortunately tended to ignore the practical computational questions which arise when a program is to be written to do deduction on an actual machine. (. . .) Yet if computational logic is to develop into a genuinely useful science it is precisely the computational issues which have to be given the most attention.

J. A. ROBINSON, 1971

Ever since the invention of machine oriented calculi the representation of formulae has been at the heart of deduction systems development. Alan Robinson's paper dating back almost thirty years (Robinson, 1971) (from which the quotation above is taken) is a good example for this interest in speeding up search by special representational techniques. While these early systems were capable of searching a clause space of some 100,000 clauses, the systems in the 1970's and 80's advanced the state of the art to clause spaces of millions of clauses (see e.g. the Argonne National Laboratory and Northern Illinois University provers WOS1 (McCharen et al., 1976) and OTTER (McCune, 1990; McCune, 1994), or the Markgraf-Karl Refutation Procedure (Ohlbach and Siekmann, 1989)). Today's systems search a clause space of several billion clauses (for example the search for the proof of the Robbins algebra problem (McCune, 1997) by the OTTER prover).

The strength of current systems, such as OTTER (McCune and Wos, 1997) or SPASS (Weidenbach, 1997) and SETHEO (Moser et al., 1997), all winners of the CADE-13 competition (see (Sutcliffe and Suttner, 1997)), therefore crucially depends on appropriate indexing techniques and memory management. The first paper in this section by P. Graf describes the current state of the art and presents one of the best known term indexing techniques, which is used inter alia by SPASS (Weidenbach, 1997) and in the ΩMEGA (Benzmüller et al., 1997) system.

A deduction system with a reasonable user-friendly interface that performs at today's level of competence is a large and structured software system, comparable say to an operating system or a compiler. Just as the structure of the latter was discovered in the late 1950's and 60's and the respec-

W. Bibel, P. H. Schmitt (eds.), Automated Deduction. A basis for applications. Vol. II
© 1998 *Kluwer Academic Publishers. Printed in the Netherlands*

tive engineering disciplines for compiler and operating systems emerged, the structure and implementational techniques for a deduction system evolved in the 1970's and 80's. Good examples for these early structural developments are the Boyer-Moore induction prover (Boyer and Moore, 1979), the LMA system of the Argonne National Laboratory (Lusk et al., 1982a; Lusk et al., 1982b) or the MKRP-system (Raph, 1984). Each of these consisted of several megabytes of source code and pioneered many of today's software engineering aspects of a deduction system.

Basic subtasks such as formula parsing, pretty-printing, implementation of the basic data structures, developing a user interface and tracing facilities etc. are time consuming but not overly demanding implementational tasks. On the other hand the implementation of many subcomponents, such as the unification algorithms, the actual inference rule(s), or the indexing techniques mentioned above are far from trivial, but nevertheless they are well understood and should not be a burden for the implementor of a new deduction system. Thus the idea of a toolbox for the development of deduction systems was born and the second paper by D. Fehrer gives a presentation of KEIM, a tool kit for deduction system development that is now in use at several sites.

A related idea is presented in the third paper by G. Neugebauer and U. Petermann, who show how an abstract specification of the inference rules for a system can be automatically translated into efficient code using a PTTP-like technique. In particular they are interested in how a given theorem proving system can be extended by additional inference rules without a reimplementation of the kernel of the system, and they demonstrate their techniques for the prover ProCom.

The proof of a difficult mathematical theorem — still within the range of present fully automated systems — may consist of several hundred steps (applications of an inference rule such as resolution) and may well represent several weeks of labour and cpu-time. An interactively generated proof in the area of program verification may actually be up to ten thousand steps long (see e.g. the KIV (Heisel et al., 1990) and VSE (Hutter et al., 1996) system) and such a proof may take months of intense work to find and patch. Thus machine generated proofs are a valuable commodity to be appropriately represented and possibly abstracted for later reuse.

An important case for reusing previously found proofs in mathematics is their use for a proof by analogy (see (Melis and Whittle, 1998) for recent work on proofs by analogy), where the earlier distinction between "derivational analogy" and "transformational analogy" (Carbonell, 1986), that was developed for the reuse of plans, becomes important for theorem proving systems as well.

Similarly the reuse of proofs in program verification as pioneered by W.

Reif (Reif and Stenzel, 1993) is becoming an important cost saving technique in software engineering. A related direction of research was opened up by an interesting observation of Ch. Walther, who noticed that techniques from explanation based learning can be applied in automated theorem proving as well (first published in (Kolbe and Walther, 1994)). The final paper in this section by Th. Kolbe and Ch. Walther on proof abstraction and reuse addresses these issues. A given proof can be analyzed and appropriately generalized to be stored in a data base of so called "proof shells". A new theorem to be shown will trigger a heuristically guided retrieval mechanism and the appropriate proof shell may then be adapted and "patched" into a proof for the theorem at hand. Automated planning techniques based on patching (Hayes-Roth, 1983) or on derivational analogy (Carbonell, 1986), proof planning by reformulation and analogy (Melis, 1995a; Melis, 1995b), proof planning and patching based on the advice of "critics" (Ireland and Bundy, 1996), proof abstraction (Giunchiglia and Villafiorita, 1996; Giunchiglia and Walsh, 1992), reuse of proofs in formal software development (Reif and Stenzel, 1993), the reuse of plans (Koehler, 1996) and finally revising proof shells as advocated in the last paper all contribute to an interesting new paradigm of reusing previously computed information.

REFERENCES

Benzmüller, C., L. Cheikhrouhou, D. Fehrer, A. Fiedler, X. Huang, M. Kerber, M. Kohlhase, K. Konrad, A. Meier, E. Melis, W. Schaarschmidt, J. Siekmann, and V. Sorge: 1997, 'ΩMEGA: Towards a Mathematical Assistant'. In: W. McCune (ed.): *Automated Deduction — CADE–14, 14th International Conference on Automated Deduction.* pp. 252–255.

Boyer, R. S. and J. S. Moore: 1979, *A Computational Logic.* London: Academic Press.

Carbonell, J. G.: 1986, 'Derivational Analogy: A Theory of Reconstructive Problem Solving and Expertise Acquisition'. In: R. S. Michalsky, J. G. Carbonell, and T. M. Mitchell (eds.): *Machine Learning: An Artificial Intelligence Approach.* Los Altos: Morgan Kaufmann, pp. 371–392.

Giunchiglia, F. and A. Villafiorita: 1996, 'ABSFOL: a Proof Checker with Abstraction'. In: *Proceedings of the 13th International Conference on Automated Deduction (CADE-13).* pp. 134–140.

Giunchiglia, F. and T. Walsh: 1992, 'A Theory of Abstraction'. *Artificial Intelligence* **57**, 323–390.

Hayes-Roth, F.: 1983, 'Using Proofs and Refutations to Learn from Experience'. In: R. S. Michalski, J. G. Carbonell, and T. M. Mitchell (eds.): *Machine Learning — An Artificial Intelligence Approach.* Palo Alto: Tioga Publishing Company.

Heisel, M., W. Reif, and W. Stephan: 1990, 'Tactical Theorem Proving in Program Verification'. In: *Proceedings of CADE 90*. pp. 117–131.

Hutter, D., B. Langenstein, C. Sengler, J. Siekmann, W. Stephan, and A. Wolpers: 1996, 'Verification Support Environment (VSE)'. *Journal of High Integrity Systems* 1(6), 523–530.

Ireland, A. and A. Bundy: 1996, 'Productive Use of Failure in Inductive Proof'. *Journal of Automated Reasoning* 16(1–2), 79–111.

Koehler, J.: 1996, 'Planning from Second Principles'. *Artificial Intelligence* 87.

Kolbe, T. and C. Walther: 1994, 'Reusing Proofs'. In: A. Cohn (ed.): *11th European Conference on Artificial Intelligence*. Amsterdam.

Lusk, E. L., W. McCune, and R. A. Overbeek: 1982a, 'Logic Machine Architecture: Inference Mechanisms'. In: *Automated Deduction — CADE*. pp. 85–108.

Lusk, E. L., W. McCune, and R. A. Overbeek: 1982b, 'Logic Machine Architecture: Kernel Functions'. In: *Automated Deduction — CADE*. pp. 70–84.

McCharen, J., R. Overbeek, and L. Wos: 1976, 'Problems and experiments for and with automated theorem-proving programs'. *IEEE Transactions on Computers* C–25(8), 773–782.

McCune, W.: 1990, 'OTTER 2.0 Users Guide'. Tech. Report ANL–90/9, Argonne National Laboratory, Argonne, IL.

McCune, W.: 1994, 'Otter 3.0 Reference Manual and Guide'. Technical Report ANL-94/6, Argonne National Laboratory.

McCune, W.: 1997, 'Solution of the Robbins problem'. *Journal of Automated Reasoning* 19(3), 263–276.

McCune, W. and L. Wos: 1997, 'Otter: The CADE-13 Competition Incarnations'. *Journal of Automated Reasoning, Special Issue on the CADE-13 Automated Theorem Proving System Competition* 18(2), 211–220.

Melis, E.: 1995a, 'A Model of Analogy-Driven Proof-Plan Construction'. In: *Proceedings of the 14th International Joint Conference on Artificial Intelligence*. Montreal, pp. 182–189.

Melis, E.: 1995b, 'Theorem Proving by Analogy — A Compelling Example'. In: C. Pinto-Ferreira and N. J. Mamede (eds.): *Progress in Artificial Intelligence. 7th Portuguese Conference on Artificial Intelligence, EPIA'95*. Madeira, pp. 261–272.

Melis, E. and J. Whittle: 1998, 'Analogy in Inductive Theorem Proving'. *Journal of Automated Reasoning* 20(3). to appear.

Moser, M., O. Ibens, R. Letz, J. Steinbach, C. Goller, J. Schumann, and K. Mayr: 1997, 'SETHEO and E-SETHEO. The CADE-13 systems'. *Journal of Automated Reasoning, Special Issue on the CADE-13 Automated Theorem Proving System Competition* 18(2), 237–246.

Ohlbach, H. J. and J. Siekmann: 1989, 'The Markgraf Karl Refutation Procedure'. In: *Computational Logic — Essays in Honor of Alan Robinson*. MIT Press, Cambridge, pp. 41–112.

Raph, K. M. G.: 1984, 'The Markgraf Karl Refutation Procedure'. SEKI-Memo MK-84-01, Universität Kaiserslautern, FB Informatik.

Reif, W. and K. Stenzel: 1993, 'Reuse of Proofs in Software Verification'. In: R. K. Shyamasundar (ed.): *Proc. 13th Conference on Foundations of Software Technology and Theoretical Computer Science*, Vol. 761 of *LNCS*. pp. 74–85.

Robinson, J. A.: 1971, 'Computational Logic: The Unification Computation'. In: B. Meltzer and D. Michie (eds.): *Machine Intelligence 6*. Edinburgh University Press, pp. 63–72.

Sutcliffe, G. and C. Suttner: 1997, 'The Results of the CADE-13 ATP System Competition'. *Journal of Automated Reasoning, Special Issue on the CADE-13 Automated Theorem Proving System Competition* **18**(2), 259–252.

Weidenbach, C.: 1997, 'SPASS: Version 0.49'. *Journal of Automated Reasoning, Special Issue on the CADE-13 Automated Theorem Proving System Competition* **18**(2), 247–252.

CHAPTER 5

TERM INDEXING

1. INTRODUCTION

Implementations of theorem provers that use generative procedures like resolution (Robinson, 1965b; Chang and Lee, 1973) or Knuth-Bendix completion (Knuth and Bendix, 1970) face the problem of *program degradation*: The theorem prover's rate of drawing conclusions falls off sharply with time due to an increasing amount of retained information (Wos, 1992).

In order to overcome this obstacle and to achieve efficient theorem proving, a designer of an automated reasoning system has several options. On the one hand, the performance of an automated reasoning system depends on the calculus, the search strategy, and heuristics. Exchanging one of these may cause an exponential difference in the performance. On the other hand, sophisticated implementation techniques may support the construction of efficient systems. Among these techniques term indexing particularly influences a system's performance by providing rapid access to terms in first-order predicate calculus (in the sequel simply first-order terms) with specific properties. Term indexing at best gives a speedup by some constant factor, but the speedup obtained is reliable. Carefully implemented indexing yields speedup for all examples, and this is the main difference to search heuristics, which sometimes perform well and sometimes poor.

Indexing techniques have first been applied in the area of databases (Date, 1991). In particular, relational databases employ B-trees (Bayer and Mc-Creight, 1997) or hashing (Morris, 1968) to efficiently access specific entities of a relation. The structure of *logical data* as used in automated reasoning programs, however, is much more complicated than the structure of data stored in a relational database. In automated reasoning systems we have to deal with first-order terms. Terms possess a more complex internal structure than objects found in relational databases. Furthermore the query types are not restricted to simple retrieval by looking for equal keys. Therefore, queries to a logical database (Butler and Overbeek, 1994) in context with theorem proving are much more complex than queries to relational databases: Given a database I containing terms (or literals) and a query term t, find all terms in I that are unifiable with, instances of, or more general than t. Standard applications of term indexing are the search for resolution partners for a given term

W. Bibel, P. H. Schmitt (eds.), Automated Deduction. A basis for applications. Vol. II
© 1998 Kluwer Academic Publishers. Printed in the Netherlands

(literal) or the retrieval of literals in clauses for both forward and backward subsumption.

Nowadays, term indexing techniques are used in various successful theorem provers and state-of-the-art performance is hard to achieve without indexing (Lusk and Overbeek, 1980). Interestingly, designers of high performance reasoning systems, like William McCune, Jim Christian, or Mark Stickel, also have published articles about indexing techniques (McCune, 1988; Christian, 1993; Stickel, 1989). McCune explicitly states that the speed of OTTER (McCune, 1990) is mainly achieved by using sophisticated term indexing methods (McCune, 1992).

2. CLASSIFICATIONS OF TERM INDEXING TECHNIQUES

2.1. *Relations on Terms*

The main purpose of indexing techniques in theorem provers is to achieve efficient access to first-order terms with specific properties. To this end, a set of terms I is represented as an indexing data structure. A retrieval in I is started for a set Q of *query terms*. The aim of the retrieval is to find tuples (s,t) with $s \in I$ and $t \in Q$ in such a way that a special relation R holds for s and t. Automated reasoning systems can profit from a retrieval based on the following relations (in the following σ always denotes a substitution):

$$\text{UNIF}(s,t) \quad \Longleftrightarrow \quad \exists \sigma.\ s\sigma = t\sigma$$
$$\text{INST}(s,t) \quad \Longleftrightarrow \quad \exists \sigma.\ s = t\sigma$$
$$\text{GEN}(s,t) \quad \Longleftrightarrow \quad \exists \sigma.\ s\sigma = t$$

Since $\text{UNIF}(s,t)$ holds if and only if the terms s and t are *unifiable*, the relation $\text{UNIF}(s,t)$ can be used for the retrieval of complementary unifiable literals in a resolution based system, for example. Moreover, possibly forward or backward subsumed clauses are found by accessing more general (GEN) resp. instance (INST) literals.

In addition to the standard relations, we also consider retrieval into the *subterms* of indexed or query terms. We therefore define two classes of relations $R^{\text{from}}(s,t)$ and $R^{\text{to}}(s,t)$ depending on whether the subterm occurs in the query or in the indexed set. The subterm of a term t at a position p is denoted by t/p.

$$R^{\text{from}}(s,t) \quad \Longleftrightarrow \quad \exists p.\ R(s,t/p)$$
$$R^{\text{to}}(s,t) \quad \Longleftrightarrow \quad \exists p.\ R(s/p,t)$$

A possible application of subterm indexing is the search for rewrite rules for the normalization of some term on the basis of the relation GEN^{from}.

Another challenging indexing task deals with *equational theories*. We generalize the relations introduced above to relations of the form $R_=$ since the tests for matching or unifiability have been performed using syntactical term equality. Hence, $R_{=_E}(s,t)$ holds if and only if the relation $R(s,t)$ holds, whereas the equality test has been performed using $=_E$. Indexing based on the relation UNIF$^{\text{to}}_{AC}$ could, for example, be valuable for paramodulation in the presence of AC function symbols.

If we are interested in retrieving indexed substitutions instead of indexed terms, a relation $R(\sigma,\tau)$ is needed. We consider relations of type $R(\sigma,\tau)$ as generalizations of relations $R(s,t)$ since indexing substitutions using the relation $R(\{x \mapsto s\}, \{x \mapsto t\})$ is equivalent to using $R(s,t)$. An application of indexed substitutions is hyperresolution. There we have to find a simultaneous unifier for the literals of the nucleus and the electrons. The retrieval of compatible substitutions can be supported by the relation

$$\text{UNIF}(\sigma,\tau) \iff \exists \lambda.\ \sigma\lambda = \tau\lambda$$

on substitutions.

In a retrieval we test for which pairs of terms or substitutions occurring in the indexed set I and the query set Q a specific relation R holds. The retrieval function is defined by

$$\text{retrieve}^R(I,Q) = \{(s,t) \mid (s,t) \in I \times Q \text{ and } R(s,t)\}$$

The result of the retrieval is a subset of the direct product of the indexed and the query set.

An ideal indexing technique supports the retrieval for all relations introduced on terms and substitutions. Unfortunately, such an indexing method has not yet been invented. Depending on the requirements of the chosen calculus we carefully have to select an appropriate indexing technique.

2.2. *Filters and Perfect Filters*

Indexing techniques can be viewed as filters for retrieving sets of terms. A filter that can be used for theorem proving must be *complete*. A complete filter retrieves at least a superset of the set in question. For example, an indexing technique used to find unifiable terms should at least retrieve the unifiable terms in the indexed set.

An indexing method which is useful for theorem provers does not need to be *sound*, i.e. not all terms that are found really need to have the desired property. If the indexing technique is not sound, the terms found have to be checked by an additional unification or matching test. A sound filter is also called perfect.

2.3. *Retrieval of Type 1:1, n:1, and n:m*

A retrieval is of *type* 1 : 1 if both sets I and Q have cardinality 1. Since both sets Q and I solely consist of one single term, the retrieval corresponds to simply testing if $R(s,t)$ holds.

Retrieval of type n : 1 is determined by a single query term t, which is used to find entries $s \in I$. The set I of n indexed terms is represented by an indexing data structure. Since the query set consists of a single entry, the retrieval is often considered as resulting in a subset I' of I rather than in a subset of $I \times Q$. Note that a very inefficient retrieval of type n : 1 could be performed by testing each entry of the index in a 1 : 1 type retrieval, because such an approach would have to consider all indexed terms explicitly.

Retrieval of type n : *m* includes all cases in which more than a single query term is involved. Exploiting n : m indexing, the query set typically is also represented by an index. Hence, we have to deal with two indexes; one of them represents the indexed and the other one represents the query set. The result of such a retrieval is a subset of the direct product of the term sets involved.

As an example, we consider the so-called *merge* operation for two sets or indexes described by the n : m retrieval function retrieve^{UNIF}. Note that a merge operation searching unifiable terms may also result in a new index representing the common instances of the unifiable terms. An application of this approach is hyperresolution, which will be discussed later.

2.4. *Attribute-, Set-, and Tree-Based Indexing*

Attribute-based indexing attaches a so-called *attribute* to every indexed term. Attributes describe complex features of terms in a compact way. Indexing is performed by comparing the precomputed attributes of indexed terms with attributes of query terms.

The basic principle of *set-based indexing* is to subdivide the indexed set of terms into not necessarily disjoint subsets. Each of the subsets contains a set of terms that share a specific property. The subsets are represented by so-called *property sets*. The complete set of property sets describing all the different properties of the indexed terms is called a *set-based* index. For a specific query term and a selected retrieval task we have to compute unions and intersections of property sets.

Tree-based indexing techniques mirror the term structures by storing sets of terms in a tree in such a way that common parts of the indexed terms are shared. Therefore, a single tree represents the structure of all indexed terms. Pointers to the entries of the index are stored at the leaves of the tree. Retrieval in a tree-based index will typically traverse both the tree and the query term, eventually finding appropriate leaves of the tree.

3. ATTRIBUTE-BASED INDEXING

An attribute describes more or less complex features of a term in a simple way. Each term's attribute is precomputed and attached to the term's data structure. Indexing is performed by comparing the attributes of two terms, thus supporting 1 : 1 type retrieval. Since the information about a term stored in its attribute typically is compressed and therefore incomplete, attribute-based indexing schemes do not provide perfect filters. In the following we shall sketch three attribute-based indexing techniques: The matching pretest, outline indexing, and the superimposed codeword method.

3.1. *Matching Pretest*

The matching pretest considers the precomputed sizes of the involved terms to detect whether or not a term can be an instance of another term. For example, we can guarantee that t is *not* an instance of s if the number of symbols in t is smaller than the number of symbols in s. When looking for instances or generalizations in a set of indexed terms, the pretest can be employed. Terms that successfully pass the matching pretest have to be checked by a regular matching routine. The retrieval of unifiable terms is not supported.

3.2. *Outline Indexing*

The outline indexing scheme (Henschen and Naqvi, 1981) exploits the fact that two terms s and t can unify only if they agree at every position where both terms do not contain a variable. The attribute for a term in this indexing scheme is called *outline*. It is a tuple (S, P), where S is an array of function symbols and P an array of Boolean values. $P[i]$ is true if and only if $S[i]$ contains a function or constant symbol. If two terms with outlines (S_1, P_1) and (S_2, P_2) have to be tested for unifiability, the symbols in S_1 and S_2 have to be identical at all the positions where P_1 and P_2 are true. Outline indexing can also be employed for the search of instances and generalizations.

3.3. *Superimposed Codewords*

For the application of superimposed codeword indexing (Wise and Powers, 1984; Ramamohanarao and Shepherd, 1986) a *codeword* for each non-variable symbol occurring in the terms involved is required. Each codeword consists of a fixed length sequence of Boolean values. The *descriptor* of a term is created by performing a bit-or operation on the codewords of the non-variable symbols occurring in the term. For example, if the codewords for the symbols f and a are 01010 and 00011, the descriptor of the term $f(x,a)$ is 01011. The retrieval of instances, for example, is achieved by computing the bit-and of the query term descriptor and the precomputed descriptors of indexed terms. Whenever the result of the bit-and is equal to the query term's descriptor, the considered term is a candidate for being an instance of the query term. Superimposed codeword indexing can also find generalizations and unifiable terms.

4. SET-BASED INDEXING

Set-based indexing techniques like top symbol hashing, coordinate indexing, and path indexing merely support $n : 1$ indexing tasks. Unfortunately, retrievals of type $n : m$ can only be simulated by performing m retrievals of type $n : 1$.

4.1. *Top Symbol Hashing*

A retrieval operation for a query term $f(t_1, \ldots, t_n)$ in a top symbol hashing index simply accesses the set of indexed terms starting with the function symbol f and applies a regular unification or matching routine to the terms that have been found. This indexing mechanism, which is used in most PROLOG systems (Warren, 1983; Ait-Kaci, 1991), merely considers the top function symbols of the terms involved.

4.2. *Coordinate Indexing*

Coordinate indexing (Hewitt, 1971) can consider *all* query term symbols. The property sets contain terms that have identical symbols at specific *positions*. For example, the set $I^b_{[1,2]}$ contains the term $g(f(a,b))$ and refers to the terms that contain the constant b as the top symbol of the second argument of the first argument of the term. Property sets are maintained in so-called FPA-lists. Thus coordinate indexing is often referred to as FPA-Indexing. For a specific query term and a selected retrieval task we compute unions and intersections

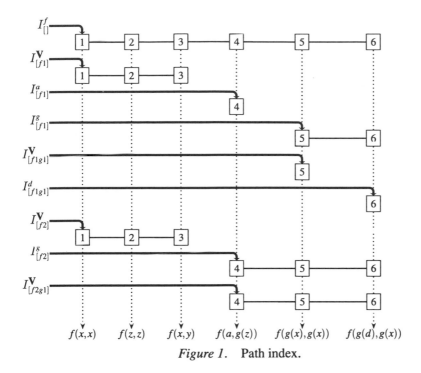

Figure 1. Path index.

of FPA-lists. Terms unifiable with $g(a)$, for example, are either variables <u>or</u> they have g as a top symbol <u>and</u> the argument of g is either a variable <u>or</u> the constant a. Thus the retrieved terms are

$$I_{[]}^{V} \cup (I_{[]}^{g} \cap (I_{[1]}^{V} \cup I_{[1]}^{a})) \subseteq \text{retrieve}^{\text{UNIF}}(I, \{g(a)\})$$

4.3. *Path Indexing*

Path indexing (Stickel, 1989; Graf, 1994) and coordinate indexing have similar basics. However, the *path lists*, which are used as property lists for path indexing, usually contain fewer terms than FPA-lists. This is because path lists store terms containing identical symbols at specific *paths*. Path indexing (PI) employs mappings from paths to symbols in order to describe subsets of indexed terms that share specific properties. For example, the set $I_{[f,2]}^{a}$ denotes all indexed terms which have the top symbol f and whose second argument is the constant a. In a *path index* sets like $I_{[f,2]}^{a}$ are represented by so-called *path lists*. Thus a path index consists of a collection of path lists. For a specific query term and a selected retrieval task we compute unions and intersections of path lists in a way similar to the one used in coordinate indexing.

In Figure 1 a path index for the indexed set

$$I = \{f(x,x), f(z,z), f(x,y), f(a,g(z)), f(g(x),g(x)), f(g(d),g(x))\}$$

is depicted. We consider an example of a very simple retrieval in this path index. Assume we would like to find instances of the *query term* $f(a,x)$. Obviously, retrieved terms should have a skeleton identical to the one of $f(a,x)$, i.e. except for the second argument the terms are identical. Therefore, the terms to be retrieved are members of the set

$$I^a_{[f,1]} \subseteq \text{retrieve}^{\text{INST}}(I, \{f(a,x)\}) \ .$$

Since the variable x occurring in $f(a,x)$ may be arbitrarily instantiated, there is no need to impose an additional restriction on the retrieved terms. A little more complicated is the search for instances if the query term is $h(a,g(b),x)$. The retrieved terms in this example are computed using the set intersection operation on property sets:

$$I^a_{[h,1]} \cap I^b_{[h,2,g,1]} \subseteq \text{retrieve}^{\text{INST}}(I, \{h(a,g(b),x)\}) \ .$$

Implementations of path indexing usually compute the set expression according to specific rules for a given query term and a selected retrieval task. This set expression is represented by a so-called *query tree*, which, when evaluated, computes unions and intersections of term sets represented by path lists.

Path indexes do not serve as perfect filters because variables in indexed terms are not treated individually. As a consequence, a path index may return a superset of the terms that were intended to be found. The retrieved set of terms is called the *candidate set* for a specific retrieval, since the found terms have to be submitted to a regular unification or matching routine.

Path indexing provides the user with a great amount of flexibility: A path index may be scanned by several queries in parallel, which allows parallel or recursive processes to work on the same index. It is possible to retrieve terms one-by-one, so we do not need to retrieve the whole candidate set at a time. Moreover, it is possible to insert entries to and delete entries from a path index even when a retrieval process is still in progress. Eventually, the strength of the filter and the memory requirements of the index can be influenced by changing the *index depth*, which determines up to which depth indexed terms are considered.

Minimizing the Size of a Path Index
Mark Stickel suggested to use *tries* for accessing path lists that correspond to a specific path (Stickel, 1989). In a trie common prefixes of paths are shared.

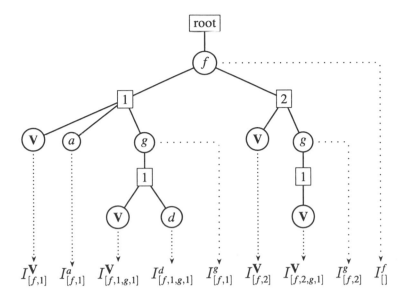

Figure 2. Accessing the path lists using a trie.

Stickel's proposal was motivated by providing efficient access to path lists. We discovered that the greatest advantage of using a trie is that *not all* paths contained in indexed terms have to be represented explicitly.

Figure 2 shows a trie that represents the paths stored for our example path index. Note that the set $I_{[]}^f$ represented by the node marked f in our example can easily be reconstructed by $I_{[f,2]}^{\mathbf{V}} \cup I_{[f,2,g,1]}^{\mathbf{V}}$ because from the structure of the trie we implicitly know that the indexed terms are either members of $I_{[f,2]}^{\mathbf{V}}$ or $I_{[f,2,g,1]}^{\mathbf{V}}$. Note that we could also have chosen the set $I_{[f,1]}^{\mathbf{V}} \cup I_{[f,1]}^a \cup I_{[f,1,g,1]}^{\mathbf{V}} \cup I_{[f,1,g,1]}^d$ to reconstruct $I_{[]}^f$. Consequently, the set $I_{[]}^f$ does not need to be maintained. Note that the sets $I_{[f,1]}^g$ and $I_{[f,2]}^g$ also do not need to be represented explicitly. Thus our path index contains path lists that correspond to leaf nodes of the trie only. The impact of this proposal on the size of a path index is strong: In experiments the size of the index was decreased to about one third of the original size with the speed of the retrieval almost being identical.

Path Indexing for AC Theories

Most indexing techniques only work for free terms. Unification within theories (Plotkin, 1972) such as associativity and commutativity (AC) usually is not supported. One approach to indexing in the presence of AC function symbols was described by Bachmair et. al. (Bachmair et al., 1993). They use a variant of discrimination trees in combination with bipartite graph matching to provide efficient retrieval of AC-generalizations of a query term. Unfortunately, their approach is not able to retrieve AC-unifiable terms.

In the following we sketch an extension of path indexing that is able to handle AC function symbols when looking for AC-unifiable terms, AC-generalizations, or AC-instances (Graf, 1996). Consider, for example, the candidate set for terms that are AC-instances of $+(a,+(g(b),x))$, where $+$ denotes an AC function symbol.

$$\texttt{instances}_{AC}(I,+(a,g(b),x)) = I^a_{[+,?]} \cap I^b_{[+,?,g,1]} \cap I^{+,\geq 3}_{[]}$$

Path indexing for AC theories considers flattened terms only, i.e. nested occurrences of AC function symbols like $+$ are changed in favor of AC function symbols with varying arities. In order to refer to *any* argument of a term, the question mark is used in paths. Therefore, the term set $I^a_{[+,?]}$ contains terms that have $+$ as a top symbol and the constant a as an argument. In addition to standard term sets, path indexing for AC theories also requires the maintenance of so-called *arity sets*. These sets contain terms that have a specific arity at a specific position. The set $I^{+,\geq 3}_{[]}$, for example, contains all terms that have an arity greater than or equal to 3 and whose top symbol is $+$. Thus AC-instances of $+(a,g(b),x)$ start with the top symbol $+$ and have two arguments a and $g(b)$. Additionally, they have an arity of at least 3 because the variable x has to be bound to the remaining arguments of retrieved terms.

In summary, the AC indexing problem can easily be reduced to ordinary indexing up to permutation of arguments of AC symbols, provided terms are flattened first.

5. TREE-BASED INDEXING

Tree-based indexes like discrimination trees, abstraction trees, and substitution trees support indexing of type $n : m$.

5.1. *Discrimination Tree Indexing*

A discrimination tree (McCune, 1988; McCune, 1992) is a variant of the trie data structure. The edges of a discrimination tree (DT) are marked with symbols in such a way that the tree represents the structure of all indexed terms and common prefixes of indexed terms are shared. Before terms are inserted, all variables are mapped to the unique symbol "$*$". Thus different variables in indexed terms are treated as equal and each path from the root of the tree to a

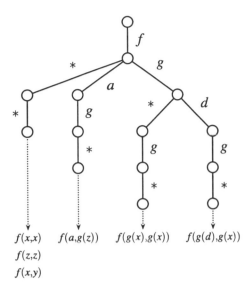

Figure 3. Discrimination tree.

leaf corresponds to a set of terms that have variables at the same positions and are syntactically equal at non-variable positions. In Figure 3 a discrimination tree is depicted.

To answer a query one has to traverse the tree using a backtracking algorithm. Insertion and deletion in discrimination trees is simple and can be accomplished very quickly. The tree's memory requirements depend on the sharing of common prefixes of indexed terms. In the example there are 3 terms which "end" in $g(*)$ and the whole tree consists of 14 nodes.

Subterm Indexing in Discrimination Trees

We distinguish retrieval *to* and retrieval *from* subterms. Retrieval *from* subterms refers to finding terms which are unifiable with, instances of, or generalizations of a subterm of the query term. This type of indexing is easy to achieve since we only have to perform a top-level retrieval for all subterms of the query term. Therefore, retrieval *from* subterms can be supported by any of the presented indexing techniques. More complicated is the retrieval *to* subterms: Here we have to find subterms of indexed terms as partners for our query term.

A straightforward solution for the retrieval *to* subterms is to simply insert all subterms of the indexed terms into the index. Obviously, a regular top-level retrieval in such an index will actually perform retrieval to subterms. The memory requirements of this method are very high; to get an index of acceptable size we employ structure sharing. Common subterms of indexed terms are shared and each of the shared subterms is inserted into the index just once.

In context with discrimination trees we still have another possibility: An additional data structure is used to index the index, i.e. to provide access to sets of subterms of the indexed set. For example, we could simply maintain a hash table for accessing lists of discrimination tree edges that are marked with a specific symbol. In order to achieve retrieval to subterms the search algorithm is started on every edge of the tree that is labeled with the top symbol of the query term.

Other indexing techniques for the retrieval of subterms have been introduced in the literature. For example, Ramesh, Ramakrishnan, and Sekar proposed a method that pre-compiles the indexed set into a string matching automaton (Ramesh et al., 1994). As with all compilation techniques, insertion in such an index is expensive since the indexed set has to be recompiled. However, the retrieval itself should be faster. A comparison of both approaches with respect to their performance in real applications should assist in deciding which technique to use in practice.

5.2. *Abstraction Tree Indexing*

The nodes of abstraction trees (Ohlbach, 1990) are labeled with lists of terms in such way that the free variables of the term list at node N and the term lists of a child of N form the domain and the codomain of a substitution. Each path from the root to a leaf of the abstraction tree (AT) represents a term. This term can be computed by applying the substitutions stored in the path to the term list attached to the root of the tree.

Due to a better grouping of the terms the abstraction tree in Figure 4 consists of 9 nodes only, which is less than the number of nodes in the discrimination tree for the same example, as shown in Figure 3. However, abstraction trees contain lots of variable renamings like $\{x_1 \mapsto x_3\}$ and $\{x_4 \mapsto x_6\}$ that are not necessary. Additionally, the original variables of indexed terms (in contrast to auxiliary variables in the substitutions at inner nodes) occur in leaf nodes of abstraction trees only. An algorithm looking for instances of a query term unfortunately cannot exploit the fact that a variable in an indexed term must not be instantiated at inner nodes of the tree, since this property does not hold for the auxiliary variables.

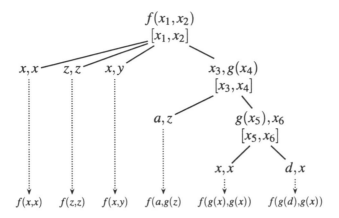

Figure 4. Abstraction tree.

5.3. *Substitution Tree Indexing*

The labels of substitution tree nodes are substitutions. Each branch in a substitution tree (Graf, 1995a) therefore represents a binding chain for variables. The substitutions of a branch from the root node down to a particular node can be composed and yield an instance of the root node's substitution. Substitution trees (ST) can index both terms and substitutions.

A term or a substitution is renamed before it is inserted into a substitution tree. This *normalization* changes all variables to so-called *indicator variables* denoted by $*_i$. For example, the normalized versions of the terms $f(x, f(y, x))$ and $f(u, f(v, u))$ both are $f(*_1, f(*_2, *_1))$. The normalization has three reasons: First, normalization results in more sharing in the index. Second, it is necessary to distinguish between variables occurring in the query and in the indexed terms for some retrieval tasks. Finally, variables of indexed terms may occur at arbitrary positions in the substitution tree because they can now be distinguished from auxiliary variables.

Figure 5 shows our standard term set. We only need 3 auxiliary variables and the whole tree contains just 9 assignments in contrast to the 16 assignments of the abstraction tree in Fig. 4. Essentially, substitution trees unify features of abstraction and discrimination trees.

In our example we stored 6 substitutions in the substitution tree. The domains of all these substitutions are identical, but this is not necessarily so. Substitution trees may also contain substitutions with different domains. We use a backtracking algorithm to find substitutions with specific properties in the tree. All retrieval algorithms are based on backtrackable variable bindings and algorithms for unification and matching that take variable bindings into

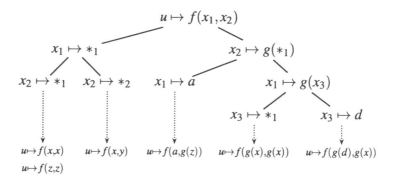

Figure 5.   Substitution tree.

account. Insertion of a substitution into the index is a complex operation. The deletion of entries is much easier and even complex deletion operations, like the deletion of all compatible substitutions in a substitution tree, can easily be accomplished. In a large number of experiments with term sets that have been created in real OTTER applications we demonstrated, that the average retrieval performance in substitution trees is faster than the average retrieval time of any other technique. In most of the experiments substitution trees even showed the best performance of all techniques that have been compared. The memory consumption of substitution trees is moderate, no matter whether terms or substitutions are indexed. When it comes to memory consumption, substitution trees are only beaten by set-based methods, which, on the other hand, perform worse in retrieval.

Indexing Substitutions with Substitution Trees
Substitutions can be represented by any index that is able to handle terms. We simply introduce a new function symbol s and store the codomain of the substitutions as arguments of s. The arity of s must equal the number of different variables that occur in the domains of the represented substitutions. In the following example, we consider the substitutions

$$\sigma = \{x \mapsto f(a,b), y \mapsto c\} \quad \text{and} \quad \tau = \{x \mapsto f(u,b), z \mapsto d\}$$

which can be represented by the two terms

$$s(x\sigma, y\sigma, z\sigma) = s(f(a,b), c, z) \quad \text{and} \quad s(x\tau, y\tau, z\tau) = s(f(u,b), y, d)$$

Figure 6 demonstrates the representation of the substitutions σ and τ in a discrimination tree, an abstraction tree, and a substitution tree.

The disadvantage of using discrimination trees is that the substitutions have to be represented by terms. There is very little structure sharing in the

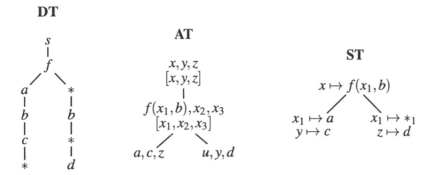

Figure 6. Representation of substitutions.

tree if these terms get more complex. The possibility of sharing further decreases if we have to deal with many different domain variables.

For the representation of substitutions the structure of abstraction trees has to be slightly modified. In abstraction trees indexed variables may only occur at leaf nodes. In order to represent the domain of the indexed substitutions, we have to allow for the occurrence of indexed variables at the root node of the tree also. An abstraction tree used to store substitutions contains many variable renamings since even domain variables that do not occur in the substitutions to be represented have to be considered (cp. variable z in the substitution σ of our example).

Substitution trees handle substitutions in a very natural manner. Even substitutions that do not have identical domains can easily be represented. In summary, substitution trees provide the better representation for substitutions compared to other indexing techniques.

6. EXPERIMENTS WITH TERM INDEXING TECHNIQUES

Unfortunately, a formal comparison of indexing techniques by complexity results will generally fail in justifying a preference for one method. The worst-case assumptions employed by most formal analyses are unrealistic and generally do not occur in practice. An average-case analysis is difficult to obtain because there is no reasonable definition of the average case. Therefore, the selection of an optimal indexing technique is possible solely on the basis of performance in the actual application. Experiments of both retrieval and maintenance operations on indexed sets that occur in practice have been considered. The experiments show that the substitution trees we introduced generally achieve a better performance than other indexing techniques. For the sake of brevity, we will in the sequel concentrate on our results for the retrieval of unifiable terms. A detailed analysis including generalizations and instances can be found in (Graf, 1995b).

Representing a major retrieval task, the search for unifiable terms is reported in detail. Table I contains the retrieval times needed for an $n : m$ type search based on m retrievals of type $n : 1$. The experiments were performed on a Sun SPARCstation ELC computer with 16 MBytes of RAM. The retrieval times include the construction of query trees for path indexing as well as the time spent for test unifications and matchings for filters that are not perfect. The retrieval times do not include the explicit construction of the unifying or matching substitutions. The term sets I and Q have been created both randomly (AVG, WIDE, GND, LIN, DEEP) and using real OTTER applications (EC+, EC-, CL+, CL-, BO+, BO-). The times in Table I refer to retrievals using no indexing at all (NONE), path indexing, discrimination tree indexing, abstraction tree indexing, and substitution tree indexing.

In the experiments discrimination trees and abstraction trees show good performance for retrieving unifiable entries. However, substitution trees really dominate the experiments. They seem to be the most suitable technique for accessing unifiable terms. The use of path indexing techniques cannot be recommended for this purpose.

6.1. *Experiments on "Average" Term Sets with Varying Cardinality*

In addition to considering completely different term sets, we also investigated how the retrieval times change if the number of terms in the indexed set is varied. The indexing techniques were tested by inserting "average" random term sets containing 1 000, 2 000, . . ., 10 000 terms. The "average" terms were created in the same way as the set AVG. The query set contained 1 000 terms in all experiments. The results of the experiments with term sets of different size are graphically illustrated in Figure 7.

Substitution trees show the best performance in the experiments on unifiable "average" terms. The performance of the other techniques is similar. Note that the time needed to find a single unifiable term in the index decreases with the number of indexed terms, regardless of the applied indexing technique. However, if none of the indexing techniques is employed, the performance of the retrieval is again very poor.

Table I. Retrieval of unifiable terms [in Seconds].

I	Q	NONE	PI	DT	AT	ST
EC+	EC+	15.0	20.8	18.5	19.0	**6.8**
EC+	EC−	25.6	35.7	29.9	34.7	**13.5**
EC−	EC+	28.2	31.9	34.8	52.7	**3.4**
EC−	EC−	100.9	109.0	112.6	125.0	**29.9**
CL+	CL+	52.3	16.8	15.2	7.5	**4.5**
CL+	CL−	68.5	23.9	27.5	10.3	**3.4**
CL−	CL+	70.6	25.7	40.9	13.4	**11.4**
CL−	CL−	27.8	12.8	8.6	4.4	**4.3**
BO+	BO+	233.0	16.8	7.7	6.4	**5.7**
BO+	BO−	214.5	14.2	3.6	3.5	**3.4**
BO−	BO+	217.0	3.7	4.6	**1.5**	**1.5**
BO−	BO−	205.9	8.3	2.2	2.2	**2.1**
AVG	AVG	136.1	20.6	16.2	14.8	**12.1**
WIDE	WIDE	155.6	39.8	**15.7**	17.7	22.7
GND	GND	124.2	2.8	1.6	1.5	**1.3**
LIN	LIN	131.9	16.3	12.6	12.8	**10.5**
DEEP	DEEP	138.6	29.9	**24.2**	25.0	26.0

7. TERM INDEXING APPLICATIONS IN THEOREM PROVERS

In this chapter we scrutinize the applicability of indexing techniques to different operations in theorem provers. Although we shall mainly concentrate on indexing for synthetic calculi, interesting applications of indexing in analytic calculi exist: In bottom-up provers identical subgoals may have to be proven over and over again, thus causing a large amount of redundancy. Nevertheless, redundant proofs for identical subgoals can be avoided by maintaining an index of lemmata. Since such an index contains all previously proven subgoals, the proof of a current goal can be skipped if a generalization of the current goal is found in the index. A version of the SETHEO prover (Letz et al., 1922), for instance, uses the dynamic path indexing approach (Graf, 1995b) to implement lemma maintenance.

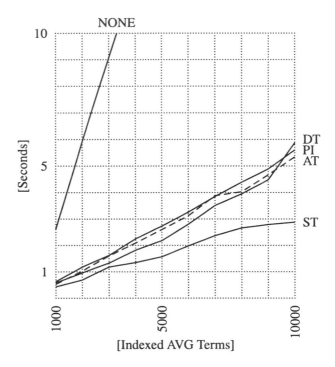

Figure 7.   Retrieval of terms unifiable with terms contained in AVG.

7.1. *Binary Resolution*

Binary resolution (Robinson, 1965b) was among the first inference rules that
have been supported by indexing techniques. One can use an index containing
all literals of kept clauses. For a given literal in a given clause we find resolution partners by retrieving complementary unifiable literals in the index.

7.2. *Binary Resolution with Theory Unification*

Plotkin proposed a method for reasoning with equality by turning a set E
of equations into an E-theory-unification algorithm (Plotkin, 1972). For two
given terms s and t such an algorithm computes all the most general substi-
tutions σ in $\mathrm{mgu}_E(s,t)$ such that E implies $s\sigma = t\sigma$, or $s\sigma =_E t\sigma$ for short.
Consequently, the modified binary resolution rule uses theory unifiers instead
of syntactical unifiers. Our version of path indexing can support the retrieval
of AC-unifiable terms, AC-instances, and AC-generalizations.

7.3. *Hyperresolution*

The search for electrons containing literals unifiable with literals of a nucleus can be supported similar to finding partners for binary resolution. However, the obstacle for efficiently implementing hyperresolution (Robinson, 1965a) is the combinatorial explosion that we face when looking for *simultaneous* unifiers for the literals of the nucleus. Let us consider an example: If the nucleus has 3 negative literals and for each of these literals 10 unifiable electrons exist, we have to test a total of $10^3 = 1000$ combinations of unifiers for compatibility. In order to master this combinatorial explosion, we propose to use the multi-merge on substitution trees: A substitution tree is maintained for every negative literal occurring in the nucleus. Each of these substitution trees contains the unifiers with electrons for a specific negative nucleus literal. Multi-merging the substitution trees results in a new substitution tree that contains all simultaneous unifiers that can be found by combining the unifiers of the negative nucleus literals and their electrons.

7.4. *Unit-Resulting Resolution*

Unit resulting resolution can be supported by indexing techniques and retrieval operations in a way similar to hyperresolution. In fact, it turns out that it is possible to design a unit resulting prover on the basis of substitution trees (instead of clauses and literals) as the primary data structure. We implemented the PURR theorem prover for testing this paradigm and for investigating the impact of indexing on a distributed application (Meyer, 1996). Compared to OTTER, the PURR system shows very competitive behavior in various experiments.

7.5. *Subsumption*

In resolution theorem provers the subsumption test for two given clauses is only a subproblem of the two bigger problems *forward* and *backward* subsumption. An index-based algorithm for forward subsumption for a given clause R aims at performing very few of the expensive subsumption tests on clauses. The main idea is to maintain all literals of kept clauses in an index. The index is searched for literals more general than those occurring in R until *all* literals of a kept clause C have been found. Only then a regular subsumption test on C and R is performed.

Backward subsumption for a clause R can be supported by indexing in the following manner: The index yields kept clauses D that contain instances of *all* literals occurring in R. A regular subsumption test on R and D is applied only for these clauses D.

Joachim Becker investigated these index-based algorithms and compared them to standard approaches (Becker, 1994): In many theorem provers a subsumption test on clauses is performed as soon as a single literal has been found in the index. Compared to this approach Becker's implementations of indexed-based subsumption showed much better average performance.

7.6. *Completion*

Most of the time spent during completion is used for rewriting terms. Thus we propose to maintain the *left-hand* sides of the rewrite rules R in an index. Retrieval of *more general* terms *from* the subterms of a term t yields rewrite rules applicable to t.

The search of critical pairs can also be supported by such an index. For every newly created rule $l \to r$ we find critical pairs by overlapping the left-hand sides of the rewrite rules in two ways: First, we retrieve all rules that contain a subterm on their *left-hand* side which is *unifiable* with l. This corresponds to indexing *into* subterms. Second, we have to retrieve *unifiable* left-hand sides *from* the subterms of l.

For both operations an index containing the left-hand sides of the rules is needed. The index has to support retrieval from and into subterms both for unifiable and more general left-hand sides. We recommend to use discrimination trees in this context.

8. CONCLUSION

We presented various indexing methods for first-order terms on the basis of a classification into attribute-based, set-based, and tree-based techniques. Moreover, we explained how indexing methods can be employed to improve the efficiency of various inference rules for theorem provers. The following conclusions can be drawn:

- The standard technique for using indexes is to maintain a set of n terms in an index and to start a retrieval for a single query term. As extensions of these $n : 1$ indexing algorithms, $n : m$ indexing techniques handle sets of query terms by also representing them in an index. For example, the merge operation supports the computation of simultaneous unifiers for hyperresolution by computing all pairs of unifiable entries occurring in two different indexes.

- A comparison of indexing techniques with respect to their functionality shows, that there is no general purpose indexing method. For example, discrimination trees can handle subterm retrieval, but they are not appropriate for accessing substitutions. Substitution trees, however, achieve the latter but cannot handle mixed retrieval and maintenance operations. Finally, path indexes can be searched and manipulated in parallel, but their performance in standard retrieval (at least for finding unifiable terms) is rather poor.

- There are many applications of indexing techniques in theorem proving that require the creation of special purpose indexing techniques. For example, a variant of the discrimination tree method provides retrieval into subterms without explicitly inserting the subterms of all indexed terms into the index. Another example is our version of path indexing for AC theories (Graf, 1996).

- Though the worst-case complexities of various unification algorithms are known, there is still no consensus on which algorithm to use in practice. When judging term indexing techniques similar problems occur. Therefore, the selection of a "good" indexing method is possible only on the basis of performance tests. In experiments with several implementations substitution trees provided a much better average retrieval performance than path and discrimination tree indexing.

The interested reader may find an exhaustive survey on up-to-date term indexing techniques and their applications in the first author's thesis (Graf, 1995b).

REFERENCES

Ait-Kaci, H.: 1991, *Warren's Abstract Machine: A Tutorial Reconstruction*, Logic Programming. Cambridge, Massachusetts: The MIT Press.

Bachmair, L., T. Chen, and I. V. Ramakrishnan: 1993, 'Associative-commutative discrimination nets'. In: *Proceedings TAPSOFT '93*. pp. 61–74.

Bayer, R. and E. McCreight: 1997, 'Organization and Maintenance of Large Ordered Indexes'. *Acta Informatica* **1**, 3.

Becker, J.: 1994, 'Effiziente Subsumption in Deduktionssystemen'. Diploma thesis, Universität des Saarlandes, Saarbrücken, Germany.

Butler, R. and R. Overbeek: 1994, 'Formula databases for high-performance resolution/paramodulation systems'. *Journal of Automated Reasoning* **12**, 139–156.

Chang, C. L. and R. C. T. Lee: 1973, *Symbolic Logic and Mechanical Theorem Proving*, Computer Science and Applied Mathematics. New York: Academic Press.

Christian, J.: 1993, 'Flatterms, discrimination trees, and fast term rewriting'. *Journal of Automated Reasoning* **10**(1), 95–113.

Date, C. J.: 1991, *An Introduction to Database Systems*, Vol. I. Addison-Wesley. 5th edition.

Graf, P.: 1994, 'Extended path-indexing'. In: *12th Conference on Automated Deduction*. pp. 514–528.

Graf, P.: 1995a, 'Substitution tree indexing'. In: *6th International Conference on Rewriting Techniques and Applications RTA-95*. pp. 117–131.

Graf, P.: 1995b, 'Term Indexing'. Phd thesis, Universität des Saarlandes, Saarbrücken, Germany. also published as Springer LNAI 1053, 1996.

Graf, P.: 1996, 'Path-indexing for AC-theories'. In: *13th Conference on Automated Deduction*. pp. 718–732.

Henschen, L. J. and S. A. Naqvi: 1981, 'An improved filter for literal indexing in resolution systems'. In: *Proceedings of the 6th International Joint Conference on Artificial Intelligence*. pp. 528–529.

Hewitt, C.: 1971, 'Description and theoretical analysis of Planner: A language for proving theorems and manipulating models in a robot'. Phd thesis, Department of Mathematics, MIT, Cambridge, Mass.

Knuth, D. and P. Bendix: 1970, 'Simple Word Problems in Universal Algebras'. In: J. Leech (ed.): *Computational Problems in Abstract Algebras*. Pergamon Press.

Letz, R., J. Schumann, S. Bayerl, and W. Bibel: 1922, 'SETHEO: A high-performance theorem prover'. *Journal of Automated Reasoning* **8**(2), 183–212.

Lusk, E. and R. Overbeek: 1980, 'Data structures and control architectures for the implementation of theorem proving programs'. In: *5th International Conference on Automated Deduction*. pp. 232–249.

McCune, W.: 1988, 'An indexing method for finding more general formulas'. *Association for Automated Reasoning Newsletter* **1**(9), 7–8.

McCune, W.: 1990, 'Otter 2.0'. In: *10th International Conference on Automated Deduction*. pp. 663–664.

McCune, W.: 1992, 'Experiments with discrimination-tree indexing and path-indexing for term retrieval'. *Journal of Automated Reasoning* **9**(2), 147–167.

Meyer, C.: 1996, 'Parallel Unit Resulting Resolution'. Diploma thesis, Universität des Saarlandes, Saarbrücken, Germany.

Morris, R.: 1968, 'Scatter storage techniques'. *Comm. ACM* **11**(1), 38–43.

Ohlbach, H. J.: 1990, 'Abstraction tree indexing for terms'. In: *Proceedings of the 9th European Conference on Artificial Intelligence*. pp. 479–484.

Plotkin, G.: 1972, 'Building in equational theories'. In: *Machine Intelligence*, Vol. 7.

Ramamohanarao, K. and J. Shepherd: 1986, 'A superimposed codeword indexing scheme for very large Prolog databases'. In: *Proceedings of the 3rd International Conference on Logic Programming*. London, England.

Ramesh, R., I. V. Ramakrishnan, and R. C. Sekar: 1994, 'Automata-driven efficient subterm unification'. In: *Proceedings of LICS-94*.

Robinson, J. A.: 1965a, 'Automated deduction with hyper-resolution'. *International Journal of Comp. Mathematics* **1**, 227–234.

Robinson, J. A.: 1965b, 'A machine-oriented logic based on the resolution principle'. *Journal of the ACM* **12**(1), 23–41.

Stickel, M.: 1989, 'The path-indexing method for indexing terms'. Technical Note 473, Artificial Intelligence Center, SRI International, Menlo Park, CA.

Warren, D. H. D.: 1983, 'An abstract prolog instruction set'. SRI Technical Note 309, SRI International.

Wise, M. and D. Powers: 1984, 'Indexing prolog clauses via superimposed code-words and field encoded words'. In: *Proceedings of the IEEE Conference on Logic Programming*. pp. 203–210.

Wos, L.: 1992, 'Note on McCune's article on discrimination trees'. *Journal of Automated Reasoning* **9**(2), 145–146.

CHAPTER 6

DEVELOPING DEDUCTION SYSTEMS:
THE TOOLBOX STYLE

1. INTRODUCTION

Implementing an automated theorem proving system poses practical problems of an order of magnitude a naive implementor hardly expects. Even an experimental version requires the solution of numerous technical problems each of which may not be too difficult (although some are), but whose mere number is frightening, such as defining basic data structures and implementing the algorithms for fundamental tasks like parsing input, or copying terms. Furthermore, the engineering discipline of building deduction systems is — after twenty years of experience — rather advanced, and many tasks such as indexing techniques for very large sets of formulae (cf. the article by Graf on indexing), fast first order unification algorithms (Paterson and Wegman, 1978; Martelli and Montanari, 1982; Corbin and Bidoit, 1983; Dowek et al., 1995) or the implementation of higher order constructs require an amount of experience and know-how that may take several years to develop. The effect is that the effort spent on the really interesting part of the experiment, namely testing a new calculus or the modification of a given one, is forced to take a back seat. Furthermore, even *moderate* efficiency — which may be enough for most experimental versions — depends on the particular encoding of the data structures and algorithms involved.

As an alternative one might want to use (sub-)systems "off the shelf". However, almost inevitably, existing systems are more often than not unsuited for the task at hand. So in order to utilize them one has to either build expensive adaptors and devices to transform in- and output, or even to modify the code itself. Even if *well documented* source code is available (which is of course not always the case for experimental software), this is not acceptable either, for it takes a trained programmer (with detailed knowledge of theorem proving) to foresee the effect of changes to foreign software.

A solution to this dilemma to be proposed here is the *toolbox approach*. A toolbox is a *library* of data structures and algorithms that is built on top of a common programming language (in our case Common Lisp and CLOS), from which a developer can choose. In contrast to earlier proposals such as LMA (Lusk et al., 1982a; Lusk et al., 1982b) the underlying programming

W. Bibel, P. H. Schmitt (eds.), Automated Deduction. A basis for applications. Vol. II
© 1998 *Kluwer Academic Publishers. Printed in the Netherlands*

language should not be encapsulated, so the user is not *forced* to use any construct of the toolbox, but free to program on his own as much as he likes. This way the developer of a new deduction system is able to concentrate on his critical, new topics, and he may use well known premanufactured parts for the rest.

In order to be really useful, a toolbox should fulfil several criteria. In particular it should

1. *be modular*, so that the user may select his favourite mix of components
2. *support different logical paradigms*
3. *contain a variety of critical algorithms*
4. *enable combinations* of as many choices as possible
5. *support* fast implementation (*rapid prototyping*)
6. *enable* at least moderate *efficiency*

In this article we describe the central aspects of the KEIM toolbox (Huang et al., 1994a; Nesmith, 1992; Richts and Nesmith, 1994) and evaluate its merits. Besides the fact that KEIM has turned out to be an essential for the implementation of the mathematical assistant ΩMEGA (Benzmüller et al., 1997; Huang et al., 1994b), it served as a prototype for the toolbox approach at several sites.

After the presentation of the main concepts of KEIM we shall discuss the experience with KEIM in the development of ΩMEGA and in several other applications. The KEIM project has been funded by the German basic research agency (Deutsche Forschungsgemeinschaft, DFG) as part of the "Schwerpunktprogramm Deduktion" (special funding scheme on deduction), grant Si 372/2–2, under the leadership of Professor Jörg Siekmann at Saarbrücken University. The system can be obtained from our ftp-server under http://www.ags.uni-sb.de/software.

2. KEIM

One of the main demands on a toolbox is its generality. This includes independence of particular approaches to theorem proving, as well as the choice of logic (and calculus). Furthermore it should support fully automated theorem provers as well as interactive systems (for which ΩMEGA is a case in point). Therefore data structures not usually found in automated theorem provers, such as e.g. partial proofs, proof plans, mathematical data bases, etc., must be handled.

Therefore we shall sometimes use the term *problem manipulation system* instead of "deduction system". A problem can be in several states, ranging from the original formulation to "proved" (possibly via several different proofs). Steps that transform one problem state into the next can be performed by automated proof procedures as well as by planning steps (i.e. a planning process), by a tactic based system, or simply by the user.

KEIM (Kernel for Enhancement of Implementation and Maintenance)[1] is written in Common Lisp (Steele jr., 1984) and relies on the object oriented features supplied by CLOS (Common Lisp Object System). It turned out that the choice of an object oriented approach was indeed critical. KEIM does not encapsulate the underlying programming language, as mentioned above, but instead adds functionality to it by supplying a hierarchy of classes defining objects related to deduction, as well as functions operating on them, thus still permitting free access to the underlying Lisp system.

The data structures for the representation of terms, types and formulae are based on a simply typed lambda calculus as the lowest representational level. Thus they are not restricted to first order languages. There exist representations for inference rules, problems and their states, as well as proof formats (currently resolution style and natural deduction). There are fundamental algorithms, such as unification, matching, substitution, representations and algorithms for inference rules like resolution, factorization and paramodulation. Due to the higher order representation language we have to deal with different forms of normalization. We have also included support routines for I/O, like parsing, pretty printing, and tools for the definition of hierarchical menus and command interpreters. A feature particularly interesting is the support for automatic documentation extraction.

Since this is not a KEIM manual (this is contained in the KEIM distribution), we will omit details and avoid KEIM-specific notation, but concentrate on those aspects and concepts of particular importance for the toolbox approach as such.

The overall module structure is shown in Figure 1. As can be seen, there is one single module that is responsible for the adaptation to different Lisp implementations. So in order to port KEIM to a new platform, changes are restricted to this distinguished location. By exploiting Lisp's feature system we have supplied code for several versions of Lisp, so they are recognized automatically. A toolbox should support numerous different calculi, the current KEIM version, however contains only modules for resolution style and natural deduction proofs: Our hope is to motivate researchers from different subareas of the field to contribute further modules to KEIM for their respective subfield.

---

[1] "Keim" is the German word for "seed", which stresses the fact that KEIM supplies the innermost ingredients of a problem manipulation system.

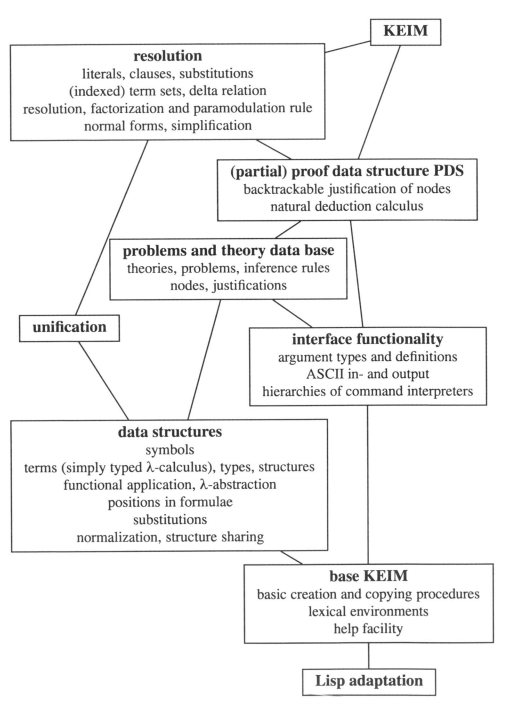

Figure 1.   The KEIM module hierarchy

3. Data Structures

Though the objects needed for the representation of various calculi differ, there exist entities common to nearly all of them. Here we present the most prominent ones and also include a discussion of elementary algorithms.

3.1. *Terms and Formulae*

Terms are the fundamental data structure. They appear as parts of *formulae* or, more specific, *clauses*, of some logical language. Most current theorem provers use some variant of first order logic. However, human mathematical knowledge is often more adequately formulated in a higher order language. Systems such as e.g. TPS (Andrews et al., 1996) or LEO (Benzmüller, 1997)) are capable of handling this. Therefore KEIM allows for the representation of formulae using a typed higher order syntax, a version of the lambda calculus, of which first order logic is a special case.

As usual in higher order logic, the difference between formulae and terms disappears, so that formulae can be seen as special forms of terms (namely terms of some particular type). The general pattern for term construction in KEIM therefore consists of building complex terms using the constructors *functional application* and *lambda abstraction* from the elementary constants and variables. This appears to be general enough to handle most representational problems, but for convenience we have added specialized constructs that supply accessors and constructors for well known subsets (such as first order formulae), since not every potential user should have to cope with the full generality (and notation) of the lambda calculus.

In addition to the representation of terms we must have a means to navigate within them, i.e. a concept of *positions*, as well as the respective accessor and constructor functions. Further important data structures are substitutions. Also included in KEIM are various algorithms for *traversing* and *labelling* structures. For efficiency reasons, and in order to keep terminology as broad as possible, we have included several special cases of formulae, such as literals or clauses, together with specialized accessor functions.

Frequently there are structures that are not terms in a strict sense, but in some respects behave as such. Generally those structures contain particular substructures that are terms. Examples are lines in a natural deduction proof, which, amongst other things like hypotheses and justifications, contain a formula. In order to make all the functions designed for terms proper accessible, a particular inheritance mechanism (called a mixin class in object oriented languages) is used. The effect is that the only thing a programmer has to do

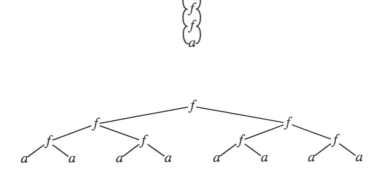

Figure 2.   An example of structure sharing.

is to define an accessor function for the term-like structure that returns the proper term. The rest is done by KEIM.

3.2. *Structure Sharing*

The concrete representation of terms can be decisive for the efficiency of the algorithms that use them. An important technique here is *structure sharing*. This issue is invisible at the calculus level and therefore should be *transparently managed* for the unskilled (or uninterested) user, but leaving the opportunity of direct access to details for the trained one if he so wishes. There are numerous variants between the extremes "never share" and "always share common subterms". Depending on the choice, all the algorithms responsible for *constructing, accessing, copying* (sub-)terms or *checking* them *for equality* differ profoundly. There are also different space requirements, as can be seen in Figure 2, where the same term is shown fully shared (top) and completely unshared (bottom). So we have to supply *optimized* versions of the algorithms for every case. This of course should be left transparent, but accessible.

   Another issue concerns equality of terms: besides pure structural equality, λ-calculus terms may be *normalized* in different ways, so there must exist normalization procedures, as well as different comparison algorithms, e.g. with respect to α-equality (which also shows up in first order theorem proving as variable renaming) or βη-equality. This touches important representational issues, such as the use of de Bruijn indices (Bruijn, 1972; Nadathur, 1994; Nadathur and Wilson, 1994). The different forms of normalization must be provided, for they all have their respective merits. C-normalized terms, e.g., where each function has exactly one argument (cf. Figure 3), allow for

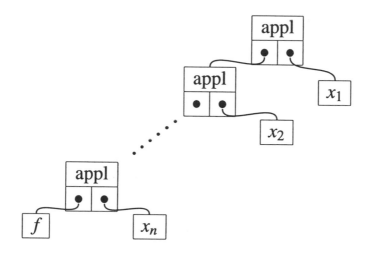

Figure 3.   The term $f(x_1,\ldots,x_n)$ in c-normal form.

simpler algorithms, at the cost of increased space requirement, as compared to n-normalized ones, where each function is expanded to maximal arity (cf. Figure 4).

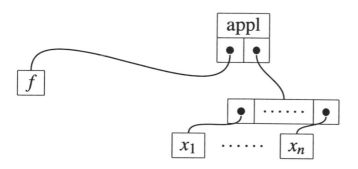

Figure 4.   The term $f(x_1,\ldots,x_n)$ in n-normal form.

The cross product of different forms of structure sharing and normalization yields a large amount of possible comparison and copying algorithms, all of which have to be supplied as *optimized* algorithms. These basic procedures are located at the bottom of any implementation and are thus very frequently called. The main reason for the failure of most naive implementations is that these algorithms are too inefficiently implemented.

3.3. *Types and Sorts*

In many applications, terms carry some additional semantic *annotation*, such as sorts, types or colours. This annotation may serve different purposes, ranging from restricting the applicability of operations (and thus effectively cutting down the search space) to introducing redundancy for error checks or plainly adding to the intuition of a human reader. The most common kind of annotation are *types*. KEIM by default supplies Church types (Church, 1940), but structurally terms and types behave very much the same, which led to the idea that both are special cases of a more general entity we call *structures*. Structures are syntactically composed as applications or abstractions from simpler structures, constants and variables being the bottom elements. In addition, every structure may have an *annotation*. The fundamental idea now is that the annotation of a structure is structurally equivalent to the structure itself, i.e. that e.g. the annotation of an abstraction is also an abstraction. As an example look at the embedding of simple Church types into this annotation scheme. Base types are the constants, and functional types can be seen as λ-abstractions: $\alpha \to \beta$ thus becomes $\lambda\alpha.\beta$. The application of a function f of type $\alpha \to \beta$ to a term t of type α has the type $(\lambda\alpha.\beta)\alpha$, which is also an application, but can be β-reduced to β. To see this, one has to slightly generalize the notion of λ-abstraction in the following way: if in $\lambda\alpha.\beta$ α is an arbitrary term, not necessarily a single variable, then β-reduction of the term $(\lambda\alpha.\beta)\gamma$ is defined as "if α and γ are unifiable with substitution σ, then $\sigma\beta$, else undefined".

By varying the kind of annotations used, different *term systems* can be defined. Further possibilities beyond Church types include *sorts* and fixed arities (for the representation of first order terms). The following example shows how sorts can be handled. It is taken from a sorted formulation of the well known Schubert's steamroller problem (Stickel, 1986). The formula

$$\forall x : \text{snail} \; \exists y : \text{plant} \quad \text{eats}(x,y)$$

can be represented (interpreting ι as the top sort) as

$$\forall_{(\iota \to o) \to o} \lambda x_{\text{snail}} \cdot \exists_{(\iota \to o) \to o} \lambda y_{\text{plant}} \cdot \text{eats}_{\text{animal} \times \iota \to o} x_{\text{snail}} y_{\text{plant}}$$

which possesses the sort

$$(\lambda(\lambda\iota.o).o)\lambda\text{snail}.$$

$$[(\lambda(\lambda\iota.o).o)\lambda\text{plant}.$$

$$\{\lambda\text{animal}.\lambda\iota.o \quad \text{snail} \quad \text{plant}\}],$$

i.e. the sort o.

Thus, sorts can be treated very similarly to types, and the difference is kept internal to the unification algorithms.

3.4. *Term Sets*

Terms usually come in sets. If these are very large, as is typically the case in theorem proving, concepts like indexing (cf. respective article in this volume) play a fundamental role. However, the description of proof procedures frequently abstracts from the particular indexing scheme. KEIM therefore supplies uniform access procedures, whose interface abstracts from these choices, so that a user may view term sets as plain unordered sets. Then the underlying indexing scheme can be changed without having to modify the rest of the algorithms built on top. Currently KEIM supports an indexing scheme based on the generalization of simple substitution tree indexing (Graf, 1995) (cf. also the article on term indexing) to higher order terms (Klein, 1997).

4. INFERENCE RULES

An important aspect of a toolbox is to support a variety of calculi. Since there are often startling differences between them we offer a rather general framework, whose view is that some piece of knowledge, is *derived* from a set of *ancestors*. Calculi differ not only in the concrete choice of inference rules used for this derivation, but also in the *direction* in which these are applied. KEIM therefore contains a mechanism to map an abstract inference rule to a concrete rule application via so called *outline patterns*. As an example consider modus ponens (MP). This inference rule can, in its propositional form, be given by the abstract rule

$$\frac{A, A \to B}{B}$$

A calculus might use the rule "forward" (given A and $A \to B$ as parents, B is derived), or in one of the "backward" directions (given B and $A \to B$, abductively derive A; or given A and B, derive $A \to B$). Of course these rules have quite a different logical status, so in addition to just generating the resulting formulae, *justifications* have to be computed and tagged to all the formulae involved. So e.g. though the direction of computation in the abductive case shown above is backward, the justification still gives the logical dependence, that is B is justified by the application of MP to A and $A \to B$.

The idea of outline patterns is that only the implementor of an inference rule should have to cope with the different behaviour of the respective directions. The user simply passes some arguments to the rule, which by itself —

driven by the particular presence or absence of arguments — determines the appropriate operation, an extreme case being a simple check, if all arguments are present.

Furthermore, there exist "layered" calculi: a sequence of inference steps of a base calculus can be condensed to yield a single step in a more abstract calculus. There are various incarnations of this idea, e.g. in tactical theorem proving (Gordon et al., 1979) or proof planning (Bundy, 1988), where the abstract inference steps show up as tacticals, tactics or methods. Also proofs at the *assertion level* (Huang, 1996) are based on this idea. For this type of reasoning, KEIM offers the possibility to define *expansion functions* that relate the levels of abstraction to one another by telling how to compute the detailed steps given a more abstract one.

In the current version of KEIM, two particular calculi are supported, namely resolution (Robinson, 1965) (including paramodulation (Robinson and Wos, 1969)) and a variant of natural deduction (Gentzen, 1935).

5. PROBLEMS AND PROOFS

Proofs may be represented at different levels of abstraction. There may even be different levels of abstraction in distinct parts of a proof. In particular in an interactive system the user may want to manipulate the abstraction level by expanding or conflating subproofs. This yields different *views* of the same proof.

It may also be the case, that one wants to look at a proof that is not yet completed (or maybe can not be completed, because the given problem is not a theorem). So KEIM supports the concept of a *partial proof*. The chronological sequence of partial proofs is another dimension of looking at a proof under development. During the proof search one might want to *backtrack* in this sequence.

The data structure representing these two dimensions of a proof (chronological completion vs. refinement) is called PDS (for plan data structure) in KEIM. The term "plan" instead of "proof" results from the insight, that plans are the more general notion, proofs being the particular instance of fully expanded (and verified) plans.

Besides these PDSs KEIM contains structures to describe problems and relate PDSs to problems, for a single problem may have several different proofs, which one might wish to store in a data base for retrieval (e.g. for analogical proofs, cf. the article on proof reuse) or educational purposes (tutoring systems).

6. MATHEMATICAL DATA BASES

Through its *theory* concept, KEIM offers a way to describe a hierarchy of mathematical theories. Whereas the theories themselves are not part of KEIM (though of course ΩMEGA contains a theory database), the means to define a theory and to efficiently maintain a theory database are included in KEIM. A KEIM theory consists of at least the *signature*, the *axioms* and the theory's parents, i.e. a theory can inherit symbols and axioms from other theories. A theory may further contain its own *inference rules* (more specifically rules, tactics and methods) as well as problems, which can be either open problems or theorems, the latter distinguished from the former by the presence of at least one valid proof.

In KEIM every problem belongs to a theory. This makes problem descriptions very brief and avoids redundant redefinition of frequently occurring signature elements. Proofs of the problem can use all the underlying theories by *import* of axioms and theorems.

7. INPUT AND OUTPUT

Programming input and output routines is one of the most tedious tasks in building systems, but of course a necessary and important one for the end user. The concrete style of I/O is very much dependent on the taste of an individual user, so we can not force a particular one, but provide some generic concepts as outlined in the sequel.

7.1. *Environments*

One very basic concept in KEIM is that of an *environment*. This is a lexical context in which symbols read from an input stream are mapped to KEIM objects, whose main use is in the parsing process. Environments can inherit from other environments, and there also exists the possibility of multiple (disjoint) inheritance. Environments play an important role in connection with theories, and every problem has its own environment, inheriting from the appropriate theory. Environments of different proofs for a problem are kept separate.

7.2. *POST*

The language *POST* is part of KEIM. *POST* is a sorted higher order language based on Church's simply typed λ-calculus. Expressions written in *POST* may not be very nice to look at, since they closely resemble Lisp

```
(problem STEAMROLLER (in base)
    (constants (wolf (o i)) (fox (o i)) (bird (o i))
    (caterpillar (o i)) (snail (o i))
    (animal (o i)) (grain (o i)) (plant (o i))
    (eats (o i i)) (smaller (o i i)))

    (assumption wolves
        (and (forall (lam (x i) (implies (wolf x)
                                          (animal x))))
            (exists (lam (x i) (wolf x)))))



        .
        .
        .




    )
```

Figure 5.   Part of Schubert's steamroller problem in $POST$.

(plain ASCII representation, prefix notation). This is, however, no coincidence, because we designed $POST$ so as to be parsable by the Lisp reader. As an example see a piece of the $POST$ formulation of Schubert's steamroller problem in Figure 5. Since this puzzle does not need much mathematical knowledge it is placed in the elementary theory base, which contains e.g. the logical connectives. After the declaration of constants local to the problem (the types are given in reverse c-normalized form, i.e. (o i i) means $\iota \times \iota \to o$) assumptions and conclusion are given[2].

The clear advantage of $POST$ is, that every KEIM object (be it terms, proofs, environments, positions, ...) has a $POST$ representation, so if the concrete style of I/O is uninteresting and the emphasis lies on quick development, very little has to be programmed in addition. Of course it is also possible to include $POST$ representations of particular KEIM objects into more sophisticated I/O languages.

---

[2] The appearance of λ-abstractions in a first order problem may appear a little irritating, but note that the type of the universal and existential quantifier (as it is defined in the base theory) is $\alpha \times o \to o$, where α is a type variable. One could hide this from the user.

7.3. *Support for User Interaction*

In an interactive system there are frequently occurring subtasks, so we supply a module that helps to design *hierarchical menus* and *command interpreters*, together with a help facility and routines for the handling of error conditions (based on the Lisp condition scheme (Steele jr., 1984)) and error messages.

8. DEVELOPMENT SUPPORT

Besides the deduction specific components, KEIM offers tools that support the development of a new system as such. For example there are rather strong *programming conventions*, including a *module concept* which replaces the package concept offered by Common Lisp by a more flexible approach to dependency management and information hiding.

Also for the purpose of accelerating the development process, KEIM contains a means for *automated documentation* of software. The KEIM style is to include documentation *directly into the source code*, from which it can be automatically extracted into written manuals in LaTeX (Lamport, 1994). The documentation of functions and classes is taken from Lisp itself, and output is generated in the well established notation introduced by (Steele jr., 1984). We supply different styles of manuals, depending on whether internal (i.e. hidden) functions are documented or not. Of course the manual extraction is coupled with the module system, thereby allowing for the extraction of documentation of subsystems including all their dependencies.

9. A NOTE ON EFFICIENCY (OR: WHY LISP?)

The main support provided by the KEIM toolbox is aimed at the *development* phase of a deduction system. Therefore aspects like modularity and fast changeability of concepts as well as quick switches between different settings are of major importance. This aspect of a quick development of a prototypical system was the reason why we chose Lisp and CLOS for the implementation of KEIM.

It has been objected that there are better choices with respect to the efficiency of the resulting software, e.g. the large Lisp runtime system seems to be an obstacle against efficiency.

In a way, these objections are true. For *top* efficiency of a final system to be employed in real world tasks we advocate a re-implementation of the *final product* using languages such as C++. KEIM, however, would then still be used for rapid prototyping. There is a way of improving the performance of such a prototypical system, yielding a smaller (though no more maintainable) program deprived of unnecessary ingredients by the method of *complete compilation to C* (Goerigk et al., 1996). This method advocates translation of the final customer's program (which must possess a single main call) into C code. The dependencies are traversed in a backward direction and code is generated only for those segments of source code that are really called, thus eliminating large parts of the Lisp runtime system and library[3].

The most important result of the KEIM project, however, is that we could isolate criteria that have to be met by *any* toolbox, as well as proving the approach valid by a prototypical implementation. The result is therefore to a large extent language independent, as long as the language meets some general demands such as containing object oriented features.

Furthermore, as it turned out, even the prototypical Lisp implementation is sufficient to satisfy moderate efficiency demands as required for experimental systems, and this certainly pays off, if the effort (in terms of time and manpower) saved during the development phase is taken into account. To test this, we have reimplemented one of the examples in favour of the PTTP approach in KEIM. One of the most frequently uttered criticism concerning KEIM, even from supporters of the toolbox idea, is that there already exist toolboxes in the guise of a programming language. The success of the PROLOG technology theorem provers (PTTP) idea (Stickel, 1988) seems to affirm this. The answer to this objection is that indeed PROLOG can be viewed as a toolbox, but as such it is too fine grained and contains a selection of items which is too narrow and thus restricts the type of system to be developed. As an example, the efficiency of PTTP based provers heavily depends on the built in *unification* and the *search strategy* (depth first with backtracking) delivered by PROLOG. But then the modularity demand is not met, since it is not that easy to exchange these parts and use a different unification algorithm, for example. Of course there are PROLOG systems such as Lambda-PROLOG (Nadathur, 1987) or ECLiPSe that are broader in some respect, by including higher order or constraint handling facilities, but none of them satisfies the strict definition of a toolbox for deduction systems in the sense given above.

Let us now look at the example: the leanTAP (Beckert and Posegga, 1995) theorem prover consists of only about ten lines of PROLOG code and nevertheless is a serious and rather efficient theorem prover. The reason for this is that it makes use of unification as well as the search strategy directly taken from PROLOG. Of course the KEIM implementation is slightly longer (the

---

[3] Unfortunately this only works for a proper subset of Common Lisp, so not every KEIM function can be translated this way.

number of rules stays the same, but some things have to be explicitly mentioned that PROLOG takes for granted). Also the result is not as fast as the PROLOG version. But, and this is the crucial point, the loss of efficiency is small in comparison to the *flexibility* gained, for it takes only a small change in KEIM to exchange e.g. standard unification for a sorted one, whereas this is simply impossible in the PROLOG implementation without changing the PROLOG version.

10. EVALUATION

KEIM has been designed to set the toolbox approach to building problem manipulation systems to work. As such, it contains the necessary data structures and fundamental algorithms as well as many tools for programming and documentation. The complete toolbox contains about 50,000 lines of Lisp code.

KEIM has been heavily used in the development of the mathematical assistant ΩMEGA. During this implementation we have not considered KEIM as closed, but frequently added further components as needed or even changed existing ones. This is exactly what we expected from the start, and what will certainly happen to continue in the future, particularly if KEIM is used for building systems of a completely different character.

Besides ΩMEGA, KEIM has been used in several quite different application areas, namely software verification (TU Wien), nonmonotonic logics with constraints (Frankfurt University), practical exercises for students in implementing unification algorithms (Bern University, Switzerland), interactive higher order theorem proving (Carnegie Mellon University, Pittsburgh, USA) and in particular in the implementation of combination algorithms for unification procedures in different theories (at CIS Munich and Aachen University) (Baader and Schulz, 1996; Kepser and Schulz, 1996).

The current KEIM system is rather biased in that it contains support for generative calculi like resolution, but lacks important concepts for the support of e.g. tableau based techniques. We hope that these components will be included by the respective community, since we intentionally gave KEIM this open architecture that allows for easy modification and extendibility.

The expertise gained by KEIM can now be easily ported to other programming languages, provided they support an object oriented programming style, as this has turned out to be an important prerequisite for the toolbox approach.

11. ACKNOWLEDGMENTS

Besides the whole current staff of the ΩMEGA group at the Universität des Saarlandes, including the students involved, I am particularly indebted to the former group members Dan Nesmith and Jörn Richts, who are essentially responsible for the success of the project. Furthermore I would like to thank Michael Kohlhase and Jörg Siekmann for helpful discussions and their readiness to repeatedly read draft versions of this article.

REFERENCES

Andrews, P. B., M. Bishop, S. Issar, D. Nesmith, F. Pfenning, and H. Xi: 1996, 'TPS: A Theorem Proving System for Classical Type Theory'. *JAR* **16**(3), 321–353.

Baader, F. and K. Schulz: 1996, 'Unification in the Union of Disjoint Equational Theories: Combining Decision Procedures'. *Journal of Symbolic Computation* **21**, 211–243.

Beckert, B. and J. Posegga: 1995, 'leanTAP: Lean Tableau-Based Deduction'. *JAR* **15**(3), 339–358.

Benzmüller, C.: 1997, 'A Calculus and a System Architecture for Extensional Higher-Order Resolution'. Research Report 97–198, Department of Mathematical Sciences, Carnegie Mellon University, Pittsburgh, USA.

Benzmüller, C., L. Cheikhrouhou, D. Fehrer, A. Fiedler, X. Huang, M. Kerber, M. Kohlhase, K. Konrad, A. Meier, E. Melis, W. Schaarschmidt, J. Siekmann, and V. Sorge: 1997, 'ΩMEGA: Towards a Mathematical Assistant'. In: W. McCune (ed.): *Automated Deduction — CADE–14, 14th International Conference on Automated Deduction*. pp. 252–255.

de Bruijn, N. D.: 1972, 'Lambda calculus notation with nameless dummies, a tool for automatic formula manipulation, with application to the Church-Rosser theorem'. *Indagationes Math.* **34**, 381–392.

Bundy, A.: 1988, 'The use of explicit plans to guide inductive proofs'. In: E. L. Lusk and R. A. Overbeek (eds.): *Proceedings of CADE 9*. Argonne, Illinois, USA, pp. 111–120.

Church, A.: 1940, 'A formulation of the simple theory of types'. *Journal of Symbolic Logic* **5**, 56–68.

Corbin, J. and M. Bidoit: 1983, 'A Rehabilitation of Robinson's Unification Algorithm'. In: R. E. A. Mason (ed.): *Information Processing*, Vol. 83. Elsevier Science Publishers B.V. (North-Holland), pp. 909–914.

Dowek, G., T. Hardin, and C. Kirchner: 1995, 'Higher Order Unification via Explicit Substitutions'. Rapport de recherche 2709, Institut National de Recherche en Informatique et Automatique.

Gentzen, G.: 1935, 'Untersuchungen über das logische Schließen I'. *Mathematische Zeitschrift* **39**.

Goerigk, W., H. Boley, U. Hoffmann, M. Perling, and M. Sintek: 1996, 'Komplettkompilation von Lisp: Eine Studie zur Übersetzung von Lisp-Software für C-Umgebungen'. *KI*.

Gordon, M., R. Milner, and W. C.: 1979, *Edinburgh LCF: A Mechanized Logic of Computation*, No. 78 in LNCS. Springer.

Graf, P.: 1995, 'Term Indexing'. Dissertation, Technische Fakultät der Universität des Saarlandes, Saarbrücken. will be published in LNAI series, 1996.

Huang, X.: 1996, *Human Oriented Proof Presentation: A Reconstructive Approach*, No. 112 in DISKI. Infix.

Huang, X., M. Kerber, M. Kohlhase, E. Melis, D. Nesmith, J. Richts, and J. Siekmann: 1994a, 'KEIM: A Toolkit for Automated Deduction'. In: A. Bundy (ed.): *Automated Deduction — CADE-12, Proceedings of the 12th International Conference on Automated Deduction, Nancy, France*. Berlin, pp. 807–810.

Huang, X., M. Kerber, M. Kohlhase, E. Melis, D. Nesmith, J. Richts, and J. Siekmann: 1994b, 'Ω-MKRP, A Proof Development Environment'. In: A. Bundy (ed.): *Automated Deduction — Cade-12, Proceedings of the 12th International Conference on Automated Deduction, Nancy, France*. Berlin, pp. 788–792.

Kepser, S. and K. Schulz: 1996, 'Combination of Constraint Solvers II: Rational Amalgamation'. In: *Proceedings of the 2nd International Conference on Constraint Programming, CP–96*.

Klein, L.: 1997, 'Indexing für Terme höherer Stufe'. Master's thesis, Universität des Saarlandes, Fachbereich Informatik, Saarbrücken.

Lumport, L.: 1994, *LaTeX User's Guide and Reference Manual*. Addison Wesley.

Lusk, E. L., W. McCune, and R. A. Overbeek: 1982b, 'Logic Machine Architecture: Inference Mechanisms'. In: *Automated Deduction — CADE*. pp. 85–108.

Lusk, E. L., W. McCune, and R. A. Overbeek: 1982a, 'Logic Machine Architecture: Kernel Functions'. In: *Automated Deduction — CADE*. pp. 70–84.

Martelli, A. and U. Montanari: 1982, 'An Efficient Unification Algorithm'. *ACM Transactions on Programming Languages and Systems* **4**(2), 258–282.

Nadathur, G.: 1987, 'A Higher-Order Logic as a Basis for Logic Programming'. Thesis, University of Pennsylvania.

Nadathur, G.: 1994, 'A Notation for Lambda Terms II: Refinements and Applications'. Technical Report CS–1994–01, Department of Computer Science, Duke University, Durham, North Carolina.

Nadathur, G. and D. S. Wilson: 1994, 'A Notation for Lambda Terms I: A Generalization of Environments'. Technical Report CS–1994–03, Department of Computer Science, Duke University, Durham, North Carolina.

Nesmith, D.: 1992, 'KEIM-Manual, Version 0'. Technical report, Universität des Saarlandes, Saarbrücken.

Paterson, M. S. and M. N. Wegman: 1978, 'Linear Unification'. *Journal of Computer and System Sciences* **16**, 158–167.

Richts, J. and D. Nesmith: 1994, 'Test-Keim Manual'. Fachbereich Informatik, Universität des Saarlandes, Saarbrücken.

Robinson, G. and L. Wos: 1969, 'Paramodulation and theorem proving in first order theories with equality'. In: *Machine Intelligence*, Vol. 4. New York: American Elsevier, pp. 135–150.

Robinson, J. A.: 1965, 'A Machine–Oriented Logic Based on the Resolution Principle'. *Journal of the ACM* **12**(1), 23–41.

Steele jr., G. L.: 1984, *Common Lisp. The Language*. Digital Equipment Corporation.

Stickel, M. E.: 1986, 'Schubert's Steamroller Problem: Formulation and Solutions'. *JAR* **2**(1), 89–101.

Stickel, M. E.: 1988, 'A Prolog Technology Theorem Prover: Implementation by an Extended Prolog Compiler'. *JAR* **4**(4), 353–380.

CHAPTER 7

SPECIFICATIONS OF INFERENCE RULES: EXTENSIONS OF THE PTTP TECHNIQUE

1. INTRODUCTION

Most theorem provers nowadays are based on a single calculus or have a few calculi hardwired. It is not so easy to experiment with other calculi. This inherent limitation can be easily overcome if you are aware of it. As a result we show how a new degree of freedom can be added to a theorem prover – in this case by extending Prolog Technology Theorem Proving (PTTP) due to Stickel (1988).

Our background is the connection method (Bibel, 1982). This method suggests that proving a theorem (or refuting its negation) means to discover a global structure within a formula — the spanning mating. An algorithmic realization of the search for a spanning mating leads to goal oriented reasoning. Goal oriented reasoning is known to be efficient with respect to memory use. Due to Prolog technology (Stickel, 1988) it is possible to achieve high inference rates already with moderate implementation effort taking advantage of the speed of present Prolog systems.

Unfortunately, like often in real life, the pure realization of a paradigm may result sometimes in poor performance. This is the case with goal oriented reasoning too. One main stream of trying to overcome this obstacle is to enhance goal oriented reasoning by dedicated forms of saturation oriented reasoning. This may be illustrated by building-in theories into theorem provers — the initial motivation of our work. Building-in a theory may be seen as a combination of saturation (of a theory) with certain forms of partial evaluation. Saturation may result in computing all consequences of a theory that are necessary for proving all theorems out of a considered class of formulas. Partial evaluation, i.e. taking into account some information about the structure of the class of the formulas to be proved, may show, just for example, that a theory may be built-in by just substituting the usual unification algorithm by a theory unification algorithm. Moreover it is quite often necessary to add some control information in order to improve the proof search.

It is not difficult to enhance a calculus with new inference rules. This may be seen in (Stickel, 1985) for enhancing resolution by theory reasoning in the

W. Bibel, P. H. Schmitt (eds.), Automated Deduction. A basis for applications. Vol. II
© 1998 *Kluwer Academic Publishers. Printed in the Netherlands*

ground case or in (Plaisted, 1990) for refining the treatment of the current path in model elimination.

In this chapter we show a way how this flexibility may be carried over to a theorem prover. Our main requirement was that the user (the experimenter) should not need to change the code of the prover in order to enhance it by new inference rules. This is difficult or even impossible for those who are not the author of the prover and at least annoying for the author. It should be sufficient to rely on the knowledge about the calculus that is behind the prover. In our case the calculus may be described by notions like current path, current goal, ancestor literals, residue etc. The user should be able to transcribe inference rules into descriptions that are formulated in a formal language. Those descriptions will then be interpreted in order to obtain a modified prover.

We applied this idea successfully to our prover ProCom (Neugebauer and Petermann, 1994) which is based on the Prolog-technology paradigm. It should be possible to apply this idea also to other prover technologies. The realization of this way of enhancing a prover is the main contribution presented in this chapter.

We present four modifications of the standard model-elimination procedure on the calculus level. After an introduction of the main features of our description language we show how these calculi may be described formally in order to form a modified prover. The first calculus is the usual model elimination procedure enhanced by using a theory unification algorithm. The second calculus is an extension of the first one towards a generalized version of constraint reasoning. Both calculi are well motivated by the translation of (multi-) modal logic into first-order logic and by constraint logics. The third calculus deals with building-in a strict order relation. The fourth calculus uses a form of equality handling — rigid E-unification (Gallier et al., 1992).

The remaining sections are devoted to the task of presenting our description language. The constructions are developed from an analysis of several calculi. The description language is sketched in Section 3. In Section 4 the calculi presented in Section 2 are formulated in this language. The translation process is finally sketched in Section 5.

2. FEATURES OF DEDUCTION RULES

In this section we present briefly our terminology. Afterwards we will have a look at some proof calculi. In further sections they will serve us as case studies for our approach to the description of calculi.

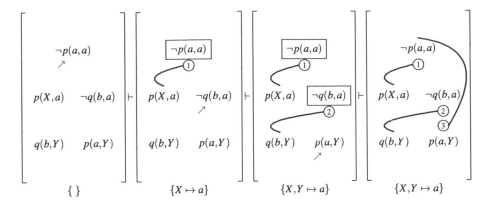

Figure 1.   A Sample Deduction

Let us consider the simple derivation shown in Figure 1 first. We have used a matrix notation dual to (Bibel, 1982). The initial matrix at the left side contains three clauses which are arranged horizontally. We are using negative representation. Clauses represent universally quantified disjunctions of literals. A path represents a conjunction.

The derivation — which is carried out with the extension procedure here — has to detect a spanning mating in an amplification of this matrix and its simultaneous unifier. Each of the matrices in Figure 1 represents a state of the derivation. A diagonal arrow pointing the *current goal* appears in every but the rightmost matrix. The *connections* forming the mating are drawn as labeled edges. The *current path* is indicated by the boxed literals.

We start in the leftmost matrix with the empty path and the goal $\neg p(a,a)$. Each complete path going through this path and literal has to be considered. The extension step 1 shown in the second matrix closes the left branch. Each complete path containing this segment is complementary with the substitution $\{X \mapsto a\}$.

Another extension step 2 leads to the substitution $\{X,Y \mapsto a\}$ and the current goal $p(a,Y)$ with the path $\{\neg p(a,a), \neg q(b,a)\}$. But all paths containing this segment are already complementary since a connection can be drawn from the current goal to a complementary literal on the current path. This step is known as reduction step 3.

A final matrix has no open goals. It completes the derivation. During the proof we have constructed a spanning mating and computed a simultaneous unifier.

Now we look at the general case. In order to describe a state of the derivation we have to represent the set of complete paths which still have to be considered. For this purpose we use hooks $(p \perp \Gamma)$. A *hook* represents the paths which continue the path fragment p via one of the literals of the clause fragment Γ. We use the sign $\perp$ in order to indicate that the partial path p and the clause Γ are in a sense orthogonal. The elements of Γ are called *goals*. The current goal of the hook has to be solved by applying an inference rule. Inference rules describe how new hooks are generated from a given hook.

A derivation starts with a single hook $(\emptyset \perp \Gamma)$ where Γ contains all literals of a given goal clause. This is called the *initial hook*. Hooks of the form $(p \perp \emptyset)$ are said to be *solved*. Solved hooks need not to be considered any more. A derivation is complete if no more unsolved hooks are left. For more details concerning the pool calculus see (Neugebauer and Schaub, 1994; Petermann, 1993) and Chapter I.2.6.

2.1. *Extension procedure with theory unification*

The extension procedure provides an extension rule and a reduction rule. In the case considered here the syntactical unification has been replaced by unification modulo some theory T. We assume that T-unifiability of pairs of terms is decidable and that complete sets of T-unifiers (Siekmann, 1989) are enumerable.

The *extension rule with unification modulo theory* T is given in (7.1). The current goal is L_0. The extension clause L_1, Γ_1 is a copy of a clause from the considered matrix. The T-unifier of L_0 and the complement of L_1 is denoted by σ. The new hooks are $(p \perp \Gamma_0)$ and $(p, L_0 \perp \Gamma_1)$. The T-connection $\sigma(L_0, L_1)$ is T-unsatisfiable and so is any path containing $\sigma(p, L_0, L_1)$. This observation justifies Rule (7.1).

$$(7.1) \quad \frac{(p \perp L_0, \Gamma_0) \qquad L_1, \Gamma_1}{(p \perp \Gamma_0),\ (p, L_0 \perp \Gamma_1)}\ \sigma$$

A *reduction inference with unification modulo theory* T has the form given in Rule (7.2). The partner L_1 for the current goal L_0 has been taken from the path p, L_1. Thus there is no extension clause. The T-unifier of L_0 and the complement of L_1 is σ. The unique new hook is $(p, L_1 \perp \Gamma_0)$. In order to justify Rule (7.2) we observe that any path which is a continuation of $\sigma(p, L_1, L_0)$ is T-unsatisfiable.

$$(7.2) \quad \frac{(p, L_1 \perp L_0, \Gamma_0)}{(p, L_1 \perp \Gamma_0)}\ \sigma$$

Indeed both rules constitute a simple generalization of the extension calculus. However it will be complete for query languages consisting only of those matrices not containing the predicate symbols used for expressing the theory T.

T may be any universal theory such that the T-unifiability of pairs of terms is decidable and complete sets of T-unifiers are enumerable (Petermann, 1993).

2.2. *Extension procedure with built-in hybrid theories*

Constraint resolution has been introduced by Bürckert (1990). As a basic feature of this calculus the signature of the underlying language is split in two sub-signatures. The built-in theory is formulated using only some of the predicate symbols of the signature. Those predicate symbols are called constraint predicates. Under certain circumstances — for example if the theory is definite and constraint literals may occur only negative — each constraint theory connection consists of a single literals. The proof search may be driven by the search for a mating spanning the non-constraint part of the matrix. For solving constraints lazy or eager strategies are possible.

At least for theories, which are given syntactically, constraint reasoning is an instance of the more general scheme of building-in hybrid theories (Petermann, 1997). Now we assume that the built-in theory consists of two sub-theories. One sub-theory may be a constraint theory $\Re$ formulated in terms of constraint predicates, whereas a second sub-theory T is formulated in terms of non-constraint predicates. Let us consider a simple case where T is an equational theory and equality does not occur in query formulas. This is the same situation as in the previous sub-section. In order to obtain a complete calculus we need both previously introduced Rules 7.1 and 7.2 and moreover the following one. A *reduction inference modulo a constraint theory* $\Re$ has the form given in Rule (7.3).

$$(7.3) \qquad \frac{(p \perp L_0, \Gamma_0)}{(p \perp \Gamma_0)} \; \sigma$$

Here the set consisting of the literal L_0 is a theory connection. Thus, there is no extension clause and even no partner from the path is needed. The $\Re$-unifier of L_0 is σ. The unique new hook is $(p \perp \Gamma_0)$. In order to justify Rule (7.3) we observe that any path which is a continuation of $\sigma(p, L_0)$ is $\Re$-unsatisfiable.

2.3. *Building-in strict order*

Theory connections do not always consist of at most two literals. An evident example is the theory of strict order (SO) of a binary relation $<$. It consists of the axioms (IR) and (T) shown below.

(IR) $\qquad \neg x < x$

(T) $\qquad x < y \wedge y < z \to x < z$

The asymmetry axiom has been omitted because it follows from irreflexivity (IR) and transitivity (T). SO-connections are $f(x) < f(a)$ as well as $a < b, b < c, c < a$. The length of a SO-connection is unpredictable.

It may be convenient to find a SO-connection by a number of smaller inference steps. This approach is known as partial theory reasoning (Stickel, 1985). With this technique a current goal, say $s < r$, will not be solved completely by a single inference step. It will be rather transformed to a new, hopefully simpler, subgoal, say $s < t$. Such a subgoal is called a *residue*. The residue causes the appearance of an additional hook in the conclusion of an inference rule (cf. Rule (7.6)). σ is a most general unifier of r and r'. The treatment of residues by descriptors of inference rules will be shown in Section 3 and Subsection 4.2.

We encourage the reader to compare the calculus introduced below with the theory inference system for strict orderings presented in Example 19 of Chapter I.2.6. Theory inference systems may be generated automatically using the method of linearizing completion presented in Chapter I.2.6 (see (Baumgartner, 1996) for details). Here we give a set of inference rules which is a slightly optimized translation of the mentioned theory inference system.

The calculus consists of three groups of inference rules. The rules of the first group are *IR-reduction* (7.4), *SO-reduction* (7.5) and *SO-extension* (7.6). The IR-reduction rule is justified by the irreflexivity of $<$, whereas the rules (7.5) and (7.6) are justified by the transitivity axiom. The mentioned three rules form a complete calculus with respect to a query language without negated literals. Substitution σ is a most general unifier of the terms s and s'.

$$(7.4) \qquad \frac{(p \perp s < s', \Gamma_0)}{(p \perp \Gamma_0)} \ \sigma$$

$$(7.5) \qquad \frac{(p, r < s \perp s' < t, \Gamma_0)}{(p, r < s \perp \Gamma_0), \ (p, r < s \perp r < t)} \ \sigma$$

$$(7.6) \qquad \frac{(p \perp s' < t, \Gamma_0) \qquad\qquad r < s, \Gamma_1}{(p \perp \Gamma_0), \ (p, s' < t \perp \Gamma_1), \ (p, s' < t \perp r < t)} \ \sigma$$

If also negative literals are admitted to occur in potential theorems, then two more groups of inference rules become necessary. The first one provides the rules *Reduction* (7.7) and *Extension* (7.8), where σ is a most general unifier of $r < r$ and $r' < s'$. The second one consists of the rules *AT1-reduction* (7.9), *AT1-extension* (7.10), *AT2-reduction* (7.11) and *AT2-extension* (7.12), where σ is a most general unifier of s and s'.

$$(7.7) \qquad \frac{(p, r' < s' \perp \neg r < s, \Gamma_0)}{(p, r' < s' \perp \Gamma_0)} \ \sigma$$

$$(7.8) \quad \frac{(p \perp \neg r < s, \Gamma_0) \qquad r' < s', \Gamma_1}{(p \perp \Gamma_0),\ (p, \neg r < s \perp \Gamma_1)} \ \sigma$$

$$(7.9) \quad \frac{(p, \neg s' < t \perp s < r, \Gamma_0)}{(p, \neg s' < t \perp \Gamma_0),\ (p, \neg s' < t \perp \neg r < t)} \ \sigma$$

$$(7.10) \quad \frac{(p \perp s < r, \Gamma_0) \qquad \neg s' < t, \Gamma_1}{(p \perp \Gamma_0),\ (p, s < r \perp \Gamma_1),\ (p, s < r \perp \neg r < t)} \ \sigma$$

$$(7.11) \quad \frac{(p, s' < t \perp \neg s < r, \Gamma_0)}{(p, s' < t \perp \Gamma_0),\ (p, s' < t \perp \neg t < r)} \ \sigma$$

$$(7.12) \quad \frac{(p \perp \neg s < r, \Gamma_0) \qquad s' < t, \Gamma_1}{(p \perp \Gamma_0),\ (p, \neg s < r \perp \Gamma_1),\ (p, \neg s < r \perp \neg t < r)} \ \sigma$$

The rules (7.9), (7.10), (7.11) and (7.12) are justified by consequences of the irreflexivity and transitivity of $<$.

2.4. *Equality handled by rigid E-unification*

We now consider a calculus with equality. The heart of this calculus is the computation of so called rigid E-unifiers. If E is a set of equations then a substitution σ is called a *rigid E-unifier* of the terms s and t modulo E iff $\sigma(s)$ and $\sigma(t)$ are congruent modulo the set of equations $\sigma(E)$ where variables occurring in $\sigma(E)$ are treated as constants. Following Bibel (1982) a set $E \cup \{s \neq t\}$ is called an eq-connection if there exists a rigid E-unifier of the terms s and t modulo E.

An eq-connection calculus consists of the rules (7.13), (7.14) and (7.15).

$$(7.13) \quad \frac{(p \perp s \neq t, \Gamma_0)}{(p \perp \Gamma_0)} \ \sigma$$

$$(7.14) \quad \frac{(p \perp s \neq t, \Gamma_0) \qquad l \doteq r, \Gamma_1}{(p \perp \Gamma_0),\ (p \perp \Gamma_1),\ (p, l \doteq r \perp s \neq t)} \ \sigma$$

$$(7.15) \quad \frac{(p \perp L_0, \Gamma_0) \qquad \Gamma_1}{(p \perp \Gamma_0),\ (p, L_0 \perp \Gamma_1)}$$

The inference rule *reduction with rigid E-unification* is given by Formula (7.13). The substitution σ is a rigid E-unifier of the terms s and t where E is the set of equations within the path p. The multi-set $E, s \neq t$ is an eq-connection.

The number of literals forming an eq-connection is unpredictable. Like in the case of building-in strict order (cf. Subsection 2.3) we use a partial theory reasoning rule, the *collection rule* (7.14). This rule is collecting subsequent equations in order to use them later in a reduction step with rigid E-unification. The extension clause $l \neq r, \Gamma_1$ is a copy of a clause from the considered matrix.

The calculus consisting of Rule (7.13) and Rule (7.14) is already complete for matrices consisting of definite clauses if equality is the unique predicate symbol (Petermann, 1994). In that case every E-unsatisfiable matrix contains a negative clause which may be chosen for forming the initial hook of a successful derivation. Then, every goal will be negative because every extension clause is definite. Therefore the path of the second resulting hook of the collection rule (7.14) does not need to contain the current goal $l \neq r$ of that inference.

Finally, in order to achieve completeness for the non-Horn case, we have to describe how to solve a positive goal. For this purpose Rule (7.15), the *restart inference rule*, can be used. L_0 is an atom, Γ_1 is a copy of a negative clause, called the *restart clause*.

Different decision procedures for rigid E-unifiability have been communicated by Gallier et al. (1992), Goubault (1993) and Becher and Petermann (1994). A procedure enumerating complete sets of rigid E-unifiers is due to Beckert (1994). The proof for the decidability of the simultaneous rigid E-unification problem of Gallier et al. (1992) has been faulty as observed by Petermann (1994). Thus, the question, whether an existing simultaneous rigid E-unifier of a eq-mating U may be actually found iteratively by rigid E-unifications of the eq-connections in U, has been open for some time. A definite negative answer has been given by Degtyarev and Voronkov (1995) with the proof of the undecidability of simultaneous rigid E-unification. Nevertheless a newer result (Degtyarev and Voronkov, 1996) shows that for every equationally refutable matrix exist an amplification M and an eq-mating U spanning M such that a simultaneous rigid E-unifier of U may be found solving independently the rigid E-unification problems induced by the eq-connections of mating U. To be more precise the method presented by Degtyarev and Voronkov (1996) allows to decide the solvability of equational and order constraints and allows to avoid the application of substitutions to the derivation state. Similarly, the pool calculus may be interpreted in such a way that the substitution which is part of an inference rule is not actually applied to the pool. The substitution may be treated rather as a set of equational constraints which is added to a set of constraints associated with the pool. The treatment of constraints is also possible on the implementation level (cf. (Rigó, 1995) for a case study).

For a more subtle co-operation between the base calculus and rigid E-unification also see (Beckert and Pape, 1996). For an overview concerning rigid E-unification and its relation to first-order theorem proving the reader is referred to Chapter I.2.8.

3. SPECIFICATION OF INFERENCE RULES

Our aim is the specification of inference rules and some control information. This specification should be used for the automatic translation of proof problems into a target language like Prolog — the PTTP paradigm.

According to the calculi, we have seen in Section 2, we can extract assumptions we will make and requirements for our envisaged specification language. The overall assumption is that we consider calculi for problems in first order predicate logic in clause form only. Some results (de la Tour, 1990) indicate that this is not really a restriction. Extensions, e.g. to typed or higher order calculi, are currently outside the scope of our interests.

We consider goal oriented calculi only. This means that at each time during the proof process we have a single current goal.

In all examples we have seen one global substitution ties together various hooks. This is the more general case than the existence of local substitutions. Local substitutions can be achieved by building new instances when appropriate. Thus we have to provide a mechanism to make such copies.

For the extension rule and its variants (Rule 7.1 and Rule 7.6) or the restart Rule (7.15) we need to have access to the matrix. This access is always combined with an instantiation. The original matrix is left unchanged.

The reduction rule and its variants need access to the current path. The current path is searched for literals or literals are added to or deleted from the current path.

The translation process we have in mind is oriented towards the procedures induced by the literals in the matrix. Thus the path plays its role in the background. This is in contrast to (Neugebauer and Schaub, 1994) where the hooks were really visible in the implementation. As a consequence of hiding the path we will have a more "procedural" view than the more "logical" point of view taken in Section 2.

Any inference rule can be given as in Section 2. When we go towards an implementation we have to consider additional properties. One main point is the sequentiality of the proof search. In this context the order of subproofs is essential for an efficient algorithm. Since this can not be detected automatically we will give the control to the user and act according to the specifi-

```
descriptor
  template(L0,goal),
  template(-L0,path).
descriptor
  template(L0,goal),
  template(-L0,extension).
```

Figure 2. Extension procedure

cation. Thus heuristic information is (more or less obvious) contained in our descriptors.

Now we will show a language to describe inference rules. Since we are aiming towards a Prolog implementation we will use Prolog terminology from now on.

Let us reconsider the extension procedure used in Section 2. Our descriptors are shown in Figure 2. There are two descriptors — one for the reduction and one for the extension rule. The order of their appearance is one place where sequentiality comes in.

Each descriptor of an inference rule starts with the keyword `descriptor`. The keyword is followed by a conjunction of instructions describing it and terminated by a point.

The first thing we need is the current goal. Since we have restricted ourselves to goal oriented calculi there is one single goal which has to be considered at a time. This is described by the construction `template(L, goal)` where L is a pattern for the goal to be solved. In an alternative description of the extension procedure in Figure 2 we have a goal template in each descriptor containing a variable as pattern. I.e. these descriptors are applicable to any subgoal.

The reduction descriptor specifies the need to search a literal on the path by the construction `template(L, path)`. Here L is the pattern of the literal to be found on the path. This corresponds to the condition for an inference rule that a certain literal has to be unified with an element of the path. In our example the literal -L0 is reduced. The minus sign acts here as negation. Positive and negative literals can have the signs ++ and - - respectively. The extension descriptor uses the construction `template(L, extension)` to express this kind of inference step. The constellation in the inference rule is as follows:

$$\frac{\ldots \qquad L, \Gamma}{\ldots (\ldots \perp \Gamma)}$$

A goal is solved by finding an associated (e.g. negated and unifiable) literal L in the matrix and all subgoals in the corresponding clause are solved.

Note that in both cases — reduction and extension step — the path has been silently managed in the background. I.e. the appropriate literals have been placed on it. This feature makes it easier to write such simple descriptors. We will see later an example where this feature has to be turned off. The path management appears there explicitly.

For this purpose we need the constructions put_on_path(L) to put the literal L onto the path, is_on_path(L) to find a literal on the path, get(path,P) to get the path, and use_path(P) to declare that the path P should be used subsequently.

The get path instruction is one element of a family of instructions giving access to various informations, or — spoken procedurally — to execute various Prolog goals. We can get the index of a literal get(index(L),I), i.e. a unique identifier, the negation of a literal get(neg(L),NL), and some more. We will see further examples in the next section.

The construction template$(L$, residue$)$ describes a new goal L not necessarily occurring in the matrix. We have seen this situation in Rule 7.6 where the residue $r < t$ had to be solved. It can be solved with all inference rules present. The general situation looks as follows:

$$\overline{\ \ \cdots\ \ \ \cdots\ \ }$$
$$\cdots (\cdots \perp L)$$

L is a literal which normally does not occur as part of an extension clause.[1] We will see a descriptor utilizing a residue template in the next section.

In addition a list of steps is admissible in a template. This describes the alternative of the elements. E.g. template(-L0,[extension,path]) could have been used in Figure 2 in order to denote the two alternatives coded in two descriptors. Thus one descriptor would have been sufficient for the extension procedure.

As we have seen with the get instructions it can be desirable to attach an external Prolog procedure and call it in a descriptor. Thus it is possible to open the full expressive power of Prolog. The call can be performed at two different times. Either at compile time or at runtime. For this purpose the constructions call(G) and constructor(G) are provided.

Usually the proof is kept as a proof tree. It can be desirable to place information in this tree. For this purpose the construction proof(P) can be used. Thus it is possible to store information which is useful for generating a presentation of the proof. As an example we can mention our experiments with

---

[1] In Figure 13 we will use a residue template on a literal taken from the matrix. This is a technical trick, not in the spirit of a residue as described in Section 2.

```
descriptor
    template(L0,goal),
    template(-L1,extension),
    constructor(t_unify(L0,L1)).
```

Figure 3.   Theory extension rule

the ILF system of Dahn and Gehne (1997) (cf. Chapter III.1.1) which generates "natural language" output describing the proof. For this purpose ILF uses
a special calculus. It also is possible to use a normal model elimination proof
as input to ILF. But it is not possible to send a proof with theory connections
unmodified to ILF.

    To overcome this barrier we have used a postprocessor which translates a
proof containing theory inference steps into a pure model elimination proof.
For this purpose we have to identify the theory inference steps such that they
can be presented appropriately. This is done with the proof primitive. As a
result we can get human readable proofs via this translation and ILF.

4. SAMPLE SPECIFICATIONS OF CALCULI

In this section we will reconsider the calculi presented in section 2. We will
see how these calculi can be expressed with the means developed in Section
3.

4.1. *Extension Procedure with Theory Unification*

Let us reconsider the extension procedure with theory unification as presented
in Section 2.1. In Figure 2 we have seen the extension rule with the usual
unification.

    Now we can consider the case with a special unification procedure. We as-
sume that the unification procedure is given as a Prolog predicate t_unify/2
which performs the unification of two literals as desired. As shown in Figure 3, the instruction constructor(t_unify(L0,L1)) allows to compile in
a call to this procedure. The definition has to be provided in a library which
is added by the linker.

    We have said before that the order of the instructions will be used to ex-
press the sequentiality of the operations to be performed. Thus we have to
read this descriptor in this sequential way. This leads us to the following interpretation. Given a goal L0 take an arbitrary literal -L1 from the matrix.
Solve the remaining goals from the clause of -L1. Afterwards try to unify

```
descriptor
    template(L0,goal),
    get(index(L1),Index),
    get(neg(L1),NotL1),
    constructor(t_unify(NotL1,L0)),
    template(L1,extension(Index)).
```

Figure 4. Another theory extension rule

```
descriptor
    template(L0,goal),
    template(L,path),
    get(neg(L),NotL),
    constructor(t_unify(NotL,L0)).
```

Figure 5. Theory reduction rule

the arbitrary result with the initial goal. One may argue that the delay of the unification after the solution of the subgoals bears some inefficiency.

The alternative would be to reverse the order of the extension template and the constructor. This would lead to an enumeration of the unifiers and afterwards searching for an appropriate literal in the matrix. Depending on the theory this might be an acceptable alternative.

Nevertheless we have in mind another solution. We want to try the unification with the literal in the matrix first and then — after its success — the remaining subgoals should be solved.

This solution is shown in Figure 4. The get-index instruction enumerates all literals in the matrix which unify with the unspecified pattern L1. This instruction specializes L1 and instantiates Index to a literal index which can be fed into the extension template to specify which extension to take.

Before this extension template is invoked the literal L1 is negated and its negation NotL1 is T-unified with the initial goal. This descriptor comes close to the description of the extension procedure as given in Section 2.

We could improve this situation even more if the theory is restricted. E.g. if the theory does not touch the predicate symbols, i.e. two literals L and L' are not unifiable if they have different predicate symbols. In this case the descriptor could be specialized to take into account this fact. Thus some additional efficiency could be gained.

The theory reduction inference rule (7.2) is much simpler to describe after we have seen the extension step already. The descriptor is given in Figure 5. The instruction constructor(t_unify(NotL,L0)) simply enumerates all

```
descriptor
    template(++(S<S),goal).
```

Figure 6.   IR-reduction rule

```
descriptor
    template(++(S<T),goal),
    template(++(R<S),[path,extension]),
    template(++(R<T),residue).
```

Figure 7.   SO-Reduction and extension rules

elements from the path and unifies them with L. Again we could improve efficiency by exploiting additional features of the underlying theory.

4.2. *Building-in strict order*

Now let us consider the calculus presented in Section 2.3. Figure 6 shows the simple descriptor for Rule (7.4). Describing the inference rules (7.5) and (7.6) we observe a phenomenon not present in the previous example. In addition to the reduction and extension template, we need a means to specify a new goal which does not appear in the matrix. The residue template introduces the specified literal as new goal. This can be seen in Figure 7. The instruction template(++(S<R),[path,extension]) of this descriptor specifies both rules (7.5) and (7.6).

The same technique is used for the specification of the AT1- and AT2-rules ((7.9), (7.10), (7.11) and (7.12)). These descriptors are shown in Figure 9 and Figure 10. With the descriptor for the reduction and extension rules for negative goals in Figure 8 we have completed the calculus specified in Section 2.3.

The translation has to take care to use a sound and complete unification algorithm. This is assumed to be provided by the underlying Prolog system.

```
descriptor
    template(--(S<R),goal),
    template(++(S<R),[path,extension]).
```

Figure 8.   Reduction and extension rules

```
descriptor
    template(++(S<R),goal),
    template(--(S<T),[path,extension]),
    template(--(R<T),residue).
```

Figure 9. AT1-Reduction and extension rules

```
descriptor
    template(--(S<R),goal),
    template(++(S<T),[path,extension]),
    template(--(T<R),residue).
```

Figure 10. AT2-Reduction and extension rules

4.3. *Equality handled by rigid E-unification*

Most of the features needed for the calculus of Section 2.4 has already been described in previous sections. First, we consider Rule (7.13). The technique used is similar to theory reduction shown in Figure 5. An external Prolog predicate re_unify/3 performs the unification. Since this predicate needs the path to do its job we use the get-path instruction to access it and give it as the third argument to re_unify/3 (cf. Figure 11).

Now we need to consider the collection rule (7.14). One attempt to describe this rule is shown in Figure 12. This example shows another possibility to manipulate the path. We assume that the automatic augmentation of the path by the current goal is turned off. The goal pattern is obvious if we assume that the predicate symbol = is used to represent the equality $\doteq$. The resulting hook $(p \perp \Gamma_0)$ does not appear explicitly.

The hook $(p \perp \Gamma_1)$ together with the premise is described by the extension template.

The hook $(p, l \doteq r \perp s \neq t)$ is described by a residue template. But in this case we need the old path — before it has been modified for the extension template. Thus we have to get this old path. This is done with the get-path in-

```
descriptor
    template(--(S=T),goal),
    get(path,P),
    constructor(re_unify(S,T,P)).
```

Figure 11. Rigid *E*-reduction rule

```
descriptor
    template(--(S=T), goal),
    get(path,P),
    template(++(L=R), extension),
    use_path(P),
    put_on_path(++(L=R)),
    template(--(S=T), residue).
```

Figure 12.   Collection rule

```
descriptor
    template(++(L=R),goal),
    put_on_path(++(L=R)),
    call(negative_clause(--(S=T),Index)),
    template(--(S=T),residue),
    template(--(S=T),extension(Index)).
```

Figure 13.   Restart rule

struction. The path is stored in P and restored after the extension is performed. This is done with the use_path instruction. Now we can put the literal (L=R) onto the path and complete the description of the last hook with the residue template.

Finally we have to treat the restart rule (7.15). This rule says that a negative clause has to be found. In the descriptor shown in Figure 13 this task is performed by calling negative_clause/2 as an external Prolog predicate. The call instruction is used for this purpose. This predicate is evaluated at compile time. Thus it can make use of all informations concerning the matrix present in the system. Especially it can access the matrix and find a negative clause.

The literal index of one literal in such a clause is returned. This index can be used to ensure that only this clause is considered for the extension step. This is done by passing the index to the extension template. Since we need to prove all goals in this negative clause we need an additional residue template. Those goals are solved with a new path which has been constructed by putting the current goal onto the old path.

5. Controlling the Translation

In the present section we describe the interpretation of descriptors. Along the line of Stickel's Prolog Technology Theorem Prover (Stickel, 1988) a given matrix is translated into a Prolog program, which performs the proof search. However, different from (Stickel, 1988), extension and reduction steps are not hard-wired into our translation process. Rather the descriptors determine the translation of Prolog procedures, which comprise the possibilities to solve a given goal.

First of all the translation determines, analyzing the matrix, which Prolog procedures to generate, and constructs a pattern of the head of those procedures. This includes coding the negation in the predicate names, adding arguments for controlling the proof search, i.e. limiting the depth or the size of a proof for iterative deepening. All those techniques have already been described in (Stickel, 1988).

For an illustration of the new capabilities of CaPrl let us analyze the descriptors given in the figures 11, 12 and 13 as a descriptions of the inference rules (7.13), (7.14) and (7.15).

In principle, the instructions of a descriptor are interpreted in the order they appear. This way the user gains finer control of the inference process. Exceptions from this rule concern the goal and the proof pattern.

Wherever the goal pattern is specified it has to be used to instantiate the head of the Prolog procedure to be constructed. Even more, it has also to be used to decide, if the descriptor is applicable for the current procedure. We have seen this effect e.g. in Section 4.3. There we have had goal templates for the negative = predicate (Figure 11, 12) and the restart rule for the positive one (Figure 13). When the Prolog procedure for the negative = predicate is constructed the restart rule is not applicable and vice versa.

The instruction proof (*Term*) can be used for collecting information about the inference steps performed which might be useful understanding the proof. The proof term is also put as additional argument to any literal.

The remaining instructions are used to construct the body of the Prolog clause. For most of the instructions it is straightforward how they influence the translation.

— The get instructions unify the second argument with the information requested. E.g. the instruction get (path, P) in Figure 12 unifies P with the current path at this place. Since no additional code is generated the effect can be seen in the next Prolog subgoals only.

The get instructions can also be used to select Prolog clauses. If a get instruction fails then the associated Prolog clause is omitted (cf. Figure 4).

```
'=/2'(S, T, Depth, Path, Proof) :-
    check_depth_bound(Depth, Depth2, 1),
    '=/3 AUX 1'(L, R, _, Depth2, Path/_, Proof1),
    put_on_path('=/2'(L, R), Path, Path2),
    '=/2'(S, T, Depth2, Path2, Proof2),
    Proof = '*proof*'(Proof1, [Proof2]).
```

Figure 14. The translated collection rule

— The use_path instruction unifies its argument with the path to be used from this time on. We have used this feature in Figure 12 to restore an old path retrieved previously.

— The put_on_path instruction calls a library predicate to place the argument on the path and use the new path from this time on.
In Figure 14 the generated code can be seen. The definition for the predicate put_on_path/3 has to be added during the linking process after the compilation.

— The constructor instruction is translated into a call of the predicate it contains as argument. Thus constructor(*Goal*) places the *Goal* as additional Prolog goal in the code generated (cf. Figure 11).

— The call instruction is simply executed when it is encountered (cf. Figure 13). This is the compile time variant of constructor.

— The residue template is translated into a call to the associated Prolog predicate. For example, the template template(--(S=T),residue) as given in Figure 12 is translated into the Prolog goal
'=/2'(S,T,Depth2,Path2,Proof2)) shown in Figure 14.

— The path template is translated into a call to a library predicate unifying its argument with some literal on the path.

— The extension template is translated to the call of an auxiliary predicate representing the possibilities to find an unifiable literal in a clause and solve the remaining subgoals.
The extension template template(++(L=R),extension) from Figure 12 has been translated into a call to '=/3 AUX 1'/6 in Figure 14. The definition of this predicate depends on the matrix.

Let us consider as example the descriptor for the collection rule in Figure 12. The translated Prolog code is shown in Figure 14.[2]

---

[2] This code has been generated with ProCom. Only a minor renaming of variables has been performed to enhance the readability.

As discussed earlier we have the sign together with the arity of the predicate coded into the translated predicate name. Three additional arguments hold the depth, the path and the proof term.

The subgoal check_depth_bound is used to manage the search strategy. In this case iterative deepening on the depth of the proof tree has been active. This first subgoal decrements the depth bound and fails if 0 is reached. Otherwise the new depth is returned in Depth2. The new depth is passed to the extension and the residue later on.

Next we see the subgoal =/3 AUX 1. This predicate is generated for the extension template. The anonymous variables denoted by _ can be used to pass back the literal index and the new path to be used afterwards. Both are not needed in this example.

Now we discuss subgoal put_on_path. Here a literal is placed on the path. But here the initial path is used. This is the effect of the get path/use path construction in the descriptor. No subgoals are generated explicitly but the effect is obvious.

The subgoal =/2 has been generated from the residue template. The final subgoal is used to construct a proof term from the subproofs P1 and P2. Since nothing has been specified in the descriptor the default operator *proof* is used.

The code as it is shown above can be generated. But it seems to be rather inefficient to use lots of intermediate predicates. It has turned out that it is inevitable to use a postprocessor to optimize this code. Especially the definitions of (most) library predicates and the auxiliary predicates can be unfolded.

The technique to use library predicates which are optimized away later gives us high flexibility. For instance we have not fixed the data type used for the path yet. The path in (Stickel, 1988) consists of two lists — the positive and the negative literals. Alternatively one might think of using one list only which contains all literals, or a more complex (maybe efficient) data structure like binary trees or discrimination trees may be desirable. In such a case only a library has to be exchanged which provides a few predicates which provide the basic operations on a path.

With the optimizing technique it is possible to get the same results as the original PTTP if the plain model elimination inference rules are used. With the auxiliary predicate introduced here it is very simple to get the flexibility we need for the general case.

6. CONCLUSION

In this paper we have presented a language to specify inference rules. It has been shown how such specifications can be used to control the translation of a matrix in a PTTP-like theorem prover.

The language together with the compilation technique have been integrated in our theorem prover ProCom. It is called CaPrI which is an acronym for *calculi programming interface*. As we have seen some standard techniques are also used. Those are — among others — the possibility to adjust the behavior of the prover (or the translation) by activation of flags, the use of an external library and a linker.

Some of the well known techniques in automated theorem proving can be formulated within the CaPrI framework. Among those are the use of lemmas or regularity constraints to prune the search space.

Others like the linearizing completion of Horn-theories ((Baumgartner, 1996), see Chapter I.2.6) may be plugged in easily. On the other hand our technique also offers a tool to include theories which need or allow specialized algorithms to handle them efficiently.

This work provides the basis for a prover kernel which combines the flexibility to specify the inference rules with the inference rates reached by translating the given matrix into a Prolog program.

The flexibility and the character of a toolkit puts ProCom/CaPrI in competition with other toolkits for theorem provers, like KEIM (Huang et al., 1994) (see also Chapter II.2.6). But in contrast to those ProCom/CaPrI covers the area of Prolog based translation techniques for goal oriented calculi.

The limitation of goal oriented calculi may be dropped. This would just mean to replace the restriction of a single goal by the restriction to an arbitrary, non-empty list of goal templates. An extended compilation technique would have to be envisaged to cope with this generalization.

Even so we have opened a new dimension of flexibility to the PTTP-like translation techniques we have not reached the limit yet.

ACKNOWLEDGMENTS

The authors would like to thank William McCune for critical and helpful remarks.

REFERENCES

Baumgartner, P.: 1996, 'Theory Reasoning in Connection Calculi and the Linearizing Completion Approach'. Ph.D. thesis, Universität Koblenz-Landau. (Accepted).

Becher, G. and U. Petermann: 1994, 'Rigid E-Unification by Completion and Rigid Paramodulation'. In: B. Nebel and L. Dreschler-Fischer (eds.): *KI-94: Advances in Artificial Intelligence*, Vol. LNAI 861. pp. 319–330, Springer Verlag.

Beckert, B.: 1994, 'A Completion-Based Method for Mixed Universal and Rigid E-Unification'. In: A. Bundy (ed.): *CADE-12*, Vol. LNCS 814. pp. 678–692, Springer Verlag.

Beckert, B. and C. Pape: 1996, 'Incremental Theory Reasoning Methods for Semantic Tableaux'. In: *Theorem Proving with Analytic Tableaux and Related Methods, 5th International Workshop, Terassini, 1996*, Vol. 1071 of *Lecture Notes in Artificial Intelligence*. pp. 93–109, Springer.

Bibel, W.: 1982, *Automated Theorem Proving*. Vieweg Verlag.

Bürckert, H.: 1990, 'A Resolution Principle for Clauses with Constraints'. In: M. E. Stickel (ed.): *Proc. CADE-10*. pp. 178–192, Springer. LNCS/LNAI 449.

Dahn, B. I. and J. Gehne: 1997, 'Integration of Automated and Interactive Theorem Proving in ILF'. In: *Proc. CADE-14*. pp. 57–60.

de la Tour, T. B.: 1990, 'Minimizing the Number of Clauses by Renaming'. In: M. E. Stickel (ed.): *CADE-10*, Vol. LNCS 449. pp. 558–572, Springer Verlag.

Degtyarev, A. and A. Voronkov: 1995, 'Simultaneous Rigid E-Unification is Undecidable'. Technical Report 105, Uppsala University.

Degtyarev, A. and A. Voronkov: 1996, 'What You Ever Wanted to Know about Rigid E-Unification'. In: *JELIA 1996, European Workshop on Logic in Artificial Intelligence*. Springer.

Gallier, J., P. Narendran, D. Plaisted, and W. Snyder: 1992, 'Theorem Proving with Equational Matings and Rigid E-Unification'. *Journal of the ACM*.

Goubault, J.: 1993, 'A Rule-Based Algorithm for Rigid E-Unification'. In: G. Gottlob, A. Leitsch, and D. Mundici (eds.): *3rd Kurt Gödel Colloquium '93*, Vol. 713 of *LNCS*. Springer Verlag.

Huang, X., M. Kerber, M. Kohlhase, et al.: 1994, 'KEIM: A Toolkit for Automated Deduction'. In: A. Bundy (ed.): *CADE-12*, Vol. LNAI 814. pp. 807–810, Springer Verlag.

Neugebauer, G. and U. Petermann: 1994, 'CaPrI: Integrating Theories into the Prolog Technology Theorem Prover ProCom'. presented at the Workshop on Theory Reasoning 12th Conference on Automated Deduction, Nancy.

Neugebauer, G. and T. Schaub: 1994, 'A Pool-Based Connection Calculus'. In: C. Bozşahin, U. Halıcı, K. Oflazar, and N. Yalabık (eds.): *Proceedings of Third Turkish Symposium on Artificial Intelligence and Neural Networks*. pp. 297–306, Middle East Technical University Press.

Petermann, U.: 1993, 'Completeness of the Pool Calculus With an Open Built In Theory'. In: G. Gottlob, A. Leitsch, and D. Mundici (eds.): *3rd Kurt Gödel Colloquium '93*, Vol. LNCS 713. Springer Verlag.

Petermann, U.: 1994, 'A Complete Connection Calculus with Rigid E-Unification'. In: *JELIA94*, Vol. LNCS 838. pp. 152–167, Springer Verlag.

Petermann, U.: 1997, 'Building-In Hybrid Theories'. In: *Workshop on First-Order Theorem Proving*.

Plaisted, D. A.: 1990, 'A Sequent-Style Model Elimination Strategy and a Positive Refinement'. *J. Automated Reasoning* **6**, 389–402.

Rigó, Z.: 1995, 'Untersuchungen zum automatischen Beweisen in Modallogiken'. Master's thesis, Universität Leipzig.

Siekmann, J. H.: 1989, 'Unification Theory'. *J. Symbolic Computation* **7**(1), 207–274.

Stickel, M.: 1985, 'Automated Deduction by Theory Resolution'. *J. Automated Reasoning* **4**(1), 333–356.

Stickel, M. E.: 1988, 'A Prolog Technology Theorem Prover: Implementation by an Extended Prolog Compiler'. *J. Automated Reasoning* **4**, 353–380.

THOMAS KOLBE, CHRISTOPH WALTHER

CHAPTER 8

PROOF ANALYSIS, GENERALIZATION AND REUSE

1. INTRODUCTION

We investigate the improvement of theorem provers by reusing previously computed proofs, cf. (Kolbe and Walther, 1994; Kolbe and Walther, 1995d; Kolbe and Walther, 1996a; Kolbe and Walther, 1996b; Kolbe and Walther, 1997). Our work has similarities with the machine learning methodologies of *explanation-based learning* (Ellman, 1989), *analogical reasoning* (Hall, 1989), and *abstraction* (Giunchiglia and Walsh, 1992).

Assume that an (interactive or automated) theorem prover shall be supplemented by a learning component which remembers already computed proofs when called for proving a new conjecture, cf. Figure 1. If the prover is asked to prove a new conjecture ψ which is *similar* to a previously proven conjecture φ, it first tries to associate each function symbol occurring in φ with a function symbol from ψ. These associations are then propagated into the *proof* of φ for obtaining a proof of ψ. But generally, there are function symbols in the *proof* of φ which do not occur in the *conjecture* φ and thus still have to be associated. If these function symbols can be associated such that (certain) axioms used in the proof of φ are mapped to provable formulas, the proof of φ is successfully *reused* for proving ψ, cf. Figure 2.

We are concerned with providing the associations for function symbols automatically such that verifiable proof obligations for ψ are obtained. Therefore we propose an indirect way of association by first *generalizing* the function symbols from (the proof of) φ to *function variables*, i.e. we use a second-order language. Then we *instantiate* the function variables occurring in the generalization of φ — the so-called *bound* function variables — with function symbols by second-order matching with ψ. For providing the remaining associations a further second-order substitution has to be found which replaces the so-called *free* function variables, i.e. function variables which occur in the generalization of the *proof* of φ, but not in the generalization of φ. We call

* This work was supported under grants no. Wa 652/4-1,2,3 by the Deutsche Forschungsgemeinschaft as part of the focus program "Deduktion". Authors' address: Fachbereich Informatik, Technische Universität Darmstadt, Alexanderstr. 10, D-64283 Darmstadt, Germany. e-mail: {kolbe,walther}@informatik.tu-darmstadt.de

W. Bibel, P. H. Schmitt (eds.), Automated Deduction. A basis for applications. Vol. II
© 1998 Kluwer Academic Publishers. Printed in the Netherlands

such a second-order substitution a *solution* (for the free function variables), if all formulas stemming from (certain) axioms in the proof of φ are provable.

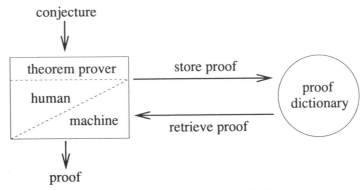

Figure 1. Theorem Proving with Reuse

Subsequently we are going to illustrate and discuss our proposal for reusing proofs which has been implemented in the PLAGIATOR system (Kolbe and Brauburger, 1997).

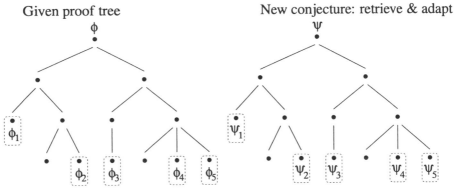

Figure 2. Reused Proof

2. REUSING PROOFS - AN EXAMPLE

Let us briefly sketch our method for reusing proofs, where our examples stem from the domain of theorem proving by *mathematical induction*: An *induction formula IH → IC* is either a *step* formula or a *base* formula in which case *IH* equals TRUE. Induction formulas are proved by modifying the induction conclusion *IC* using given axioms until the induction hypothesis *IH* is applicable. For instance, let the functions plus, sum and app be defined by the following equations where 0 and s(x) (resp. empty and add(n,x)) are the constructors of the sort number (resp. list):[1]

(plus-1)	plus$(0,y)$	$\equiv y$
(plus-2)	plus$(\mathsf{s}(x),y)$	$\equiv \mathsf{s}(\mathsf{plus}(x,y))$
(sum-1)	sum(empty)	$\equiv 0$
(sum-2)	sum$(\mathsf{add}(n,x))$	$\equiv \mathsf{plus}(n,\mathsf{sum}(x))$
(app-1)	app(empty,y)	$\equiv y$
(app-2)	app$(\mathsf{add}(n,x),y)$	$\equiv \mathsf{add}(n,\mathsf{app}(x,y))$

These defining equations form a theory which may contain *lemmata*, i.e. statements which were (inductively) inferred from the defining equations and other already proved lemmata. For instance

$$(\text{lem-1}) \quad \mathsf{plus}(\mathsf{plus}(x,y),z) \equiv \mathsf{plus}(x,\mathsf{plus}(y,z))$$

can be easily proved and therefore may be used like any defining equation in subsequent deductions. We aim to optimize proving such conjectures by reusing previously computed proofs of other conjectures. For instance consider the statement

$$\varphi[x,y] := \mathsf{plus}(\mathsf{sum}(x),\mathsf{sum}(y)) \equiv \mathsf{sum}(\mathsf{app}(x,y))$$

We prove the conjecture $\forall x,y \; \varphi[x,y]$ by induction upon the list-variable x and obtain two induction formulas, viz. the base formula φ_b and the step formula φ_S as

$$\varphi_b \quad := \quad \forall y \; \varphi[\mathsf{empty},y]$$
$$\varphi_S \quad := \quad \forall n,x,y \; (\forall u \; \varphi[x,u]) \rightarrow \varphi[\mathsf{add}(n,x),y].$$

The following proof of the step formula φ_S is obtained by modifying the induction conclusion $\text{IC} := \varphi[\mathsf{add}(n,x),y] =$

$$\mathsf{plus}(\mathsf{sum}(\mathsf{add}(n,x)),\mathsf{sum}(y)) \equiv \mathsf{sum}(\mathsf{app}(\mathsf{add}(n,x),y))$$

in a backward chaining style, i.e. each statement is implied by the statement in the line below, where terms are underlined if they have been changed in the corresponding proof step:[2]

---

[1] We usually omit universal quantifiers at the top level of formulas as well as the sort information for variables.

[2] We omit a proof for the base formula φ_b as there are no particularities compared to the step case.

$$
\begin{array}{ll}
\text{plus}(\text{sum}(\text{add}(n,x)),\text{sum}(y)) \equiv \text{sum}(\text{app}(\text{add}(n,x),y)) & \text{IC} \\
\text{plus}(\underline{\text{plus}(n,\text{sum}(x))},\text{sum}(y)) \equiv \text{sum}(\text{app}(\text{add}(n,x),y)) & \text{(sum-2)} \\
\text{plus}(\text{plus}(n,\text{sum}(x)),\text{sum}(y)) \equiv \text{sum}(\underline{\text{add}(n,\text{app}(x,y))}) & \text{(app-2)} \\
\text{plus}(\text{plus}(n,\text{sum}(x)),\text{sum}(y)) \equiv \underline{\text{plus}(n,\text{sum}(\text{app}(x,y)))} & \text{(sum-2)} \\
\text{plus}(\text{plus}(n,\text{sum}(x)),\text{sum}(y)) \equiv \underline{\text{plus}(n,\text{plus}(\text{sum}(x),\text{sum}(y)))} & \text{IH} \\
\underline{\text{plus}(n,\text{plus}(\text{sum}(x),\text{sum}(y)))} \equiv \text{plus}(n,\overline{\text{plus}(\text{sum}(x),\text{sum}(y))}) & \text{(lem-1)} \\
\hline
\quad\quad\quad\quad\quad\quad\quad\quad \text{TRUE} & X \equiv X
\end{array}
$$

Given such a proof, it is *analyzed* to distinguish its *relevant* features from its *irrelevant* parts. Relevant features are specific to the proof and are collected in a *proof catch* because "similar" requirements must be satisfied if this proof is to be reused later on. We consider features like the positions where equations are applied, induction conclusions and hypotheses, general laws like $X \equiv X$ etc. as irrelevant because they can always be satisfied. Formally the catch of a proof is a *subset* of the set of leaves of the corresponding proof tree.

Analysis of the above proof yields (sum-2), (app-2) and (lem-1) as the catch. E.g. all we have to know about plus for proving $\forall x, y\; \varphi[x,y]$ is its associativity, but not its semantics or *how* plus is computed. We then *generalize*[3] the conjecture, the induction formula and the catch for obtaining a so-called *proof shell*. This is achieved by replacing function *symbols* by function *variables* denoted by capital letters F, G, H etc., yielding the *schematic conjecture* $\Phi := F(G(x), G(y)) \equiv G(H(x,y))$ with the corresponding *schematic induction formula* Φ_S as well as the *schematic catch* C_S:

$$
\begin{aligned}
\Phi_S \quad &:= \quad (\forall u\; F(G(x), G(u)) \equiv G(H(x,u))) \rightarrow \\
&\qquad F(G(D(n,x)), G(y)) \equiv G(H(D(n,x),y))
\end{aligned}
$$

$$
C_S \quad := \quad
\left\{
\begin{array}{lll}
(8.1) & G(D(n,x)) \equiv F(n, G(x)) \\
(8.2) & H(D(n,x),y) \equiv D(n, H(x,y)) \\
(8.3) & F(F(x,y),z) \equiv F(x, F(y,z))
\end{array}
\right\}
$$

Figure 3. The proof shell PS_S for the proof of φ_S (simple analysis)

If a new statement ψ shall be proved, a suitable induction axiom is selected by well-known automated methods, cf. (Walther, 1994), from which a set of induction formulas I_ψ is computed for ψ. Then for proving an induction formula $\psi_i \in I_\psi$ by reuse, it is tested whether some proof shell PS *applies for* ψ_i, i.e. whether ψ_i is a (second-order) instance of the schematic induction formula of PS. If the test succeeds, the obtained (second-order) matcher is

---

[3] Not to be confused with *generalization* of a formula φ as a preprocessing for proving φ by induction.

applied to the schematic catch of *PS*, and if all formulas of the instantiated schematic catch can be proved (which may necessitate further proof reuses), ψ_i is verified by reuse since the truth of an instantiated schematic catch implies the truth of its instantiated schematic induction formula.

For instance, assume that the new conjecture $\forall x, y \; \psi[x, y]$ shall be proved, where

$$\psi[x, y] := \text{times}(\text{prod}(x), \text{prod}(y)) \equiv \text{prod}(\text{app}(x, y))$$

and times and prod are defined by the axioms

(times-1)	$\text{times}(0, y)$	$\equiv 0$
(times-2)	$\text{times}(s(x), y)$	$\equiv \text{plus}(y, \text{times}(x, y))$
(prod-1)	$\text{prod}(\text{empty})$	$\equiv s(0)$
(prod-2)	$\text{prod}(\text{add}(n, x))$	$\equiv \text{times}(n, \text{prod}(x)).$

The induction formulas computed for ψ are

$$\psi_b \quad := \quad \forall y \; \psi[\text{empty}, y]$$
$$\psi_s \quad := \quad \forall n, x, y \; (\forall u \; \psi[x, u]) \to \psi[\text{add}(n, x), y].$$

Obviously ψ is an instance of Φ and ψ_s is an instance of Φ_s w.r.t. the matcher $\pi := \{F/\text{times}, G/\text{prod}, H/\text{app}, D/\text{add}\}$. Hence (only considering the step case) we may reuse the given proof by instantiating the schematic catch C_s and subsequent verification of the resulting proof obligations:

$$\pi(C_s) = \left\{ \begin{array}{lr} (8.4) & \text{prod}(\text{add}(n, x)) \equiv \text{times}(n, \text{prod}(x)) \\ (8.5) & \text{app}(\text{add}(n, x), y) \equiv \text{add}(n, \text{app}(x, y)) \\ (8.6) & \text{times}(\text{times}(x, y), z) \equiv \text{times}(x, \text{times}(y, z)) \end{array} \right\}$$

Formulas (8.4) and (8.5) are axioms, viz. (prod-2) and (app-2), and therefore are obviously true. So it only remains to prove the associativity of times (8.6) and, if successful, ψ_s is proved. Compared to a direct proof of ψ_s we have saved the user interactions necessary to apply the right axioms in the right place (where the associativity of times must be verified in either case). Additionally, conjecture (8.6) has been *speculated as a lemma* which is required for proving conjecture ψ.

3. THE PHASES OF THE REUSE PROCESS

Although our examples stem from the domain of mathematical induction, our method is based on *first-order* reasoning only. Roughly speaking, a conjecture φ is proved by induction, if the *validity* of certain formulas, viz. the induction formulas, can be verified, where these formulas entail the truth of φ by means of some *induction axiom*. Since we do not consider induction axioms by our method, but only the proofs of first-order (induction) formulas, our method applies to any first-order reasoning procedure.

Our approach for reusing proofs is organized into the following steps, cf. Figure 4:

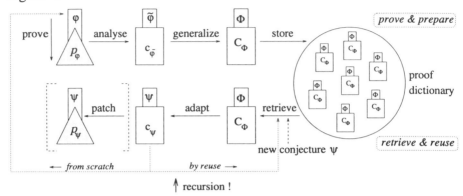

Figure 4. The Phases of the Reuse Process

Prove: [cf. Sections 1, 2] If required, a *direct* proof p for (an induction formula) φ from a set of axioms AX is given by the human advisor or an automated theorem prover. The set of axioms AX consists of *defining equations*, previously proved *lemmata*, and logical axioms like $x \equiv x$, $\varphi \rightarrow \varphi$ etc.

Analyze: (Kolbe and Walther, 1994) The *simple proof analysis* which was illustrated in Section 2 analyzes a proof p of φ, yielding a *proof catch* c. Formally, the catch c is a finite subset of the non-logical axioms of AX such that c logically implies φ. For increasing the applicability of proof shells and the reusability of proofs, we have developed the *refined proof analysis* which also distinguishes different *occurrences* of function symbols in the conjecture and in the catch of a proof, cf. Section 4. For instance the (step formula of) statement $\psi_2 := \text{plus}(\text{len}(x), \text{len}(y)) \equiv \text{len}(\text{app}(x,y))$ cannot be proved by reusing the proof shell from Figure 3, because one formula of the instantiated catch does not hold. However, the reuse succeeds if refined analysis is applied (see below).

Generalize: (Kolbe and Walther, 1994) Both φ and c are *generalized* by replacing (different occurrences of) function *symbols* with (different) function *variables*. This yields a *schematic conjecture* Φ and a *schematic catch C*, where the latter is a set of schematic formulas which — considered as a set of first-order hypotheses — logically implies the schematic conjecture Φ. Such a pair $PS := \langle \Phi, C \rangle$ is called a *proof shell* and serves as the data structure for reusing the proof p. E.g. after the refined analysis of the proof of φ_S from Section 2, generalization yields $\Phi' := F^1(G^1(x), G^2(y)) \equiv G^3(H^1(x,y))$ and the proof shell of Figure 5. Here e.g. the function variables F^1, F^2, F^3 correspond to different occurrences of the function symbol plus, i.e. the schematic equation (8.10) stems from generalizing (lem-1), cf. also Section 4.

$$\Phi'_S := (\forall u\, F^1(G^1(x), G^2(u)) \equiv G^3(H^1(x,u))) \rightarrow$$
$$F^1(G^1(D^1(n,x)), G^2(y)) \equiv G^3(H^1(D^1(n,x),y))$$

$$C'_S := \left\{ \begin{array}{ll} (8.7) & G^1(D^1(n,x)) \equiv F^2(n, G^1(x)) \\ (8.8) & H^1(D^1(n,x),y) \equiv D^4(n, H^1(x,y)) \\ (8.9) & G^3(D^4(n,x)) \equiv F^3(n, G^3(x)) \\ (8.10) & F^1(F^2(x,y),z) \equiv F^3(x, F^1(y,z)) \end{array} \right\}$$

Figure 5. The proof shell PS'_S for the proof of φ_S (refined analysis)[4]

Store: (Kolbe and Walther, 1995c) Proofs shells $\langle \Phi, C_1 \rangle, \ldots, \langle \Phi, C_n \rangle$ (sharing a common schematic goal formula Φ) are merged into a *proof volume* $PV := \langle \Phi, \{C_1, \ldots, C_n\} \rangle$ which then is stored in the *proof dictionary PD*, i.e. a library of "proof ideas" organized as a set of proof volumes, cf. Section 5.

Retrieve: (Kolbe and Walther, 1995c) If a new conjecture ψ is to be proved, the proof dictionary is searched for a proof volume $PV := \langle \Phi, \{C_1, \ldots, C_n\} \rangle$ such that $\psi = \pi(\Phi)$ for some second-order matcher π. For instance, $\pi_2 := \{F^1/\text{plus}, G^{1,2,3}/\text{len}, H^1/\text{app}, D^{1,2}/\text{add}\}$ is obtained by matching Φ'_S from Figure 5 with ψ_2 above. If successful, the schematic conjecture Φ and in turn also the proof volume *PV applies for* ψ (*via the matcher* π), cf. Section 5. Then a catch C_i is selected by heuristic support from the proof volume *PV* and the *partially instantiated catch* $\pi(C_i)$ serves as a candidate for proving ψ by reuse. For our example, the partially instantiated catch is obtained as

Adapt: (Kolbe and Walther, 1995a; Kolbe and Walther, 1995d) Since a partially instantiated catch $\pi(C_i)$ may contain *free function variables*, i.e. function variables which occur in C_i but not in Φ, these function variables have to be instantiated by known functions. Free function variables, as F^2, F^3, and D^4 in $\pi_2(C'_S)$ resulting from the *refined* analysis, provide an increased flexibility of the approach, because different instantiations correspond to different proofs. Hence a further second-

---

[4] Note that corresponding function variables in the induction hypothesis resp. the induction conclusion are provided with the same indices.

order substitution ρ is required for replacing these function variables such that the resulting proof obligations, i.e. all formulas in the *totally instantiated catch* $\rho(\pi(C_i))$, are provable from *AX*. Such a second-order substitution ρ is called a *solution* (for the free function variables), and upon existence of ρ conjecture ψ is proved by reuse because semantical entailment is invariant w.r.t. (second-order) instantiation (Kolbe, 1997). Solutions ρ are computed by *second-order matching modulo symbolical evaluation*, cf. Section 6. For the example, the solution $\rho_2 :=$ $\{F^{2,3}/s(w_2), D^4/add\}$ is obtained which e.g. instantiates (8.11) to the axiom $len(add(n,x)) \equiv s(len(x))$. [5]

$$\pi_2(C_S') = \left\{ \begin{array}{rl} (8.11) & len(add(n,x)) \equiv F^2(n, len(x)) \\ (8.12) & app(add(n,x),y) \equiv D^4(n, app(x,y)) \\ (8.13) & len(D^4(n,x)) \equiv F^3(n, len(x)) \\ (8.14) & plus(F^2(x,y),z) \equiv F^3(x, plus(y,z)) \end{array} \right\}.$$

Figure 6. A partially instantiated catch.

Patch: (Kolbe and Walther, 1995a; Kolbe and Walther, 1995b) Often one is not only interested in the *provability* of ψ, but also in a *proof* of ψ which can be presented to a human or can be processed subsequently. In this case it is not sufficient just to *instantiate* the schematic proof P of Φ (which is obtained by generalizing the proof p of φ) with the computed substitution $\tau := \rho \circ \pi$ because τ might destroy the structure of P. Therefore the instantiated proof $\tau(P)$ is *patched* (which always succeeds) by removing void resp. inserting additional inference steps for obtaining a proof p' of ψ, cf. Section 6.

4. PROOF ANALYSIS AND GENERALIZATION

We formalize the different styles of proof analysis which were illustrated in Sections 2 and 3. In general, these methods have to be augmented by a technique for treating also *sort informations* since we consider formulas in a many-sorted logic, see (Kolbe, 1997; Kolbe and Walther, 1997) for details.

---

[5] Function variables may also be replaced by functional terms, i.e. (first-order) terms where special *argument variables* $w_1, ..., w_n$ serve as formal parameters. The instantiations of F^1 and F^2 are different here, viz. plus and s, and this is why reuse fails for the *simple* analysis, cf. Figure 3.

4.1. *Simple Proof Analysis*

To formalize the *simple* proof analysis, the calculus in which we prove (base and step) formulas is augmented by stipulating in addition which formula has to be remembered as a "relevant feature" if a particular inference rule is used in a proof. Hence the rules are applied to expressions of the form $\langle \varphi, A \rangle$, where φ is a formula and A, called the *accumulator*, holds the catch collected so far. Thus e.g. replacement rules for applying an equation (stemming from the axioms or from the hypotheses) differ in updating the accumulator component: If the applied equation is an *axiom* it has to be recorded in A for obtaining the catch, but if the applied equation is a *hypothesis* it is irrelevant for the learning step and A remains unchanged.

If $\langle \varphi, \emptyset \rangle \vdash \langle \mathrm{TRUE}, A \rangle$ can be established, then φ is proved, i.e. $AX \models \varphi$, and A contains the catch of the proof, i.e. $A \models \varphi$. Now given a formula φ and a set of induction formulas $\{\varphi_0, \dots, \varphi_n\}$ for φ, we try to infer $\langle \varphi_i, \emptyset \rangle \vdash \langle \mathrm{TRUE}, A_i \rangle$ for each i in our calculus. If successful, φ is proved by means of some induction axioms and for each i an accumulator A_i is obtained which holds the catch of the proof of φ_i. Then we replace each function symbol f occurring in $\varphi, \varphi_0, \dots, \varphi_n, A_0, \dots, A_n$ by a function variable F, and obtain the proof shells $PS_i := \langle \Phi_i, C_i \rangle$ with the schematic induction formulas Φ_i and the schematic catches C_i (for the common schematic conjecture Φ).

4.2. *Refined Proof Analysis*

The success of proof reuses directly depends on the generality of what has been learned from a given proof. Reconsider the statement

$$\psi_2[x,y] := \mathsf{plus}(\mathsf{len}(x), \mathsf{len}(y)) \equiv \mathsf{len}(\mathsf{app}(x,y))$$

from Section 3 where the length $\mathsf{len}(x)$ of a list x is defined by

(len-1)	$\mathsf{len}(\mathsf{empty})$	$\equiv$	0
(len-2)	$\mathsf{len}(\mathsf{add}(n,x))$	$\equiv$	$\mathsf{s}(\mathsf{len}(x))$.

Since PS_S (cf. Figure 3) applies for the step formula of ψ_2, we instantiate the schematic catch C_S and obtain the proof obligations

(8.15)	$\mathsf{len}(\mathsf{add}(n,x))$	$\equiv$	$\mathsf{plus}(n, \mathsf{len}(x))$
(8.16)	$\mathsf{app}(\mathsf{add}(n,x),y)$	$\equiv$	$\mathsf{add}(n, \mathsf{app}(x,y))$
(8.17)	$\mathsf{plus}(\mathsf{plus}(x,y), z)$	$\equiv$	$\mathsf{plus}(x, \mathsf{plus}(y,z))$.

But statement (8.15) does not hold, hence ψ_2 cannot be verified by reuse. One idea is that ψ_2 requires an original proof which differs in its structure from all previously computed proofs, and therefore the proof reuse fails. But it may also be the case, that what has been learned is simply not enough, so the reuse fails only for this reason. The latter is true for statement ψ_2 and we improve our technique so that the reuse eventually is successful. The key idea for the improvement is to distinguish different *occurrences* of function symbols in the conjecture and in the catch of a proof, cf. Section 3. This necessitates modifications of the proof analysis yielding the technique of *refined proof analysis*. To distinguish the different occurrences of function symbols we label function symbols with an *index*: For a function symbol f and $i \in \mathbb{N}$ we call f^i an *indexed* function symbol with *index* i and *root* f. We use a mapping *index* to supply all occurrences of all function symbols in a term with new unique indices,[6] e.g. $index(s(plus(s(x),y))) \equiv s^1(plus^1(s^2(x),y))$. We demand *index* to yield indices in ascending order starting with 1.

We call a pair $\langle f^i, f^j \rangle$ of indexed function symbols with *different indices* and *common root* an *(index-) collision*, and the analysis calculus from Section 4.1 is modified to incorporate the bookkeeping of the index collisions which appear during a proof. Hence the modified calculus operates on triples $\langle \varphi, A, K \rangle$, where all function symbols in φ and A are indexed now and K is a set of index collisions.

Now e.g. replacement rules not only differ in updating the accumulator component (as they already do in the simple analysis approach), but they also differ in the treatment of indices: Only *freshly indexed* axioms are applied, whereas (already indexed) hypotheses are applied without index modifications. In both cases the indexed function symbols with common root which have to be identified in the replacement step are recorded in the collision component. Therefore only function symbols with common root are identified which *must* be identified and this guarantees that a *most general* catch will be obtained subsequently.

For instance, let $AX = \{f(a) \equiv a, g(a) \equiv b, g(b) \equiv a\}$ and consider the conjecture $\varphi = f(f(a)) \equiv a$. Then

$$\langle f^1(f^2(a^1)) \equiv a^2, \emptyset, \emptyset \rangle$$
$$\vdash \quad \langle f^1(a^4) \equiv a^2, \{f^3(a^3) \equiv a^4\}, \{\langle f^2, f^3 \rangle, \langle a^1, a^3 \rangle\} \rangle$$
$$\vdash \quad \langle a^6 \equiv a^2, \{f^3(a^3) \equiv a^4, f^4(a^5 \equiv a^6\}, $$
$$\{\langle f^2, f^3 \rangle, \langle a^1, a^3 \rangle, \langle f^1, f^4 \rangle, \langle a^4, a^5 \rangle\} \rangle$$
$$\vdash \quad \langle \text{TRUE}, \{f^3(a^3) \equiv a^4, f^4(a^5) \equiv a^6\}, $$
$$\{\langle f^2, f^3 \rangle, \langle a^1, a^3 \rangle, \langle f^1, f^4 \rangle, \langle a^4, a^5 \rangle, \langle a^2, a^6 \rangle\} \rangle$$

is computed with refined proof analysis, the accumulator holds $\{f^3(a^3) \equiv a^4, f^4(a^5) \equiv a^6\}$ as the indexed catch, and the collision component consists of $\{\langle f^2, f^3 \rangle, \langle a^1, a^3 \rangle, \langle f^1, f^4 \rangle, \langle a^4, a^5 \rangle, \langle a^2, a^6 \rangle\}$.

---

[6] Strictly speaking, *index* is an operation having a side effect since we demand that indices obtained by *index* have been "never used before".

For proving a formula φ we establish $\langle index(\varphi), \emptyset, \emptyset \rangle \vdash \langle \text{TRUE}, A, K \rangle$ in the refined calculus. The collision set K contains pairs $\langle f^i, f^j \rangle$ of indexed function symbols which have been identified in the proof. This information must be propagated into the accumulator A and the indexed conjecture when building the proof catch, because the proof is based on the assumption that f^i and f^j denote *identical* functions: Either (i) f^i and f^j are identified *syntactically* by replacing f^i with f^j (or vice versa) in $index(\varphi)$ and the equations of A or (ii) f^i and f^j are identified *semantically* by insertion of $[\forall x_1, ..., x_n \; f^i(x_1, ..., x_n) \equiv f^j(x_1, ..., x_n)]$ into A. We use the pragmatical criterion of identifying *bound* function symbol from the proven conjecture semantically, while the other *free* function symbols are identified syntactically. In this way a proof shell is obtained where both the applicability is maximized and the reuse effort is minimized, since the function symbols from the conjecture are propagated into the proof catch as far as possible (and necessary). For the example above, identification yields e.g. the proof catch $\{f^2(a^1) \equiv a^4, f^1(a^4) \equiv a^2\}$.

Now the equations in the modified accumulator A are generalized as before yielding the schematic catch, where however function symbols with *different indices* are replaced by *different function variables*. This yields the proof shell $PS^* = \langle F(G(A)) \equiv B, \{G(A) \equiv C, F(C) \equiv B\} \rangle$ for the example above, and e.g. the conjectures $\psi_1 = f(g(b)) = a$ and $\psi_2 = g(g(a)) \equiv a$ can be proved by reuse based on PS^*. However, if only simple proof analysis is used, the proof shell $PS = \langle F(F(A)) \equiv A, \{F(A) \equiv A\} \rangle$ is obtained from the analyzed derivation of φ and an attempt to verify the conjectures ψ_1 and ψ_2 by reuse based on PS fails, either because the proof shell does not apply or unprovable proof obligations are obtained.

For instance, after assigning indices to the conjecture φ from Section 2, we prove the induction formulas for the indexed conjecture

$$\text{plus}^1(\text{sum}^1(x), \text{sum}^2(y)) \equiv \text{sum}^3(\text{app}^1(x, y))$$

and obtain the proof shell PS_S' from Figure 5 with the schematic conjecture Φ_S' and the schematic catch C_S' for the step formula:[7] Now reconsider statement ψ_2 from the beginning of this section. Since PS_S' applies for ψ_2, the instantiated catch $\pi_2(C_S')$ given in Section 3 is computed. This schematic catch is only *partially* instantiated because the function variables F^2, F^3 and D^4 stemming from the function symbols plus, plus and add in the catch of the original proof are not replaced by concrete function symbols. This is because these function variables do not occur in the schematic induction formula of the proof shell.

---

[7] Dropping the indices in these equations yields the schematic catch C_S from Figure 3.

We call such function variables of a proof shell *free* and we call all function variables occurring in the schematic induction formulas *bound* function
variables.

A formula φ with a free function variable F is *true* iff some function exists such that the formula obtained from φ by replacing F with this function
can be proved. Thus e.g. a provable formula, viz. the axiom (app-2), is obtained from (8.12) if D^4 is replaced by add, and provable formulas are also
obtained from (8.11), (8.13), and (8.14), viz. the axioms (len-2) and (plus-2),
if $\rho_2 := \{F^{2,3}/\mathsf{s}(w_2), D^4/\mathrm{add}\}$ is used as in Section 3. Hence ψ_2 is proved
by reuse only. The goal-directed computation of useful instantiations like ρ_2
is discussed in Section 6.

Since also more general *schematic conjectures* are obtained when using
the refined proof analysis, proofs more often can be reused only because a
proof shell applies more often. Consider e.g. the associativity of multiplication and the statement $z^{xy} = (z^y)^x$, i.e.

$$\varphi_3[x,y,z] :\equiv \quad \mathrm{times}(\mathrm{times}(x,y),z) \quad \equiv \mathrm{times}(x,\mathrm{times}(y,z))$$
$$\psi_3[x,y,z] :\equiv \quad \exp(\mathrm{times}(x,y),z) \quad \equiv \exp(x,\exp(y,z))$$

and suppose that φ_3 is already proved. With the *simple* proof analysis, the
proof of φ_3 cannot be reused for verifying ψ_3 because ψ_3 is not an instance
of the schematic conjecture

$$\Phi_3 := \quad F(F(x,y),z) \equiv F(x,F(y,z))$$

obtained from φ_3. But with the *refined* proof analysis, φ_3 is generalized to the
schematic conjecture

$$\Phi_4 := \quad F^1(F^2(x,y),z) \equiv F^3(x,F^4(y,z))$$

which matches ψ_3, so now the proof of φ_3 can be reused (Kolbe, 1997). Here
the occurrences of times in φ_3 correspond to the *distinct* function symbols
exp and times in ψ_3, i.e. the artificially generalized schematic catch generated
from the proof of φ_3 is sufficient to prove the *more general* conjecture ψ_3. A
further increase of reusability is obtained by miscellaneous modifications of
the matching algorithm concerning the order and presence of arguments in
function applications, see Section 7 for examples.

5. PROOF MANAGEMENT

The concept of similarity using second-order matching as presented in the
previous sections is very broad. But since there may be different proofs for

different instances of a schematic conjecture Φ, we have to select a reusable proof shell among the applicable proof shells for a new conjecture ψ which shall be proved. For supporting an efficient retrieval of applicable proof shells we organize the set of computed proof shells by so-called *proof volumes* and a *proof dictionary*. A proof volume represents a set of proof shells sharing the same schematic conjecture, and a proof dictionary is a set of proof volumes:

Definition 5.1 (Proof volume, Proof dictionary) A *proof volume PV* is a pair $PV = \langle \Phi, \{C_1, ..., C_n\} \rangle$ consisting of a schematic conjecture $root(PV) := \Phi$ and a non-empty ($n \geq 1$) set of schematic catches $leaves(PV) := \{C_1, ..., C_n\}$, such that $\langle \Phi, C_i \rangle$ is a proof shell for $1 \leq i \leq n$. The proof volume *applies* for a conjecture ψ iff Φ applies for ψ. A *proof dictionary* is a set $PD := \{PV_1, ..., PV_m\}$ of proof volumes PV_i, $1 \leq i \leq m$, $m \geq 0$. $\qquad\square$

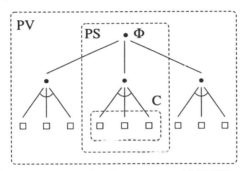

Figure 7. A proof volume as an AND/OR-tree

Example 5.2 (Proof Management) Assume that the first conjecture φ_1 to be proven is the associativity of the multiplication function times, i.e.

$$\varphi_1 := \text{times}(x, \text{times}(y, z)) \equiv \text{times}(\text{times}(x, y), z).$$

As we are not provided with any proof shells we have to compute direct proofs of the induction formulas

$$\varphi_{1b} := \text{times}(0, \text{times}(y, z)) \equiv \text{times}(\text{times}(0, y), z),$$

$$\varphi_{1s} :=$$
$$(\forall u, v \, \text{times}(x, \text{times}(u, v)) \equiv \text{times}(\text{times}(x, u), v))$$
$$\rightarrow \text{times}(\text{s}(x), \text{times}(y, z)) \equiv \text{times}(\text{times}(\text{s}(x), y), z).$$

These proofs are analyzed and generalized subsequently, resulting in the proof shells $PS_{1b} := \langle \Phi_{1b}, C_{1b} \rangle$ and $PS_{1s} := \langle \Phi_{1s}, C_{1s} \rangle$, where e.g. in

$$\Phi_{1b} := F^1(D^1, F^2(y, z)) \equiv F^3(F^4(D^2, y), z)$$

the function variables $F^1, ..., F^4$ correspond to the different occurrences of the function symbol times in φ_{1b}.

If the next conjecture is the associativity of plus, i.e.

$$\varphi_2 := \mathsf{plus}(x, \mathsf{plus}(y, z)) \equiv \mathsf{plus}(\mathsf{plus}(x, y), z),$$

the proof shell PS_{1b} applies for the base formula φ_{2b}, and PS_{1s} applies for the step formula φ_{2s}. But while the reuse succeeds for φ_{2s}, it fails for the base case such that a direct proof has to be computed for φ_{2b}, resulting in the proof shell $PS_{2b} := \langle \Phi_{2b}, C_{2b} \rangle$.

Now as both proof shells PS_{1b} and PS_{2b} are constructed by generalizing (the proofs of) structurally similar formulas, viz. base cases for a law of associativity, their schematic conjectures Φ_{1b} and Φ_{2b} are identical. Thus PS_{1b} and PS_{2b} apply for exactly the same formulas, while e.g. PS_{1b} and PS_{1s} never apply for the same formulas.

Hence, after finding a proof shell which applies for a given conjecture ψ, all other applicable proof shells can be accessed (without further matching operations) if we organize our proof shells in the following way: We collect all proof shells with the same schematic conjecture Φ in a proof volume for Φ, e.g. the proof volume $PV_{2b} := \langle \Phi_{1b}, \{C_{1b}, C_{2b}\} \rangle$ covers the proof shells PS_{1b} and PS_{2b}. The proof volumes PV_{2b} and $PV_{1s} := \langle \Phi_{1s}, \{C_{1s}\} \rangle$ then form a proof dictionary $PD_2 := \{PV_{2b}, PV_{1s}\}$ providing all proof shells which are available in the present state.   □

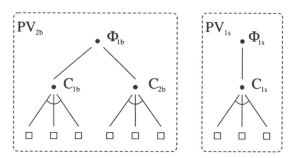

Figure 8. The proof dictionary $PD_2 = \{PV_{2b}, PV_{1s}\}$

The example reveals that sometimes a *partial* reuse of induction proofs is possible, viz. if only some of the several induction formulas of a conjecture are provable by reuse (of the currently available proofs).

Example 5.3 (generalization) Reconsider the proof dictionary $PD_2 = \{PV_{2b}, PV_{1s}\}$ from Figure 8 in Example 5.2. Let the third conjecture be given as

$$\varphi_3 := \mathsf{minus}(x, \mathsf{plus}(y, z)) \equiv \mathsf{minus}(\mathsf{minus}(x, y), z).$$

The proof dictionary is searched for a proof volume which applies for the base formula φ_{3b} and PV_{2b} is obtained. Now the contained proof shells are tried for verifying φ_{3b}, where the reuse succeeds for PS_{1b} (and fails for PS_{2b}). In the same way PV_{1s} is determined for the step formula φ_{3s}, but the reuse fails here and consequently a direct proof of φ_{3s} has to be computed.

Now we could generalize the found proof as usual, which however would yield a proof shell that had to be adapted subsequently for an insertion into PV_{1s}. Hence we prefer to combine the generalization with the necessary adaptation by generalizing the proof with respect to the schematic conjecture Φ_{1s} from the proof volume PV_{1s}. The key idea is to use the *inverse* of the matcher π between Φ_{1s} and φ_{3s} as a "permutative generalization function". [8] Thus the resulting proof shell $PS_{3s} = \langle \Phi_{1s}, C_{3s} \rangle$ can be inserted in the proof dictionary yielding $PD_3 := \{PV_{2b}, PV_{3s}\}$ with $PV_{3s} := \langle \Phi_{1s}, \{C_{1s}, C_{3s}\} \rangle$.

Let the conjecture φ_4 be the associativity of the maximum function max, i.e.

$$\varphi_4 := \mathsf{max}(x, \mathsf{max}(y, z)) \equiv \mathsf{max}(\mathsf{max}(x, y), z).$$

The base formula φ_{4b} can be proved using the proof catch $C_{2b} \in leaves(PV_{2b})$ obtained from the associativity of plus, and the step formula φ_{4s} is provable using the catch $C_{3s} \in leaves(PV_{3s})$ from the conjecture φ_3. $\square$

This example reveals that sometimes a "mixed" reuse of induction proofs is possible, i.e. the several induction formulas which are computed for one conjecture may be verified by reusing the proofs of induction formulas belonging to several conjectures.

Using proof volumes the set of available proof shells is divided into disjoint subsets (each represented by a proof volume) such that for each conjecture ψ either *none* or *all* proof shells of a proof volume apply for ψ, thus reducing the matching costs when checking for applicability. For supporting an efficient retrieval of proofs we demand that there is *at most one* proof volume in a proof dictionary which applies for ψ, i.e. each pair of different schematic conjectures (at the roots of the proof volumes) which both apply for *some* conjecture must apply for exactly the *same* conjectures. This uniqueness, however, can only be achieved if we restrict the class of matchers which determine the applicability of a proof shell for a conjecture, cf. (Kolbe and Walther, 1995c).

---

[8] More precisely, we have to consider the *indexed* conjecture φ'_{3s} which is obtained from φ_{3s} by distinguishing different occurrences of function symbols. Note that the refined proof analysis yields a proof (catch) for this indexed conjecture.

Generally, there is a trade-off between *high reusability* and *efficient retrieval* because the latter imposes some restrictions on the similarity criterion which spoils the former. Hence the proof management component should be designed according to the reuse system's domain and task: If a machine working in a real-time environment should be improved by reuse, the system has to be tailored to optimize retrieval and adaptation time, even if the degree of reusability is decreased. If user interactions in a hybrid system should be saved, the system has to be designed to achieve *high reusability*, because the savings of (even a few) human interactions will outweigh the costs of (even a large amount of) machine resources (wrt. time and memory) spent for retrieval and adaptation, see also Section 9.

6. ADAPTATION OF PROOFS

After a proof shell is retrieved for a new conjecture, the computed matcher is applied for partially instantiating the schematic catch, leaving only the free function variables uninstantiated, cf. Section 3. As the solvability of free function variables is defined w.r.t. a set of axioms, it is undecidable in general whether a solution exists, and in the positive case the solution is not unique. Here we develop an efficient procedure for computing solutions by *second-order matching "modulo evaluation"*. Our algorithm incorporates the underlying axioms by heuristically combining the second-order matching algorithm of (Huet and Lang, 1978) with the technique of *symbolic evaluation*, cf. e.g. (Walther, 1994).

6.1. *Second-Order Matching modulo Evaluation*

We assume that each function symbol either is a *constructor* of some data structure or a *defined function symbol* specified by some axioms, cf. (Walther, 1994). For instance the declaration

$$structure \ \mathsf{empty} \ \mathsf{add(number, list)} \ : \ \mathsf{list}$$

defines a data structure for linear lists of natural numbers with the constructors empty and add. The axioms (len-1), (len-2) define the function len computing the length of a list. A term t is *evaluated* by applying the defining axioms for the function symbols contained in t. Thus e.g. $\mathsf{len(app(add}(n,k),l))$ is evaluated via $\mathsf{len(add}(n, \mathsf{app}(k,l)))$ to $\mathsf{s(app}(k,l))$ by applying the defining axioms for len and app. This is called *symbolic* evaluation because the term to be evaluated may contain variables as e.g. z (which are not evaluated).

Reconsider the partially instantiated catch $C_2 := \pi_2(C'_S)$ from Figure 6. Our goal is to *compute* a second-order substitution ρ_S for these free function variables which *solves* C_2, i.e. we are looking for some ρ_s such that all for-mulas of $\rho_s(C_2)$ are provable. The attempt of solving F^2 in the first equation (8.11) of C_2 by (syntactical) second-order *matching* fails, but we may *evaluate* the left hand side (lhs) of (8.11) by the defining axiom (len-2) yielding $\mathsf{s}(\mathsf{len}(x)) \equiv F^2(n, \mathsf{len}(x))$. Now the lhs matches the rhs via the unique matcher $\rho_1 := \{F^2/\mathsf{s}(w_2)\}$ which means that (8.11) is solved because $\rho_1((8.11)) = $ (len-2).

A *second-order substitution* is a finite set of pairs $\{..., F/s, ...\}$ where F is a function variable and s is a functional term. A functional term may contain special *argument variables* $w_1, w_2, ...$ where e.g. w_2 denotes the second argument of the binary function variable F^2, i.e. ρ_1 replaces F^2 with the function symbol s, but the first argument w_1 of F^2 is ignored.[9] A target term t *matches* a pattern P iff there is a second-order substitution π with $t = \pi(P)$. Second-order matching is decidable and we can adapt the algorithm from (Huet and Lang, 1978) for computing a finite set $\Pi := match(t, P)$ of matchers which is consistent, complete, and minimal.

We resume our example and apply ρ_1 to the remaining schematic formulas obtaining

$$
\rho_1(C_2 \setminus \{(8.11)\}) = \left\{ \begin{array}{ll} (8.12) & \mathsf{app}(\mathsf{add}(n,x),y) \equiv D^4(n, \mathsf{app}(x,y)) \\ (8.13) & \mathsf{len}(D^4(n,x)) = F^3(n, \mathsf{len}(x)) \\ (8.18) & \mathsf{plus}(\mathsf{s}(y),z) \equiv F^3(x, \mathsf{plus}(y,z)) \end{array} \right\}.
$$

We continue with the next schematic equation whose (say) lhs does not contain function variables, i.e. with (8.12) or with (8.18). Here we choose (8.12) and evaluating its lhs using a defining axiom for app yields $\mathsf{add}(n, \mathsf{app}(x,y))$ which matches the rhs via the unique matcher $\rho_2 := \{D^4/\mathsf{add}\}$. Now the lhs of $\rho_2((8.13))$ is purely first-order and can be evaluated, i.e. $\rho_3 := \{F^3/\mathsf{s}(w_2)\}$ is obtained by matching. Finally all free function variables are instantiated and the remaining proof obligation $\rho_3((8.18))$ can be passed to the prover. As $\rho_3((8.18))$ is a defining axiom for plus, the prover is success-ful, i.e. the proof reuse is completed. Thus the alternation of *evaluation* and *second-order matching* allows us to *compute* the solution $\rho_s = \{F^2/\mathsf{s}(w_2), F^3/\mathsf{s}(w_2), D^4/\mathsf{add}\}$.[10]

---

[9] Note that neither first-order variables or function variables are introduced by our (restricted) second-order substitutions nor are (first-order) variables replaced.

[10] The proposed method is more flexible than simply matching the schematic equations from C_2 with the defining axioms.

6.2. *Solving a Proof Catch*

We discuss more precisely the algorithm *solve_catch* which implements the second-order matching "modulo evaluation" illustrated in the previous section. However, first we present the auxiliary functions *choose_eq*, *eval*, *match*, and *choose_matcher* which are used by *solve_catch*:

The function *choose_eq*(C') selects some equation $l \equiv R$ from a set C' of partially instantiated schematic equations, such that the (say) lhs l does *not* contain function variables, but the rhs R *does* contain function variables. If there is no such equation then *choose_eq* fails, and otherwise a heuristic for guiding the selection of one of the maybe several candidate equations $l \equiv R \in C'$ is used. This heuristic prefers the schematic equation which is most "constrained", where we consider an equation with a high number of function symbols and a low number of function variables as "highly constrained", because both criteria limit the number of possible matchers, cf. (Kolbe and Walther, 1995d).

The operation *eval*(l) implements the *terminating* symbolic evaluator which computes the normal form $l' := eval(l)$ of a term l by applying one or more defining equations of the involved function symbols, such that $AX \models l \equiv l'$ is satisfied for all terms l and a set AX of axioms, cf. (Walther, 1994). The second-order matching algorithm *match* already mentioned in Section 6.1 is called for l' and R and collects all computed matchers in a set Π.

The function *choose_matcher*(Π) fails if Π is empty, and otherwise selects one of the computed matchers due to the following heuristic:[11] The "simplest" matcher π with the minimal number of introduced function symbols is selected, because such a matcher preserves the given structure of the proof being reused and it is more likely to find a valid instance of the schematic proof in this case.

Now the algorithm *solve_catch* has two sets C', X of partially instantiated schematic equations as input arguments, where initially C' holds a partially instantiated schematic catch and X is empty. *solve_catch* operates by successively (1) selecting the next equation $e \in C'$ to be solved by the auxiliary function *choose_eq*, (2) evaluating the totally instantiated part of e by the auxiliary function *eval*, (3) computing the set Π of solutions for e by the auxiliary function *match*, (4) selecting a matcher $\pi \in \Pi$ by the auxiliary function *choose_matcher*, and (5) repeating from (1) with $\pi(C'), \pi(X)$ until no function variables remain in C'.

---

[11] For improved efficiency, selecting the matcher can also be built into the matching algorithm *match*.

If $\Pi = \emptyset$ in (3), then instead of (4) in (4') the unsolvable equation e is removed from C' and inserted into the *remainder set X*. We will not try to solve the equations collected in X by matching again (as this must fail), but we hope that the function variables occurring in X are instantiated while solving the remaining part of the catch. If *choose_eq* fails in (1) i.e. all non-first-order equations in C' (if any) contain function variables on *both* sides, then the alternating process of matching modulo evaluation and instantiating the catch fails. Now if no function variables occur in C' and X, then *solve_catch* is successful and $C' \cup X$ is returned as the set of first-order proof obligations together with (the composition of) the matchers computed as solutions. Otherwise there is an equation in C' with function variables on both sides (which has not yet been processed) or there is an equation in X with function variables on one side (where the matching failed). We regard the catch as unsolvable then and *solve_catch* fails, although one can think of several alternative heuristically motivated attempts of solving the remaining function variables, cf. (Kolbe and Walther, 1995d). We summarize these considerations with the following theorem:

Theorem 6.1 (soundness of *solve_catch*) *If AX is a set of axioms, C' is a schematic catch and X is a finite set of schematic equations, then solve_catch(C',X) either fails or returns $\langle E, \rho \rangle$ such that $AX \cup E \models \rho(C' \cup X)$.*

By this theorem, a proof reuse for $\psi = \pi(\Phi)$ with a partially instantiated catch $\pi(C)$ succeeds if for $\langle E, \rho \rangle := solve\_catch(\pi(C), \emptyset)$ all equations in E, i.e. the lemmata speculated for the reuse, are provable from AX, because then $AX \models E$ and consequently $AX \models \rho(\pi(C))$, i.e. $AX \models \psi$ is implied. For instance $solve\_catch(C_2, \emptyset) = \langle \rho_S(C_2), \rho_S \rangle$ follows for ψ_{2S} from Section 6.1 and therefore ψ_{2S} is proved by reuse. The general usefulness of *solve_catch* is demonstrated with more examples in Section 7.

6.3. *Patching Proofs*

Often one is not only interested in the *provability* of ψ, but also in a specific *proof* of ψ which can be presented to a human user or which can be processed subsequently, e.g. by translation into natural language to obtain proofs like those found in mathematical textbooks (Huang, 1994). Furthermore proofs can be worked up for planning or synthesis tasks if plans or programs should be extracted form proofs (Bates and Constable, 1985). These applications require a *specific* proof, i.e. it is not enough to know that *some* proof exists.

Reconsider the proof shell PS'_S from Figure 5. The schematic catch C'_S provides the relevant features for performing a schematic proof P of the schematic conjecture $\Phi'_S =: IH \rightarrow IC$, i.e. P is the generalization the original proof p of φ_S. In P the schematic conclusion IC is modified in a backward chaining style by successively applying the schematic equations $(8.7) - (8.10)$ and the hypothesis IH:

$$
\begin{aligned}
F^1(G^1(D^1(n,x)),G^2(y)) &\equiv G^3(H^1(D^1(n,x),y)) \\
F^1(F^2(n,G^1(x)),G^2(y)) &\equiv G^3(H^1(D^1(n,x),y)) \\
F^1(\overline{F^2(n,G^1(x))},G^2(y)) &\equiv G^3(D^4(n,H^1(x,y))) \\
F^1(F^2(n,G^1(x)),G^2(y)) &\equiv F^3(\overline{n,G^3(H^1(x,y))}) \\
F^1(F^2(n,G^1(x)),G^2(y)) &\equiv F^3(n,\overline{F^1(G^1(x),G^2(y))}) \\
F^3(n,F^1(G^1(x),G^2(y))) &\equiv F^3(n,\overline{F^1(G^1(x),G^2(y))})
\end{aligned}
$$
$$\overline{}$$
$$\text{TRUE}$$

Such a schematic proof is *instantiated* with a second-order substitution $\tau = \rho \circ \pi$ which composes a matcher π of the schematic conjecture Φ'_S and a new conjecture ψ_S with a solution ρ of the partially instantiated catch $\pi(C'_S)$ computed by *solve_catch*. As long as the matcher τ only replaces function variables with function symbols, the instantiated schematic proof $\tau(P)$ is a proof of ψ_S from $\tau(C'_S)$, i.e. $\tau(C'_S) \vdash_{\tau(P)} \psi_S$, because the structure of P is preserved. However, function variables are also replaced using general second-order substitutions like $\tau := \{F^{1,2,3}/w_2,\ G^{1,2,3}/minus(w_1,w_1),\ H^1/plus(w_1,w_2),\ D^{1,4}/s(w_2)\}$ (where minus denotes subtraction) obtained by matching Φ'_S with a new conjecture $\psi_S := \tau(\Phi'_S) = \tau(IH \to IC) =$

$$
\begin{aligned}
(\forall u\ minus(u,u) &\equiv minus(plus(x,u),plus(x,u))) \\
\to minus(y,y) &\equiv minus(plus(s(x),y),plus(s(x),y))
\end{aligned}
$$

and solving the resulting partially instantiated catch. C'_S is instantiated yielding the set of proof obligations

$$
\tau(C'_S) = \left\{
\begin{array}{ll}
\tau(8.7) & minus(s(x),s(x)) \equiv minus(x,x) \\
\tau(8.8) & plus(s(x),y) \equiv s(plus(x,y)) \\
\tau(8.9) & minus(s(x),s(x)) \equiv minus(x,x) \\
\tau(8.10) & z \equiv z
\end{array}
\right\}.
$$

By instantiating the schematic proof P with τ, we obtain $\tau(P)$ as

$$
\begin{array}{lll}
minus(y,y) &\equiv minus(plus(s(x),y),plus(s(x),y)) & \tau(IC) \\
minus(y,y) &\equiv minus(plus(s(x),y),plus(s(x),y)) & \tau(8.7) \\
minus(y,y) &\equiv minus(s(plus(x,y)),s(plus(x,y))) & \tau(8.8) \\
minus(y,y) &\equiv minus(\overline{plus(x,y),plus(x,y)}) & \tau(8.9) \\
minus(y,y) &\equiv \overline{minus(y,y)} & \tau(IH) \\
minus(y,y) &\equiv \overline{minus(y,y)} & \tau(8.10)
\end{array}
$$
$$\overline{}$$
$$\text{TRUE} \qquad\qquad X \equiv X$$

Figure 9. An instantiated schematic proof

But $\tau(P)$ is *not* a proof, i.e. $\tau(C'_S) \not\vdash_{\tau(P)} \psi_S$. Although each statement in Figure 9 is implied by the statement in the line below, the *justifications* of the inference steps are not valid. E.g. the *replace*($\tau(8.7)$)-step is illegal because the *position* of the replacement (the former first argument of F^1) does not exist in $\tau(IC)$. Also the *replace*($\tau(8.8)$)-step is illegal, as it actually consists of *two* replacements which must be performed separately at different positions. Finally the redundant *replace*($\tau(8.10)$)-step should be omitted. Thus $\tau(P)$ has to be *patched* for obtaining a proof of ψ_S.

Patching a proof is performed by successively patching the single replacement steps in a proof, cf. (Kolbe and Walther, 1995b): An algorithm *patch_positions*(t, p, τ) yields for arbitrary terms t, positions p, and second-order substitutions τ a list of independent positions $[p_1, ..., p_k]$ such that instantiating the modification of t at p with τ is equivalent to modifying the instance $\tau(t)$ at $p_1, ..., p_k$. With this auxiliary algorithm we can compute $P_\tau :=$ *patch_proof*(P, τ) to obtain a patched proof for the conjecture $\psi_S = \tau(IH) \rightarrow \tau(IC)$ from above:

$$
\begin{array}{lll}
\text{minus}(y,y) & \equiv \ \text{minus}(\text{plus}(\text{s}(x),y), \text{plus}(\text{s}(x),y)) & \tau(IC) \\
\text{minus}(y,y) & \equiv \ \text{minus}(\text{s}(\text{plus}(x,y)), \text{plus}(\text{s}(x),y)) & \tau(8.8) \\
\text{minus}(y,y) & \equiv \ \text{minus}(\text{s}(\text{plus}(x,y)), \text{s}(\text{plus}(x,y))) & \tau(8.8) \\
\text{minus}(y,y) & \equiv \ \text{minus}(\text{plus}(x,y), \text{plus}(x,y)) & \tau(8.9) \\
\text{minus}(y,y) & \equiv \ \underline{\text{minus}(y,y)} & \tau(IH) \\
& \overline{\text{TRUE}} & X \equiv X
\end{array}
$$

Figure 10. A patched proof

7. EXAMPLES FOR THE REUSE OF PROOFS

In the previous sections we have described our approach to prove reuse w.r.t. the steps illustrated in Figure 4. Apart from initial proofs provided by human advisor or an automated theorem prover in the "prove" step, none of these steps necessitates human support. Thus the proof shell from Figure 5 can be automatically reused for proving the step formulas φ_i^s of the *apparently different* conjectures φ_i given in Table I below.[12]

This table illustrates a typical session with the PLAGIATOR-system (Kolbe and Brauburger, 1997): At the beginning of the session the human advisor submits statement $\varphi_{8.0}$ (in the first row) and a proof p of φ to the system.

---

[12] $<>$ concatenates two lists, $\prod$ multiplies the numbers in a list, and rev reverses a list. For $\varphi_{8.6}$ the instance $\sigma_{8.6}(\varphi_{8.12})$ is speculated where $\sigma_{8.6} = \{z/\text{add}(n, \text{empty})\}$.

Then the statements $\varphi_{8.1}$, $\varphi_{8.2}$, ... , $\varphi_{8.6}$ are presented to the PLAGIATOR, which proves each statement (and also statements $\varphi_{8.7}$, ... , $\varphi_{8.12}$ obtained as speculated lemmata) only by reuse of p such that no user interactions are required. The third column shows the subgoals speculated by *solve_catch* when proving a statement by reuse. E.g. statement $\varphi_{8.7}$ is speculated when verifying $\varphi_{8.3}$, which leads to speculating $\varphi_{8.8}$ which in turn entails speculation of conjecture $\varphi_{8.9}$, for which eventually $\varphi_{8.10}$ is speculated.

$\varphi_{8.0}$ Φ	$\sum x + \sum y \equiv \sum (x <> y)$ $F^1(G^1(x), G^2(y)) \equiv G^3(H^1(x,y))$	$\varphi_{8.10}$
	conjectures proved by reuse	subgoal
$\varphi_{8.1}$	$\lvert x \rvert + \lvert y \rvert \equiv \lvert x <> y \rvert$	–
$\varphi_{8.2}$	$2x + 2y \equiv 2(x+y)$	–
$\varphi_{8.3}$	$(z^y)^x \equiv z^{x \times y}$	$\varphi_{8.7}$
$\varphi_{8.4}$	$\prod x \times \prod y \equiv \prod (x <> y)$	$\varphi_{8.8}$
$\varphi_{8.5}$	$x + y \equiv y + x$	$\varphi_{8.11}$
$\varphi_{8.6}$	$\mathrm{rev}(y) <> \mathrm{rev}(x) \equiv \mathrm{rev}(x <> y)$	$\sigma_{8.6}(\varphi_{8.12})$
$\varphi_{8.7}$	$z^x \times z^y \equiv z^{x+y}$	$\varphi_{8.8}$
$\varphi_{8.8}$	$x \times (y \times z) \equiv (x \times y) \times z$	$\varphi_{8.9}$
$\varphi_{8.9}$	$x \times z + y \times z \equiv (x+y) \times z$	$\varphi_{8.10}$
$\varphi_{8.10}$	$x + (y+z) \equiv (x+y) + z$	–
$\varphi_{8.11}$	$x + s(y) \equiv s(x+y)$	–
$\varphi_{8.12}$	$x <> (y <> z) \equiv (x <> y) <> z$	–

Table I. Conjectures proved by reuse

8. TERMINATION OF THEOREM PROVING BY REUSE

Table I reveals that the subgoals which are speculated when proving some conjecture φ by reuse often also can be proved by reuse, which introduces recursion and consequently the problem of termination into the reuse procedure. For preventing *infinite* reuse sequences, we impose a *termination requirement* on the reuse procedure: We demand $\varphi >_F \varphi_i$ for each conjecture φ and each *reducible* member φ_i of a totally instantiated catch, where $>_F$ is a well-founded relation on formulas and a formula is *reducible* iff it cannot be evaluated to a tautology and some proof shell applies for it. Since proof reuse never is attempted for an irreducible conjecture, $\varphi >_F \varphi_i$ is not required for guaranteeing termination if φ_i is irreducible. Thus e.g. proving $\varphi_{8.1} := \lvert k \rvert + \lvert l \rvert \equiv \lvert k <> l \rvert$ by reuse terminates vacuously as all formulas from the totally instantiated catch $\pi_2(\rho_2(C_S'))$ are instances of axioms, cf.

Section 6.1. But when proving e.g. $\varphi_{8.3}$ by reuse, $\varphi_{8.3} >_F \varphi_{8.7}$ is required, cf. Table I.

For developing a well-founded relation $>_F$ on the set of formulas, we start by separating function symbols from the signature Σ into the set Σ^c of constructor function symbols, as 0, s, empty, add, etc., and the set Σ^d of defined function symbols, e.g. exp, prod, times, sum, plus etc. Then the *defined-by relation* $>_{def}$ is a relation on Σ^d defined as:

$f >_{def} g$ iff (1) g occurs in one of the defining axioms for f and $f \neq g$

    or (2) $f >_{def} h >_{def} g$ for some $h \in \Sigma^d$.

Obviously, $>_{def}$ is *transitive* and by the requirements for the introduction of function symbols which in particular exclude mutual recursion, cf. (Walther, 1994), $>_{def}$ is *well-founded*. We have, for instance, exp $>_{def}$ times $>_{def}$ plus and sum $>_{def}$ plus as well as prod $>_{def}$ times $>_{def}$ plus.

Now let $\gg_{def}$ be the strict multiset order imposed by $>_{def}$ on multisets of Σ^d. Then $\gg_{def}$ is well-founded because $>_{def}$ is, cf. (Dershowitz and Manna, 1979), and since each finite *set* also is a *multiset*, finite sets $S_1, S_2 \subseteq \Sigma^d$ can be compared by the multiset order $\gg_{def}$. We find e.g. $\{$exp, prod, sum$\} \gg_{def}$ $\{$times, sum$\} \gg_{def} \{$times, plus$\} \gg_{def} \{$plus$\}$ and thus $\{$exp, prod, sum$\}$ $\gg_{def} \{$plus$\}$.

We use the well-founded order $\gg_{def}$ on sets of defined function symbols for defining an order $>_F$ on formulas. The idea underlying the development of $>_F$ is to model the difficulty of a proof, i.e. $\varphi >_F \psi$ *should* hold if φ is (expected to be) more difficult to prove than ψ.

For realizing this idea, we consider sets of (defined) *maximal symbols* (w.r.t. $>_{def}$), since their presence have a substantial influence on the difficulty of a proof: A finite subset $S \subseteq \Sigma^d$ is called *pure* iff $s_1 \not>_{def} s_2$ for all $s_1, s_2 \in S$. For instance, $\{$exp, prod, sum$\}$ is pure whereas $\{$exp, times, plus$\}$ is not. We let *purifyS* denote the *maximal pure subset* of $S \subseteq \Sigma^d$, i.e. *purifyS* is the set of $>_{def}$ -maximal elements of S. For instance, *purify*$\{$exp, times, plus, prod, sum$\} = \{$exp, prod, sum$\}$.

We let $\Sigma^d(\phi)$ denote the set of all defined function symbols in a formula ϕ, i.e. if a formula ϕ contains one defined function symbol at least, then *purify*$\Sigma^d(\phi)$ is the set of all maximal defined function symbols occurring in ϕ, with which the difficulty of (proving) ϕ is estimated.

Using *purify* and $\gg_{def}$, a relation $>_F$ on formulas now can be defined, where we use the number of occurrences $\#_f(\phi) \in \mathbb{N}$ of a function symbol $f \in \Sigma^d$ in a formula ϕ as an additional criterion: $\varphi >_F \psi$ iff

(a) $purify\Sigma^d(\varphi) \gg_{def} purify\Sigma^d(\psi)$, or

(b) $purify\Sigma^d(\varphi) = purify\Sigma^d(\psi)$ and

$$\Sigma_{f \in purify\Sigma^d(\varphi)} \#_f(\varphi) > \Sigma_{f \in purify\Sigma^d(\psi)} \#_f(\psi).$$

$>_F$ is well-founded, because it is formed as a lexicographic combination of well-founded relations. The restriction on *maximal* defined function symbols models the observation that for proving a statement φ about some function f, quite inevitably also properties of functions $g <_{def} f$ have to be considered, *independent of whether g already occurs in φ or not.*

Criterion (b) is a simple refinement regarding the number of occurrences of maximal symbols. Note that although (a) compares *sets* of maximal symbols with the *multiset*-order $\gg_{def}$ and (b) compares the *number of occurrences* of maximal symbols, we do not merge these criteria such that the *multisets* of occurrences of maximal symbols would be compared. This is because e.g. for φ containing one occurrence of len as well as rev and for ψ containing only two occurrences of len, criterion (a) succeeds with $purify\Sigma^d(\varphi) = \{len, rev\} \gg_{def} \{len\} = purify\Sigma^d(\psi)$ while the combined criterion would fail due to $\{len, rev\} \not\gg_{def} \{len, len\}$, where $\{...\}$ denotes a multiset.

For instance, reconsider the examples from Table I: Proving $\varphi_{8.4}$ by reuse yields the speculated lemmata $\varphi_{8.8}$, $\varphi_{8.9}$, and $\varphi_{8.10}$ in turn. We find $purify\Sigma^d(\varphi_{8.4}) = \{prod, app\} \gg_{def} \{times\} = purify\Sigma^d(\varphi_{8.8}) = purify\Sigma^d(\varphi_{8.9}) \gg_{def} \{plus\} = purify\Sigma^d(\varphi_{8.10})$ and $\#_{times}(\varphi_{8.8}) = 4 > 3 = \#_{times}(\varphi_{8.9})$, and therefore $\varphi_{8.4} >_F \varphi_{8.8} >_F \varphi_{8.9} >_F \varphi_{8.10}$. Also $\varphi_{8.6} >_F \sigma_{8.6}(\varphi_{8.12})$ because $purify\Sigma^d(\varphi_{8.6}) = \{rev\} \gg_{def} \{app\} = purify\Sigma^d(\sigma_{8.6}(\varphi_{8.12}))$ for the instance $\sigma8.6(\varphi_{8.12})$ of $\varphi_{8.12}$ with $\sigma_{8.6} = \{p/m :: \varepsilon\}$ which is speculated when proving $\varphi_{8.6}$ by reuse.

The well-founded termination order $>_F$ is refined by additional criteria, cf. (Kolbe and Walther, 1996b; Kolbe and Walther, 1998), and this refined order has proved useful for guaranteeing the termination of the reuse procedure without spoiling the system's performance: So far, we were not faced with a conjecture which can be proved by reuse *without* the termination requirement, but *cannot* if the termination requirement is obeyed. Furthermore, examples reveal that by the termination requirement, unsuccessful reuse attempts can be avoided and also the termination of *lemma speculation* for induction theorem proving in general can often be guaranteed, cf. (Kolbe and Walther, 1996b; Kolbe and Walther, 1998).

9. Why Reusing Proofs ?

Incorporating reuse in problem solving is motivated by the expected benefits of exploiting past problem solutions. We identify and analyze two potential benefits of reuse which are claimed to justify the additional effort to be spent for reuse, cf. e.g. (Etzioni, 1993; Segre and Elkan, 1994).

In the *problem reduction paradigm*, a problem p is mapped to a finite set of subproblems $\{p_1, ..., p_n\}$ by some *(problem-)reduction operators*, and each of the subproblems p_i is mapped to a finite set of subproblems in turn, etc., cf. (Nilsson, 1971). The reduction process stops successfully, if *each* subproblem eventually is reduced to a *primitive* problem p' where primitiveness is a *syntactical* notion depending on the particular problem solving domain. The only requirement is that "primitive" problems are trivially solvable indeed and that a solution is obvious. Since it is demanded in addition that each reduction operator only yields a set P of subproblems for a given problem p such that the solvability of *all* subproblems in P imply the solvability of p, successful termination of the reduction process entails the solvability of the original problem.

Problem solving within this paradigm creates a search space which is organized as an AND/OR-tree: Several reduction operators may be applicable for a problem, which creates an OR-branch in the search tree. On solving *all* subproblems obtained by the application of one reduction operator, an AND-branch is created. A subtree of such a *search tree* is a *solution tree* iff each OR-node has exactly one successor and each leaf is labeled with a primitive problem.

9.1. *Saving Resources*

An obvious potential benefit of reuse is *speedup* which is gained by spending additional memory for storing solutions: Resources, i.e. problem solving time, might be saved if restricting the search space by guidance through an old solution outweighs the additional costs for retrieval and adaptation. For the above model of problem solving we make the simplifying assumption that for each reducible node n of a search tree exactly s reductions can be applied for expanding n, each of which has exactly b successor nodes. Then the worst case complexity for computing a closed search tree T in breadth first search is in $O(s^N b^N)$, where $2N$ is the depth of T, i.e. T has N layers of OR-nodes and N layers of AND-nodes. Since the solution tree of T is in $O(b^N)$, an *optimal* reuse saves the amount of time s^N caused by the search factor at most, i.e. the worst case complexity of constructing the solution is decreased from

$O(s^N b^N)$ to $O(b^N)$ by a successful reuse (without considering the additional costs necessitated by the reuse procedure).[13]

In general, however, reuse cannot always be successful, and problems still must be solved from scratch from time to time. Thus for sake of a stronger statement about the savings by reuse, a *sequence* of problems (and their solutions) has to be considered where we assume a search depth fixed to N for all problems ("fixed depth assumption"): If m denotes the length of a problem-sequence, then the worst case complexity of constructing solutions for *all* m problems *from scratch* is $O(m s^N b^N)$. The partial success of reuse is represented by a *reuse factor* $r_i \in [0, 1)$ which denotes the probability that problem $i \in \{1, ..., m\}$ can be solved by reuse if an average sequence of problems is considered. Since the (initially empty) solution base is extended whenever a problem cannot be solved by reuse, its size $|SB_i| \in [0, m]$ grows with i and therefore also r_i grows with i as more and more reusable solutions exists. For enabling further considerations, we assume the reuse factor to be of a schematic form $r_i := 1 - 1/i^a$, where the *reuse gradient* $a \in [0, \infty)$ models different successes of reuse, e.g. $a = 0$ means that reuse always fails and $a > 1$ implies that reuse almost always succeeds. Figure 11 shows the courses of r_i for $i \in (0, 10000]$ and $a \in \{0.1, 0.2, 0.5, 1, 1.5\}$:

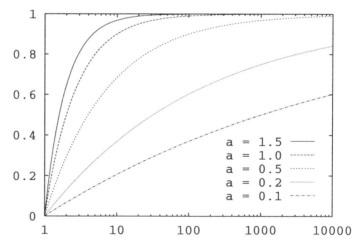

Figure 11. Reuse factors r_i for different gradients a

For the reuse gradients of Figure 11, a problem can be solved by reuse with e.g. a probability of 90% if 10^{10}, 10^5, 10^2, 10^1, or $10^{0.66}$ problems,

---

[13] The amount of time $O(b^N)$ caused by the branching factor b is not saved as it merely measures the solution's size.

respectively, have been considered before. This means that a reuse gradient $a > 1$ models a quite unrealistic assumption about the degree of problem reusability.

The worst case complexity of constructing solutions for *all* m problems with reuse is given as follows: When considering the ith problem p_i in the sequence, the costs for proving p_i *from scratch* is proportional to $(1 - r_i)b^N s^N$ plus $(1 - r_i)b^N$ for the costs of updating the solution base. The retrieval and update costs when proving p_i by *reusing* a solution is proportional to $b^N|SB_i|$, where $|SB_i| \approx \sum_{j=1}^{i}(1 - r_i)$ is the size of the solution base after i problems have been considered. For computing the costs for the whole sequence $\langle p_1 \dots p_m \rangle$ of problems we sum up these costs for $i = 1 \dots m$, and the values given in Table II are obtained.

a	*from scratch*	*retrieval*	$|SB_m|$
$\in [0, 1)$	$O(b^N s^N m^{1-a})$	$O(b^N m^{2-a})$	$O(m^{1-a})$
$= 1$	$O(b^N s^N \ln(m))$	$O(b^N m \ln(m))$	$O(\ln(m))$
$\in (1, \infty)$	$O(b^N s^N)$	$O(b^N m)$	$O(1)$

Table II. Search costs for different reuse gradients

The last row of Table II shows (as already recognized) that $a > 1$ is an unrealistic assumption as a finite solution base would suffice for solving all problems by reuse. Therefore we focus on reuse gradients $a \in \{0.1, 0.2, 0.5, 1\}$ subsequently.

Figure 12 shows the relative savings computed as $|SB_m|(1/m + 1/s^N)$ obtained by reusing past problem solutions as compared to pure problem solving from scratch (where a "search part" $s^N = 500$ is assumed). E.g. for $a = 0.2$, only 64% of the costs for pure problem solving from scratch must be spent after 10 problems have been considered, only 47% after 100 problems and only 76% after 1000 problems. However, problem solving by reuse is more than 3 times costly after 10000 problems have been considered.

Figure 12 shows the typical shape of a firstly increasing and then permanently decreasing efficiency of problem solving with reuse as compared to pure problem solving from scratch, cf. (Holder, 1992). This effect is due to the increasing retrieval costs as the solution base SB grows with the system's lifetime, i.e. $|SB_m| \to \infty$ for $m \to \infty$. In our model the system efficiency is increased as long as the size of the solution base is bound by the search part s^N of problem solving, i.e. for $|SB_m| < s^N$.

If a solution base of *fixed size* is used, then the retrieval costs reduce to $O(b^N m)$. However, if the solution base remains unchanged once it is filled up, we have to assume a constant reuse factor r_i from this point such that from-scratch costs of $O(b^N s^N m)$ like in the pure from-scratch case apply and nothing is saved. But if the solution base's content is "improved" by replacing solutions, then the previous approach with $r_i := 1 - 1/i^a$ is justified and relative costs of $O(|SB_m|/m)$ are obtained which monotonely *decrease* for growing m, compared to $O(|SB_m|)$ which does *not* decrease with growing m.

Our analysis clarifies the role of the "utility problem" and also reveals that a solution base of limited size should be used, where sophisticated techniques for utility assessment (Minton, 1990), self-limitizing the solution base at its optimal size (Holder, 1992), and/or case deletion (Smyth and Keane, 1995) should be applied.

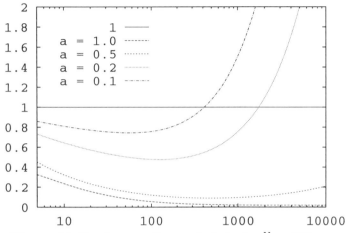

Figure 12. Total costs of reuse (relative); $s^N = 500$

This consideration must of course be relativized as computing the savings in *real applications* often imposes severe simplifications of the regarded domains for obtaining necessary probabilistic distributions for the problem space. Hence in most applications one relies on empirical studies for measuring the average benefits of reuse, and present research reports savings from only a few percent up to speedups of factor 20 and more, cf. e.g. (Etzioni, 1993; Segre and Elkan, 1994).

9.2. *System Performance*

Another potential benefit of reuse is the improvement of the system's *performance*, i.e. more problems are solvable as compared to pure problem solving from scratch. Of course this only applies for *incomplete* problem solvers which cannot solve every solvable problem. This feature presupposes sophisticated adaptation techniques for reuse which *apply the problem solver for adapting* a retrieved candidate solution s' for a new problem p. This means that a set of problems $\{p_1, ..., p_n\}$ is generated by adaptation such that a solution s of p can be obtained uniformly from s' and the solutions $s_1, ..., s_n$ of $p_1, ..., p_n$ respectively. By incorporation of the problem solver, powerful adaptation techniques can be obtained, cf. e.g. Section 6. Since adaptation then works like problem reduction, the rule base of the overall system is *dynamically extended* by reuse. Thus problems become solvable which are unsolvable if only the *fixed* initial rule base of the problem solver is considered.

Example 9.1 (Reuse increases Performance)
We reconsider the reuse of induction proofs, cf. Table I. Assume a simple induction prover IP which verifies $\varphi_{8.10}$, the associativity of addition. Using this property as lemma, IP verifies also the statement $\varphi_{8.0}$ stipulating that summation over finite lists of numbers distributes over list concatenation. However, IP fails to prove the conjecture $\varphi_{8.6}$ since the induction proof gets blocked and no useful lemma can be speculated by IP. Now if IP is provided with the ability of reusing proofs, then $\varphi_{8.6}$ can be verified by reusing the proof of $\varphi_{8.0}$ because the lemma $\sigma_{8.6}(\varphi_{8.12})$ (corresponding to $\varphi_{8.10}$ in the original proof) is speculated by reuse. Since IP can prove this lemma, the performance of IP is increased by reuse.[14]

The effect of increased performance is strengthened if the subproblems generated by adaptation are considered by the reuse mechanism before the problem solver is called. Now for each of these retrieved problems, the solution base is searched again for reusable solutions and so on, until eventually a retrieved problem either is primitive or the retrieval fails. For instance also the speculated lemma $\varphi_{8.12}$ from Example 9.1 can be proved by reusing the proof of the previously verified statement $\varphi_{8.10}$.[15]

9.3. *Summary of Benefits*

Our analysis reveals that *speedup* is not the only resp. not the most important benefit of reusing past problem solutions (if it is obtained at all). The benefits rather depend on the application domain and the kind of problem solver which is available. Therefore the design of a reuse system should be influenced by this analysis: An *interactive* theorem prover should be designed to achieve *high reusability*, because the costs of employing even a large amount

---

[14] Induction provers more sophisticated than *IP* might perform a *generalization* for obtaining $\varphi_{8.12}$, but due to the inherent incompleteness of induction (more complicated) examples can be found also for these systems.

[15] The complexity-theoretic considerations of Section 9.1 must be refined to capture the recursion in the reuse procedure because the "fixed depth assumption" now is violated.

of machine resources (time and memory) will be outweighed by the savings of even few human interactions. For an *automated* theorem prover the reuse system has to be tailored to optimize retrieval and adaptation time, even if the degree of reusability is decreased. Consequently reuse in automated theorem provers pays only in domains where *drastic* savings are achievable, or where an *incomplete* system is improved by providing additional solutions by reuse. Similar observations have been made if analogical reasoning is applied to inductive theorem proving (Melis and Whittle, 1998).

10. Conclusion

We have presented an approach for reusing proofs which is implemented with the PLAGIATOR-system (Kolbe and Brauburger, 1997). Experiments with the system reveal that *automated* proof reuse is successful even for conjectures which differ considerably in both their structure and their domain. Therefore we claim that the complex task of automated reasoning sometimes can be learned from a human advisor. For achieving high reusability, sophisticated techniques with powerful heuristics have been developed to gain the high flexibility which is required for retrieval and adaptation of known proofs.

References

Bates, J. L. and R. L. Constable: 1985, 'Proofs as Programs'. *ACM Transactions on Programming Languages and Systems* **7**(1), 113–136.

Dershowitz, N. and Z. Manna: 1979, 'Proving Termination with Multiset Orderings'. *Communications of the ACM* **22**(8), 465–476.

Ellman, T.: 1989, 'Explanation-Based Learning: A Survey of Programs and Perspectives'. *ACM Computing Surveys* **21**(2), 163–221.

Etzioni, O.: 1993, 'A Structural Theory of Explanation-Based Learning'. *Artificial Intelligence* **60**, 93–139.

Giunchiglia, F. and T. Walsh: 1992, 'A Theory of Abstraction'. *Artificial Intelligence* **57**, 323–389.

Hall, R. P.: 1989, 'Computational Approaches to Analogical Reasoning: A Comparative Analysis'. *Artificial Intelligence* **39**, 39–120.

Holder, L. B.: 1992, 'Empirical Analysis of the General Utility Problem in Machine Learning'. In: *Proceedings AAAI-92*.

Huang, X.: 1994, 'PROVERB: A System Explanining Machine-Found Proofs'. In: *Proc. of 16th Annual Conf. of the Cognitive Science Society*. Atlanta, Georgia.

Huet, G. and B. Lang: 1978, 'Proving and Applying Program Transformations Expressed with Second-Order Patterns'. *Acta Informatica* **11**, 31–55.

Kolbe, T.: 1997, 'Optimizing Proof Search by Machine Learning Techniques'. Ph.D. thesis, Technische Hochschule Darmstadt.

Kolbe, T. and J. Brauburger: 1997, 'PLAGIATOR — A Learning Prover'. In: W. McCune (ed.): *Proceedings of the 14th International Conference on Automated Deduction (CADE-97)*, Townsville, Australia. pp. 256–259, Springer.

Kolbe, T. and C. Walther: 1994, 'Reusing Proofs'. In: A. Cohn (ed.): *Proceedings of the 11th European Conference on Artificial Intelligence (ECAI-94)*, Amsterdam, The Netherlands. John Wiley & Sons, Ltd., pp. 80–84.

Kolbe, T. and C. Walther: 1995a, 'Adaptation of Proofs for Reuse'. In: D. W. Aha and A. Ram (eds.): *Adaptation of Knowledge for Reuse. Papers from the 1995 AAAI Fall Symposium*, Cambridge, MA, USA. pp. 61–67, The AAAI Press.

Kolbe, T. and C. Walther: 1995b, 'Patching Proofs for Reuse'. In: N. Lavrac and S. Wrobel (eds.): *Proceedings of the European Conference on Machine Learning (ECML-95)*, Heraklion, Greece. pp. 303–306, Springer LNAI 912.

Kolbe, T. and C. Walther: 1995c, 'Proof Management and Retrieval'. In: *Proceedings of the IJCAI'95 Workshop on Formal Approaches to the Reuse of Plans, Proofs, and Programs*. pp. 16–20.

Kolbe, T. and C. Walther: 1995d, 'Second-Order Matching modulo Evaluation — A Technique for Reusing Proofs'. In: *Proceedings of the 14th International Joint Conference on Artificial Intelligence (IJCAI-95)*, Montreal, Canada. pp. 190–195.

Kolbe, T. and C. Walther: 1996a, 'Proving Theorems by Mimicking a Human's Skill'. In: Y. Gil (ed.): *Acquisition, Learning & Demonstration: Automating Tasks for Users. 1996 AAAI Spring Symposium*, Menlo Park, CA, USA. pp. 50–56.

Kolbe, T. and C. Walther: 1996b, 'Termination of Theorem Proving by Reuse'. In: M. McRobbie and J. Slaney (eds.): *Proceedings of the 13th International Conference on Automated Deduction (CADE-96)*, New Brunswick, NJ, USA. pp. 106–120.

Kolbe, T. and C. Walther: 1997, 'Proving Theorems by Reuse'. Research Report IBN 97/52, TU Darmstadt.

Kolbe, T. and C. Walther: 1998, 'On Terminating Lemma Speculations'. Journal *Information and Computation.* (to appear).

Melis, E. and J. Whittle: 1998, 'Analogy in Inductive Theorem Proving'. *Journal of Automated Reasoning* 20(3). To appear.

Minton, S.: 1990, 'Quantitative Results Concerning the Utility of Explanation-Based Learning'. *Artificial Intelligence* 42, 363–391.

Nilsson, N. J.: 1971, *Problem Solving Methods in Artificial Intelligence.* McGraw Hill, New York.

Segre, A. and C. Elkan: 1994, 'A High-Performance Explanation-Based Learning Algorithm'. *Artificial Intelligence* 69, 1–50.

Smyth, B. and M. T. Keane: 1995, 'Remembering to Forget'. In: *Proc. 14th Intern. Joint Conf. on Artif. Intell. (IJCAI-95).* Morgan Kaufmann, pp. 377–382.

Walther, C.: 1994, 'Mathematical Induction'. In: D. M. Gabbay, C. J. Hogger, and J. A. Robinson (eds.): *Handbook of Logic in Artificial Intelligence and Logic Programming*, Vol. 2. Oxford University Press, pp. 127–227.

Part 3

Parallel Inference Systems

Editor: Wolfgang W. Küchlin

WOLFGANG KÜCHLIN

INTRODUCTION

Symbolic Computation in general, and Automated Theorem Proving in particular, place extremely high demands on computational resources. Currently, parallelism appears to be the most promising kind of architectural support for Symbolic Computation. Parallel computation is a vast and diverse field offering many different architectures and programming paradigms. Today, it is still open which combination(s) will ultimately turn out to be the winner(s). However, a number of major questions have emerged that need to be investigated.

1. Parallel inference algorithms must be developed.
2. A suitable parallel hardware and middleware architecture must be selected.
3. The algorithmic parallelism must be efficiently mapped to the real architecture.
4. Significant speedups must be achieved on significant problems.

The three articles in this part are concerned with several of these questions. Bündgen *et al.* present their PaReDuX parallel term-rewriting completion system for networks of shared-memory parallel workstations. Schumann *et al.* describe several approaches to parallel first order theorem proving based on their SETHEO model elimination prover. Bornscheuer *et al.* develop new connectionist approaches to inference problems with the aim of exhibiting massive amounts of parallelism.

Before we give a more detailed introduction to these articles in Section 4 below, we shall first discuss some of the fundamental problems of parallel Symbolic Computation.

1. GENERAL MOTIVATION AND PROBLEMS

Parallelization is one of the most promising approaches for speeding up time- and memory-intensive computations. Multi-processor machines have already arrived at the desk-side, and computer networks are abundant. Even by the conservative approach (taken both by Bündgen *et al.* and Schumann *et al.*) of using such a general purpose network, processing power and, equally important, memory capacity can be multiplied by a factor of ten to several hundred.

W. Bibel, P. H. Schmitt (eds.), Automated Deduction. A basis for applications. Vol. II
© 1998 *Kluwer Academic Publishers. Printed in the Netherlands*

However, parallel programming is still extremely complex. Theoretically, changing the parallel architecture model may even change the asymptotic complexity of algorithms. In practice, we also see vast performance changes—from slowdowns to significant speedups by orders of magnitude—when going from one architecture to another. Parallel code is still difficult to develop and extremely difficult to debug. The question of an appropriate programming paradigm and environment is therefore of great practical importance. Without a powerful but simple programming paradigm there can be no sustainable development of a complex parallel inference system. Since as of now there is no parallel operating system standard, the programming environment must currently be formed by a middleware layer of software between the operating system and the application.

Add to these general problems the high complexity of Symbolic Computation algorithms with highly irregular loads and extreme data dependence, and it becomes clear that parallel Symbolic Computation is probably one of the "worst cases" of software development. Symbolic Computation occurs on a very high level of abstraction, and hence it is comparatively complex and slow. It is characterized by dynamic data structures combined with highly data-dependent control flow and execution time. In Computer Science, apart from developing better algorithms one gains efficiency by tailoring hardware and software to the particular computation domain, exploiting regular execution patterns. In numeric computation, vector operations on floating point numbers may be supported directly by hardware, because the execution time for each operation is independent of the size of the number. In Computer Algebra, another field of Symbolic Computation, coefficient operations on polynomials are already data-dependent because coefficients may be bignums or again polynomials, but at least algorithms can still be tailored to particular domains such as polynomial algebras. In Automated Deduction, however, we rise to an even higher level of abstraction, because even particular domains are given to us only at run-time, in the form of axiomatizations. Computations are typically no longer algorithms computing some algebraic value, but they are search procedures looking for proofs in extremely large and poorly structured search spaces. The performance of the prover depends heavily on the search heuristics, and different examples may well require different heuristics, precluding dedicated architectural support. Finally, Symbolic Computation needs garbage collection, whose parallelization is surprisingly difficult and forms a research subject in its own right.

Nevertheless, it is of great promise to investigate parallelism in Automated Deduction. Logical inference systems are naturally parallel, and the significant increases in computing power and main memory on parallel architectures will likely expand the range of practically solvable applications.

2. PARALLEL ARCHITECTURES

A *multicomputer* is a parallel machine consisting of *loosely coupled* processors; a *multiprocessor* consists of several *tightly coupled* processors. Typically, a multicomputer will be a *distributed memory (DM)* machine, and a multiprocessor will be a *shared memory (SM)* machine. We may think of a network of workstations and a single multi-processor workstation, respectively, consisting of multiple independent standard microprocessors. This class of computers is known as MIMD (multiple instruction streams, multiple data streams); the SIMD class of machines requires that all processors execute a single instruction stream, and hence SIMD machines do not execute data-dependent programs efficiently. SIMD machines, which had been mostly dedicated towards numerical parallel linear algebra, have declined in importance in favor of the more generally useful MIMD machines.

Distributed memory architectures, as e. g. a workstation network or the Transputer system used by Schumann *et al.*, achieve communication by passing messages, and they potentially scale to many thousands of processors. Because of the comparatively high overhead and low bandwidth of message passing, only relatively large (coarse-grain) tasks can be profitably executed in parallel. Shared memory architectures either use a common bus (e. g. Multiprocessor SPARCstation20), or a more powerful network switch such as a crossbar (e. g. Multiprocessor SPARC Ultra), for fast communication. Due to the low overhead and high bandwidth, smaller (more fine-grained) tasks can be profitably executed in parallel, but the architecture scales only to a few dozen processors. A modern parallel workstation (or high-end PC), such as the one used by Bündgen *et al.*, combines both forms of parallelism: there is SM parallelism over the processors of the workstation itself, and there is DM parallelism over the network to which it is connected.

Although scalable DM parallel Multicomputers are attractive machines, they are relatively hard to program. It is difficult to balance the computation load, and to organize the communication and synchronization, between hundreds or thousands of processors. In single machines such as Transputer based systems, it is also hard to break the computation down into small enough tasks that will fit into the small local memories of each node. In general, it is much easier to work with a hierarchical multicomputer of k n-processor SM machines than with one layer of $k * n$ single processors. Therefore, parallel inference techniques must be developed for both DM and SM machines and should ultimately be able to exploit a hierarchical combination of both.

3. PARALLEL PROGRAMMING ENVIRONMENTS

In order to cooperate towards a common goal, parallel tasks must communicate data and synchronize operations. It is the purpose of a parallel programming environment—ideally provided by a parallel operating system—to support this in a perspicuous, efficient, reliable, and portable way. More information on this issue can be found in any modern book on operating systems, such as (Tanenbaum, 1992; Tanenbaum, 1995).

In the DM case, major issues are the placement of tasks across the network so that the system load is evenly balanced, and the creation of communication primitives which must be high-level enough to provide abstraction and low-level enough to provide efficiency. In the SM case, main goals are to provide (logically) unrestricted parallelism, to map the parallel tasks efficiently onto the available real parallelism (processors), and to integrate the parallelism smoothly both into the application programming environment above and into the operating system below.

DM communication is often built upon message passing using the standard low level TCP/IP protocols whose services are accessible through *sockets* in the operating system. On a higher level, the *remote procedure call (RPC)* protocol allows a client task on one machine to call a procedure on a remote server. The PVM system (Sunderam, 1990) provides facilities for placing tasks, passing messages, and synchronizing across an inhomogeneous net. PVM is already available on a wide variety of architectures, so that it is possible to develop a distributed application on a workstation net and later to run it on a parallel supercomputer. Schumann *et al.* have used both RPC and PVM to build parallel theorem provers on a large network of workstations.

On SM machines, modern operating systems provide lightweight processes *(threads of control)* as the fundamental support for parallel programming. A thread is an execution context for a procedure much as a traditional process is an execution context for a program. However, their overhead is greatly reduced and they can be used to execute tasks on a much finer level of granularity. Threads are part of the POSIX operating system standard (POSIX 1003.4a), and they are now provided by the operating system kernels of all manufacturers of parallel workstations.

Bündgen *et al.* have used the *S-threads* system (POSIX threads extended by a parallel list processing capability) for parallelizing the completion procedure in term-rewriting. *S-threads* have also been used in the PARSAC parallel Computer Algebra system (Küchlin, 1995). The DTS system (cf. (Küchlin, 1995)) extends the threads abstraction across the network, effectively providing an *asynchronous remote procedure call*. DTS also provides automatic load balancing and node failure recovery, and has been implemented both on top of PVM and on top of the ACE distributed programming environment. Bündgen *et al.* have used DTS in their implementation of parallel unfailing completion, combining two multiprocessor SPARCstations.

Dedicated parallel programming languages can bridge the architectural gap between DM and SM parallelism, providing implementations on both architectures. One can think of them as providing a special language binding for the (raw) operating system and network communication primitives. Parallel Prolog dialects (Shapiro, 1987) enable the parallel solution of subgoals in Prolog clauses by the parallel application of rules. Functional languages also lend themselves well for the expression of parallelism. The Scheme dialect MultiLisp (Halstead, 1985) provides the important concept of *futures*. If a computation is designated as a *future*, then the caller immediately receives a token (the future) which can be incorporated into list data-structures in lieu of the actual result while the future is evaluated concurrently.

Using a parallel programming language for parallel Symbolic Computation has several advantages, such as convenience of coding and integrated garbage collection. However, parallel Prolog and Lisp dialects have different semantics than their sequential forms and thus require the construction of completely new parallel provers, whereas more conservative approaches allow parallel constructs to be inserted into existing sequential programs. The higher programming abstraction may also create a problem in that it becomes far too easy to express parallelism which cannot be efficiently executed by the hardware. Both Bündgen *et al.* and Schumann *et al.* take the conservative approach combining traditional languages like C and Prolog with operating system and network primitives for parallel programming.

4. PARALLELISM IN SYMBOLIC COMPUTATION

As areas of Parallel Symbolic Computation we briefly mention the fields of Parallel AI in general, Parallel Automated Theorem Proving, Parallel Computer Algebra, and Parallel Symbolic Programming. In order to provide a few initial literature pointers with survey character, let us mention (Suttner and Schumann, 1993; Küchlin, 1995; Halstead, 1986; Shapiro, 1987). The PASCO conference sequence (Hong, 1994; Hitz and Kaltofen, 1997) and the Symbolic Computation track of the EUROPAR conference series (Lengauer et al., 1997) attempt to bring together researchers of all these fields.

The three articles included in this part of the present work are concerned respectively with the practical parallelization of completion based term rewriting and of model elimination based theorem proving, and with the theoretical development of connectionist approaches to theorem proving in order to exploit massive amounts of parallelism.

The article by Bündgen, Göbel, Küchlin, and Weber, *Parallel Term Rewriting with PaReDuX*, describes the current state of parallel term rewriting on the example of the parallel rewrite laboratory PaReDuX. The core PaReDuX system is designed for shared memory parallel architectures, such as multiprocessor workstations or PC's where it shows good performance on a variety of examples. In a first part they discuss in detail the parallelization of three term completion procedures: Knuth-Bendix completion, completion modulo associativity and commutativity, and unfailing completion. Those parallelizations are strategy-compliant, i.e., the parallel code performs exactly the same work as the sequential code, yet the work load is shared by many processors. In a second part they examine the more general situation of a hierarchical multiprocessor as it occurs in a network of shared-memory multi-processors. PaReDuX was extended by a network component based on the distributed threads system DTS. They show on several examples how coordinated search parallelism combined with the parallel execution of each search algorithm on a multi-processor results in superior overall speedups.

The article by Schumann, Wolf, and Suttner, *Parallel Theorem Provers Based on SETHEO*, describes the concept and implementation of seven parallel theorem provers which are based on the sequential inference machine SETHEO. Nondeterministic, uninformed search algorithms (which are the basis of each automated theorem prover) are in particular suited for a variety of different parallel execution models. The research sets a focus on the development of parallel models and their implementation which go beyond the classical parallelization paradigms of AND/OR-parallelism. This allowed for a comprehensive evaluation and comparison of the different parallel execution models: static and dynamic search-space partitioning, random competition, strategy-parallelism, and parallel cooperation.

In accordance with these articles, current research seems to indicate that in general there is enough theoretical parallelism in current inference systems to keep today's parallel machines busy. The problem is, however, to organize and master the theoretical concurrency so that a practically efficient parallel system results. In particular, both articles describe various forms of parallelism such as AND-parallelism, OR-parallelism, and cooperation based approaches. The latter are necessary in order to remove unneccessary redundancy from OR-parallel approaches; cooperative theorem provers are discussed in more detail in Part 3 of this volume.

The article by Bornscheuer, Hölldobler, Kalinke, and Strohmaier, *Massively Parallel Reasoning*, is concerned with new connectionist approaches to theorem proving that will exhibit truly massive amounts of parallelism for future parallel machines. Their article begins with an introduction to logic programs, metrics, and connectionist models. Then they present multi-flip networks for propositional logic together with simulation results. Finally they look at normal programs and establish a strong relationship between logic programming and connectionist models of computation. This enables the distributed processing of interpretations of normal logic programs within a fixed massively parallel architecture.

REFERENCES

Halstead, Jr., R. H.: 1985, 'Multilisp: A Language for Concurrent Symbolic Computation'. *ACM Trans. Programming Languages and Systems* **7**(4), 501–538.

Halstead, Jr., R. H.: 1986, 'Parallel Symbolic Computing'. *Computer* **19**(8), 35–43.

Hitz, M. and E. Kaltofen (eds.): 1997, 'Second Intl. Symp. Parallel Symbolic Computation PASCO'97'. Wailea, Maui, HI:, ACM Press.

Hong, H. (ed.): 1994, 'First Intl. Symp. Parallel Symbolic Computation PASCO'94', Vol. 5 of *Lecture Notes Series in Computing*. Linz, Austria:, World Scientific.

Küchlin, W. W.: 1995, 'PARSAC-2: Parallel Computer Algebra on the Desk-Top'. In: J. Fleischer, J. Grabmeier, F. Hehl, and W. Küchlin (eds.): *Computer Algebra in Science and Engineering*. Singapore, pp. 24–43, World Scientific. Proc. of a ZiF workshop, Bielefeld, Germany, Aug. 28–31, 1994.

Lengauer, C., M. Griebl, and S. Gorlatch (eds.): 1997, 'EUROPAR'97 Parallel Processing', Vol. 1300 of *LNCS*. Passau, Germany:, Springer-Verlag. (3rd Intl. Euro-Par Conf.).

Shapiro, E. (ed.): 1987, *Concurrent Prolog*, MIT Press series in Logic Programming. Cambridge, MA: MIT Press.

Sunderam, V. S.: 1990, 'PVM: A Framework for Parallel Distributed Computing'. *Concurrency: Practice and Experience* **2**(4), 315–339.

Suttner, C. and J. Schumann: 1993, 'Parallel Automated Theorem Proving'. In: L. Kanal, V. Kumar, H. Kitano, and C. Suttner (eds.): *Parallel Processing for Artificial Intelligence I*. pp. 209–257, Elsevier, Amsterdam.

Tanenbaum, A. S.: 1992, *Modern Operating Systems*. Englewood Cliffs, NJ: Prentice Hall.

Tanenbaum, A. S.: 1995, *Distributed Operating Systems*. Englewood Cliffs, NJ: Prentice Hall.

CHAPTER 9

PARALLEL TERM REWRITING WITH PAREDUX

1. INTRODUCTION

Nowadays high performance computer architectures are parallel architectures. They can be subdivided into *shared memory multiprocessors* and *distributed memory multiprocessors*. Their most popular and commonly available representatives are shared-memory parallel workstations, and networks of workstations, respectively. Combining both kinds of architectures in the form of a network of shared memory multiprocessors results in a *hierarchical multiprocessor* with non-uniform memory access.

There are also two major approaches to parallelizing software for symbolic computation. The first approach to parallelizing symbolic computation software relies on *work parallelism* (*AND-parallelism*), which spreads a given amount of work over a number of processors. If only work parallelism is used, then the parallel completion algorithm executes exactly the same strategy no matter how many processors it runs on, and therefore we also speak of a *strategy compliant parallelization*. Due to the high synchronization requirement of a single completion loop with a fixed strategy, and the fine grain of its parallel components, strategy compliant parallelization is typically only profitable on shared memory multiprocessors.

In a first part of this chapter we focus on this approach to parallelizing term rewriting systems. Specifically, we demonstrate the approach taken in PaReDuX. Our results show that the PaReDuX system can be used for parallel completion with an efficiency of 75 %–95 % on a desktop shared memory multiprocessor.

The second approach to parallelizing software for symbolic computation uses *search parallelism* (*OR-parallelism*). It exploits the fact that commonly there are several strategies for finding the solution, the best of which is not known a priori. In this case, executing several different strategies in parallel, or randomizing strategies by parallelization effects until the first solution is found by any of them, often speeds up the search. This kind of parallelization may even yield super-linear speed-ups, if the best strategy in the pool executes less work than the sequential default strategy. Due to the minimal synchronization requirements and the coarse granularity of search parallelism, it is an ideal candidate for distribution on a network of workstations.

W. Bibel, P. H. Schmitt (eds.), Automated Deduction. A basis for applications. Vol. II

However, its effect vanishes as the known sequential strategies approach an optimal strategy. In particular, all strategy dependent speed-ups are limited by the performance of an optimal strategy. Therefore it is interesting to speed up each individual strategy by performing its work in parallel (but preserving the strategy), as is done in the core PaReDuX system and is described in the first part.

The goal of the second part of this chapter is twofold. Algorithmically, we want to show that the *modular* combination of search parallelism and work parallelism on an hierarchical memory processor can yield aggregate speed-ups exceeding those of each isolated scheme and can overcome the limits that both approaches to parallel completion have. These limits are given by the facts that the speed-ups of search parallelism are limited by the number and relative quality of different selection strategies; work parallelism is limited by the amount of parallel work within each strategy and by the architectural constraints of today's shared memory processors that allow at most a few dozen processors. Our second goal is to demonstrate that a uniform programming technique can be used for the network part and the shared memory part. Our DTS parallel system environment supports the notion of a lightweight process fork across the network. Hence we can employ a common fork/join paradigm for parallelization on uniprocessors (for virtual parallelism), multiprocessors, and across the network.

For an empirical validation of our concept we extended the core PaReDuX system to *distributed PaReDuX*. This system implements an equational theorem prover, based on an unfailing completion procedure, on a network of multiprocessor SPARC-stations. The prover combines a top-level master-slave scheme distributing search parallelism over the net with the strategy compliant parallel completion scheme of the core PaReDuX system that works on each multi-processor node. The architecture of the distributed PaReDuX system is sketched in Fig. 1.

1.1. *Organization of the Chapter*

In Sec. 2 we give a short historical overview of the system and sketch the general system concepts that are the basis of our efficient implementations. The first of these concepts is that of *Virtual S-threads*, a quite general concept that has proven to be useful in other areas of parallel symbolic computation, too. Second, we recall the theoretical basis of term rewriting. Specifically, we examine this basis towards its potentials in parallelization.

After this preliminary part, we investigate the parallelization of various completion procedures on a shared memory multiprocessor in Sec. 3. Specifically we will focus on their realization in the core PaReDuX system.

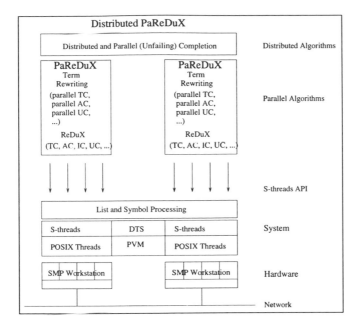

Figure 1. The *Distributed PaReDuX* system.

Section 4 is devoted to parallel symbolic computation on a hierarchical multiprocessor. In Sec. 4.1 the master-slave approach to unfailing completion, which is implemented in distributed PaReDuX, is described. Empirical results of strategy compliant vs. master-slave unfailing completion are the content of Sec. 4.2. A comparison with related work in Sec. 4.3 concludes this second part. Some concluding remarks are given in Sec. 5.

2. BACKGROUND

2.1. *History and Overview of the PaReDuX System*

Our PaReDuX system (Bündgen et al., 1996b; Bündgen et al., 1995; Bündgen et al., 1994c; Bündgen et al., 1994b) is the parallel form of ReDuX (Bündgen, 1993). The origins of ReDuX go back to the early eighties (Küchlin, 1982), and it was originally developed within the framework of the ALDES/SAC-2 computer algebra library (Collins, 1980), using the list processing and symbol packages of SAC-2. In the early nineties, SAC-2, and later ReDuX, were translated from ALDES to C; the C version of SAC-2 is known as SACLIB (Buchberger et al., 1993). At the same time, parallelization of SACLIB programs began; historically, the first multi-threaded parallel form is PARSAC (Küchlin, 1990b; Küchlin, 1995), and a similar more recent variant is PACLIB (Schreiner and Hong, 1993).

PaReDuX was created within the PARSAC framework using its parallel system levels S-threads (Küchlin, 1990a) and DTS (Bubeck et al., 1995). The S-threads system extends a given threads kernel by a light-weight user level system with synchronized concurrent access to a heap of list cells. For our earlier work on parallel Knuth-Bendix and AC completion we used PCR threads on UNIX System V, our later work with unfailing completion used the threads of Solaris 2.x. The system level of PARSAC includes DTS, an environment which carries the essence of the threads concept—without shared memory—over to the network, allowing a parallelization concept based on asynchronous remote procedure calls. The DTS version we used for parallel unfailing completion is implemented on top of the PVM system. It uses PVM to distribute copies of a single multi-threaded program across the network and implements the asynchronous remote procedure call abstraction with load balancing and error recovery. The use of PVM is in no way fundamentally connected with DTS, and recent versions of DTS implement this functionality without relying on PVM.

2.2. *Parallel Symbolic Computation with Virtual S-Threads*

In traditional operating systems, each process has an address space and a single *thread of control*. A thread of control is an execution context for a procedure, much as a process is an execution context for a complete program. A *threads system* allows several threads, i. e., procedures, to be active concurrently within a process. Hence multi-threading achieves finer grained parallelism than multi-processing and allows shared access to the memory of the surrounding process. A threads system can be implemented at the user level, but most modern operating systems (Tanenbaum, 1992), such as Mach, Solaris 2.x, OS/2, or Windows NT, provide kernel threads.

New threads can be created by a *fork operation* and collected by a *join operation*. Forking a new thread is similar to calling a procedure, except that the caller does not wait for the procedure to return. Instead, the parent continues to execute concurrently with the newly forked child; on a multiprocessor system this may result in true parallelism. At some later time, the parent may rendezvous with the child by means of a *join operation* and retrieve its results.

A thread provides a procedure with a private register file and a private stack. However, no separate heap is assigned—in contrast to a fork operation of a new process in UNIX. For efficient parallel *symbolic* computation heap access is needed in addition to a private register file and a private stack, but a fully separate heap would prohibit efficient parallelization in many cases. A solution is to have a private portion of the heap together with an appropriate (parallel) garbage collection facility.

In PARSAC this is provided by the *S-threads* system (Küchlin, 1990a). S-threads was originally modeled after C Threads (Cooper and Draves, 1988), the interface to the Mach operating system. It assumes that a minimal standard threads interface, such as POSIX threads, is provided by some kernel below. Each kernel thread is then extended to an S-thread capable of concurrent list processing. In this way, virtually all sequential C programs of the SACLIB library, on which ReDuX is built, will execute unmodified as a single S-thread, with a slight execution penalty imposed by the parallel list-processing context. S-threads runs on the Mach and Solaris 2.x kernel threads, but also on the user-level threads provided by PCR (Weiser et al., 1989), which in turn runs on UNIX System V.

The original S-threads memory management scheme distributes the SACLIB heap to threads as paged segments of cells, and it uses *preventive garbage collection* (Küchlin and Nevin, 1991). If a side-effect free functional programming style is used, then the result of a function can be copied into its parent's heap segment, and its own heap segment, containing all garbage, can be recycled in bulk. This scheme is efficient, naturally concurrent, and does not assume that first all threads be stopped by a user.

During Knuth-Bendix completion, however, partial modifications are made to very large critical pair queues. For reasons of efficiency, these queues are updated via side-effects, and preventive garbage collection is not applicable in this case. Therefore, the memory management of S-threads was changed to use the PACLIB scheme (Schreiner and Hong, 1993). Since it works by synchronized access to a shared available cell list, cells allocated by one thread can be woven into a global data-structure via side-effects. The PACLIB scheme assumes however that all threads can be stopped by a user. Since we did not parallelize nor optimize our implementation of this scheme for the work reported here, garbage collection times are always excluded from our measurements.

S-threads has been successfully employed to parallelize a number of algebraic algorithms (Küchlin, 1995). It was found, however, that kernel threads do not cope well with large amounts of dynamically created parallelism. Since each S-thread is mapped one-to-one onto a kernel thread, limitations of kernel implementations may show up in S-threads. This might include high fork/join times or a strict limitation on the number of concurrently active threads. It has been observed elsewhere that differences in kernel thread efficiency may destroy all parallel speed-ups on a new machine, making the system effectively non-portable (Morisse and Oevel, 1995). As a consequence, either the application must become involved in thread management decisions, or the threads system must be improved.

S-threads was enhanced to *Virtual S-threads* (Küchlin and Ward, 1992). The role of a *virtual* thread is to document *logical* concurrency, i.e., it represents a task which can possibly run in parallel on a separate processor. Roughly speaking it is the thread analogue to a Multilisp *future* (Halstead, 1985; Ito and Halstead, 1990). It is then up to the underlying thread scheduler to decide whether a virtual thread is executed as a subroutine call, or whether it is executed on a separate kernel thread leading to *real* concurrency. *Virtual S-threads* not only add application-level efficiency but also insulate the application from the idiosyncrasies and limitations of the underlying threads implementation. The Virtual S-threads system manages fork requests using *lazy task creation* (Mohr et al., 1991), handling most threads system calls itself and passing only a tiny fraction to the OS kernel. It keeps its own run-queues of micro-tasks, and it manages a small pool of kernel threads which it employs as *workers*. On each fork, a record containing the fork parameters is put in the run-queue. These tasks can be asynchronously stolen by idle workers and executed as S-threads. However, if a join finds that the task was not yet stolen, the parent S-thread executes the task as a procedure call.

Thus the virtually unbounded logical concurrency of the application is dynamically reduced to the bounded amount of real parallelism that the kernel can support. At the same time there is a significant reduction in the number of kernel thread context switches, and the grain-size of the remaining threads is increased by executing child tasks as procedures. Virtual threads also have a much lower overhead than kernel threads, so that on average thread overhead drops and becomes easier to handle.

The system now manages the tasks at a small cost in execution overhead and a great savings in programming complexity. Virtual S-threads allow us to routinely generate tens to hundreds of thousands of parallel tasks while maintaining execution efficiency. Thus, to a great extent, virtual threads take the pain out of parallel symbolic programming in practice.

2.3. *Middleware for a Hierarchical Multiprocessor*

As was explained in Sec. 2.2, a thread of control is an execution context for a procedure on a SMP.[1] A fork will cause the procedure to run concurrently, and a join will wait for its termination and retrieve its result. All threads live within the same process with shared memory. Standard threads can therefore

---

[1] In the following we use the abbreviations SMP for shared memory multiprocessors, DMP for distributed memory multiprocessors, and HMP for hierarchical multiprocessors.

not migrate across the nodes of a DMP. However, if a thread does not exploit shared memory during its execution, then it is functionally equivalent to an *asynchronous remote procedure call.*

The DTS system (Bubeck et al., 1995) implements the asynchronous remote procedure call abstraction together with some dynamic load balancing and failure recovery on the network. DTS in its current implementation uses PVM to place a copy of a program on each (SMP) node of the network, together with a local DTS representative, the *node manager.* The node manager intercepts thread forks which have been marked *networkable* by the user, and communicates them together with their parameters to other nodes on the net with spare capacity. If a remote node manager accepts such a remote procedure call, it immediately forks the computation as a local thread. This thread can in turn fork more threads, both local and networkable. If the thread terminates, the server node manager sends its result back to the client. The client node manager accepts the result and delivers it to a joining thread.

The combined S-threads and DTS system forms the middle-ware between operating system and algorithms. The high-level theorem proving algorithms are all written for its application programming interface which provides functionality, abstraction, and portability. The outcome is a heterogeneous multithreaded rewrite laboratory running on a non-uniform memory access machine (cf. Fig. 1).

Robust middleware that allows the programmer a high level of abstraction is critical for successful parallelization of symbolic software. Of course, the abstractions given by the combined S-threads and DTS system are not the only possibility. Another middle-ware that has been used successfully in connection with distributed term rewriting is the TWlib (Denzinger and Lind, 1996), a C++ library that implements abstractions of the *Teamwork* method (Denzinger, 1993; Denzinger and Schulz, 1996).

2.4. *Term rewriting*

Unfailing Completion
We assume that the reader is familiar with term rewriting systems (TRS) and completion procedures. Our notation follows Dershowitz and Jouannaud (1990). Completion without failure is an enhancement of the Knuth-Bendix method (Knuth and Bendix, 1970). Whereas general completion wishes to construct a complete set of reductions from a set of equational axioms, unfailing completion attempts to find a ground Church-Rosser system, in which every ground term has a unique normal form (Hsiang and Rusinowitch, 1987; Bachmair et al., 1989; Bonacina and Hsiang, 1991). Unfailing com-

Delete	$\dfrac{(P \cup \{s \leftrightarrow s\};\ E;\ R)}{(P;\ E;\ R)}$	
Simplify	$\dfrac{(P \cup \{s \leftrightarrow t\};\ E;\ R)}{(P \cup \{s \leftrightarrow u\};\ E;\ R)}$	if $t \to_{R \cup E_{\succ}} u$
Orient	$\dfrac{(P \cup \{s \leftrightarrow t\};\ E;\ R)}{(P;\ E;\ R \cup \{s \to t\})}$	if $s \succ t$
Unfail	$\dfrac{(P \cup \{s \leftrightarrow t\};\ E;\ R)}{(P;\ E \cup \{s \leftrightarrow t\};\ R)}$	if $s \not\succ t$ and $t \not\succ s$
Compose	$\dfrac{(P;\ E;\ R \cup \{s \to t\})}{(P;\ E;\ R \cup \{s \to u\})}$	if $t \to_{R \cup E_{\succ}} u$
Collapse$_1$	$\dfrac{(P;\ E;\ R \cup \{s \to t\})}{(P \cup \{u \leftrightarrow t\};\ E;\ R)}$	if $s \to_{\{l \to r\}} u$ where $l \to r \in R \cup E_{\succ}$ and l is not an instance of s.
Collapse$_2$	$\dfrac{(P;\ E \cup \{s \leftrightarrow t\};\ R)}{(P \cup \{u \leftrightarrow t\};\ E;\ R)}$	if $s \to_{\{l \to r\}} u$ where $l \to r \in R \cup E_{\succ}$ and l is not an instance of s.
Deduce	$\dfrac{(P;\ E;\ R)}{(P \cup \{s \leftrightarrow t\};\ E;\ R)}$	if $(s,t) \in EP_{\succ}(R \cup E)$

Figure 2. Inference rules for unfailing completion.

pletion is usually used to prove an equation (*goal*) $a \leftrightarrow b$, and it stops with success as soon as a and b have a common $R \cup E$-normal form.

A *term ordering* $\succ$ is a terminating ordering on terms such that if $l \succ r$ for all $l \to r \in R$ then the reduction relation is terminating. Term orderings can also be used to define a terminating reduction relation by a set of equations E: Let $E_{\succ} = \{s\sigma \to t\sigma \mid s \leftrightarrow t \in E \text{ or } t \leftrightarrow s \in E \text{ and } s\sigma \succ t\sigma\}$ be the set of all orientable instances of E where $s\sigma \leftrightarrow t\sigma$ is an instance of $s \leftrightarrow t$. Then $\to_{E_{\succ}}$ is the *ordered rewrite relation* defined by $E_{\succ}$. Note for $s \leftrightarrow t \in E$, neither $s \to t$ nor $t \to s$ need to be in $E_{\succ}$. Such equations are called *unorientable*. Since it is much easier to rewrite w.r.t. a rule set than w.r.t. a set of equations, orientable equations will in general be turned to a set of rules R. Thus we will in general reduce w.r.t. $R \cup E_{\succ}$.

Given a set of equations P, an unfailing completion procedure attempts to compute a terminating and (ground) confluent reduction relation $\to_{R \cup E_{\succ}}$ defined by a set of unorientable equations E and a set of rewrite rules R. This is done by repeatedly transforming a triple $(P_i; E_i; R_i)$ of two equation sets P_i and E_i and one rule set R_i into a triple $(P_{i+1}; E_{i+1}; R_{i+1})$ using the inference rules in Fig. 2. In the *Deduce*-step the set of *extended critical pairs* $EP_{\succ}(R \cup E)$ is computed. See (Bachmair et al., 1989) for a precise definition of $EP_{\succ}(R \cup E)$. Our implementation gives the highest priority to reductions (Compose, Collapse$_i$, Simplify) and deletions (Delete). Thus all equations

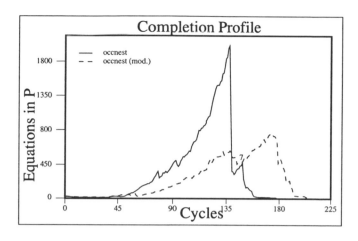

Figure 3. Completion profiles for different selection functions.

and rules are always kept completely normalized[2]. Several reductions on a single term are bundled in a normalization procedure that exploits marking techniques to avoid redundant reduction trials. The next priority is given to the computation of extended critical pairs (Deduce) and least priority is granted to the orientation steps (Orient, Unfail). The work performed between two orientation steps is called a *completion cycle*. To guarantee *fairness* of the completion procedure we must ensure that no equation stays in P forever.

Let $l \leftrightarrow r$, $l' \leftrightarrow r' \in R \cup E$ and let p be a position in l such that l' and $l|_p$ unify with most general unifier μ. Then $(r\mu, \tilde{l}\mu)$ is a *proper critical pair* between $l \leftrightarrow r$ and $l' \leftrightarrow r'$ if $r\mu \not\succ l\mu$ and $\tilde{l}\mu \not\succ l\mu$, where $\tilde{l} = l[p/r']$. The *extended critical pairs* $EP_\succ(R \cup E)$ is the set of all proper critical pairs in $R \cup E$.

Completion Strategies
The exact order in which equations from P are turned into rules in R, or into equations in E, respectively, is called the *completion strategy*. The completion strategy is determined by a selection function choosing the *best* equation in P. To yield a fair completion the best equation must be minimal in P w. r. t. some terminating ordering on equations. Small changes in this strategy can have huge effects (positive or negative) on the duration and work profile of a completion or proof task. Fig. 3 shows the completion profile obtained when completing the same input with two different strategies. The x-axis counts the

---

[2] Experiments with a version that does not completely normalize P did not perform uniformly better than the version described here. Cf. (Maier et al., 1995) for these experiments.

completion cycles and the y-axis shows $|P|$ measured before the i-th orientation operation.

Let F be the set of operators and let Var be the set of variables. The selection function is typically based on a weight function $\phi: F \cup Var \to \mathbb{N}$ defined by $\phi(x) = c_v$ for each variable $x \in Var$ and $\phi(f) = c_f$ for each function symbol $f \in F$. The weight function can be homomorphically extended to terms $t \in T(F, Var)$, e.g.:

$$\phi(t) = \begin{cases} c_v & \text{if } t \equiv x \in Var \\ c_f + \sum_{i=1}^{n} \phi(t_i) & \text{if } t \equiv f(t_1,\ldots,t_n), f \in F \end{cases}$$

Then for an equation $c \leftrightarrow d$, $\phi(c) + \phi(d)$ (add-weight) or $max\{\phi(c), \phi(d)\}$ (max-weight) define typical selection functions, which have proven to perform well (Küchlin, 1982).

Besides weight based selection functions that predominate in standard completion procedures, goal oriented selection functions have been proven to be very successful in unfailing completion procedures (Denzinger and Fuchs, 1994). Goal orientation is achieved by comparing measures acquired from a critical pair $c \leftrightarrow d$ with the corresponding measures obtained from the goal.

In our experiments—cf. Sec. 4.2—we used the following selection functions. The standard selection function was the function ON_G defined as follows. Let $occ(f,t)$ and $nest(f,t)$ be the occurrence and nesting of the symbol f in a term t—cf. (Fuchs, 1995)—, let $O_{f,t_1,t_2}^G = G(occ(f,t_1), occ(f,t_2))$ and let

$$N_{f,t_1,t_2}^G = G(nest(f,t_1), nest(f,t_2))$$

for some operation G. Then ON_G is defined as

$$ON_G(c \leftrightarrow d) = (\phi(c) + \phi(d)) \prod_{f \in F} ((O_{f,c,d}^G \dot- O_{f,a,b}^G) + 1)((N_{f,c,d}^G \dot- N_{f,a,b}^G) + 1)$$

where ϕ is a weight function, and $a \leftrightarrow b$ is the goal. Then $occnest = ON_{max}$ is the instance of ON_G defined in (Anantharaman and Andrianarievelo, 1990; Fuchs, 1995). We also use the instance $occnest\ (mod.) = ON_\times$ (i.e., G is the product) in our experiments (cf. Sec. 4.2). $occnest\ (mod.)$ takes into account operator specific information of both terms of an equation and in contrast to $occnest$ it gives strong penalties if none of the two terms of a pair is close to the goal.

Additionally, the following selection functions were used: The function $cp2siz$ is the square of the sizes of c and d. The function $cpmsiz$ gives the maximum of the sizes of c and d, and $cpnmsiz$ gives the absolute value of the difference between the sizes of c and d.

3. PARALLELIZATION ON A SHARED MEMORY MULTI-PROCESSOR

In this section we will first discuss some key decisions in constructing the core PaReDuX system, since they had a substantial influence on the shape of the parallel system: We always attempted to parallelize existing well proven sequential code before radically altering the system. Thus for the PaReDuX system we started from the sequential ReDuX system (Bündgen, 1993) (translated to C) which in turn uses SACLIB. We programmed in C and parallelized our code by including calls to Virtual S-threads. Thus our core PaReDuX code is compatible with the existing sequential ReDuX code, and it contains many calls to the sequential system.

The Virtual S-threads system supports a programming style that focuses on concurrent procedures, by making sure that they can be used very much like sequential procedures. They can be forked in parallel largely (but within reason) where program logic permits rather than where system load or application grain-size dictate. They can be called in sequential or parallel code, sequentially or in parallel. They can be put into libraries and linked and combined in the usual way, without worry of overloading the system.

Usually we followed the divide and conquer paradigm when parallelizing code. Where ReDuX normalizes a list of terms in greedy fashion, PaReDuX splits the list and forks two threads with recursive calls to the reduction procedure. Of course, as an optimization, we usually do stop dividing when the list is short. The point is that with the low overhead of virtual threads it is often possible to determine by intuition rather than extensive tests when short is short enough. Some of this intuition must be acquired for the parallel case, so that testing is necessary in a first phase for a first application. We built this intuition for plain completion first (Bündgen et al., 1994a), and then used it for AC and unfailing completion.

An abstract rendition of our parallelization methodology is roughly as follows: A given list C of uniform data, e. g., a list of critical pairs or rewrite rules, is either processed sequentially if it is too short and the work it represents falls below a predetermined grain-size, or it is split into two equal parts C_1, C_2 with $C = C_1 \circ C_2$. In the latter case, one recursive call is forked in parallel for the list C_2, and one recursive call is done by the parent thread itself for the list C_1. After computing the result for C_1, the result for C_2 is joined and both results are merged.

In combination with the Virtual S-threads environment, this divide-and-conquer approach to parallelization has a most desirable effect. Tasks generated early on have large grain-sizes and these are the tasks that are stolen by initially idle workers. Tasks generated later on have smaller grain-sizes, but those tasks are likely to be executed as procedure calls, with a substan-

---

$$R \leftarrow \text{COMPLETE}(P, \prec)$$

Input parameters: a set of equations $P = \{s_i = t_i \mid 1 \leq i \leq m\}$,
and a term ordering $\prec$

Output parameter: a complete TRS $R = \{l_i \to r_i \mid 1 \leq i \leq n\}$

$R := \emptyset$;

while $P \neq \emptyset$ **do**

   [**Orient.**] Select the best equation from P; remove it from P and add it as
   a rule $l \to r$ to R, provided that $r \prec l$, else stop with failure;

   [**Extend.**] If needed (in AC-completion) attach its extension rule to $l \to r$;

   [**Collapse.**] **for each** $l' \to r' \in R$ **do**
   if l' is reducible **then** remove $l' \to r'$ from R and put $l' = r'$ back to
   P;

   [**Compose.**] **for each** $l' \to r' \in R$ **do** normalize r' w. r. t. R;

   [**Deduce.**] **for each** $l' \to r' \in R$ **do** add critical pairs of $l \to r$ and $l' \to r'$
   to P;

   [**Simplify.**] **for each** $s = t \in P$ **do** normalize s and t w. r. t. R;

   [**Delete.**] **for each** $s = t \in P$ **do** **if** $s = t$ is trivial **then** remove it from P;

od

Figure 4.   The Procedure COMPLETE.

tially lower overhead. Thus we enjoy a dynamic adjustment of grain-size,
with mostly large-grain Virtual S-threads executing in parallel when there is
much work to do, and fine-grain Virtual S-threads executing in parallel only
when workers would remain idle otherwise. Note well that task scheduling
is done automatically within Virtual S-threads and remains transparent to the
application programmer.

3.1. *Parallel Completion in the Core PaReDuX System*

In the core PaReDuX system we parallelized the Knuth-Bendix completion
procedure, the Peterson-Stickel completion procedure for term rewriting systems with associative-commutative operators, and the unfailing completion
procedure of the ReDuX system. In all three cases we exploited medium
grained parallelism by parallelizing the inner loops of the completion procedure. A first abstraction of the one used in ReDuX and PaReDuX is given
in Figure 4. This leaves us with a sequential synchronization phase in each

cycle of the outer loop comprising the orientation, extension, collapse, and composition steps.

The methodology used in the procedure COMPLETE is certainly only one out of several possible choices, cf. (Huet, 1981). The reasons why we decided to choose a parallelization based on this methodology are manifold. First it turned out that the time spent in the synchronization phases makes up only a small portion of the total completion time. Second, our completion scheme keeps all critical pairs in normalized form. This allows to avoid unnecessary reductions (see below) and to use better heuristics in selecting the best equation in the orientation step. A further advantage of our approach is that it allows to keep the overall completion strategy fixed by fixing the selection function and the normalization strategy only. This is important for several reasons: The measurements are reliable, i. e., timings do not change if an experiment is repeated. Thus improvements during the development of the parallel program can be contributed to better parallelization techniques rather than to haphazard strategy effects. Similar arguments hold for the search of good non-problem specific grain size settings. Comparisons between sequential and parallel code is fair if both use the same strategy. For fixed strategies parallelization is scalable. That is, for problems that contain enough parallelism, timings improve if processors are added. In addition the correctness proof of a parallelization comes for free if it uses the same strategy as a correct sequential program. A last very practical argument in favor of our completion scheme is that it minimizes ordering decisions which is important for research in completing new equational theories. In these cases an appropriate term ordering is rarely known a priori and completion is run in semi-automatic mode where ordering decisions are made interactively, controlled by the intuition of the researcher. In this setting it is of utmost importance that repeated experiments yield the same results, so that a wrong decision at one point can be corrected in a new trial and that the (human idle) time between two ordering decisions is as short as possible.

We will describe the completion procedure implemented in ReDuX more precisely, before we describe our parallelization of COMPLETE. As mentioned before, the procedure COMPLETE maintains all rules in R and equations in P in fully (inter)reduced form. Thus *collapse*, *compose*, and *simplify*, apply only to those rules and old equations which are reducible by the newly oriented rule $l \rightarrow r$. Also, *delete* need only be tried on newly simplified equations. A further improvement which is also included in ReDuX and PaReDuX is a special case of the subconnectedness criterion (Küchlin, 1985; Küchlin, 1986) which allows to remove all equations in P which were derived from a collapsed rule.

Now we will investigate the synchronization requirements for the access

$$R \leftarrow \text{COMPLETE}(P, \prec)$$

Input parameters: a set of equations $P = \{s_i = t_i \mid 1 \leq i \leq m\}$,
and a term ordering $\prec$

Output parameter: a complete TRS $R = \{l_i \rightarrow r_i \mid 1 \leq i \leq n\}$

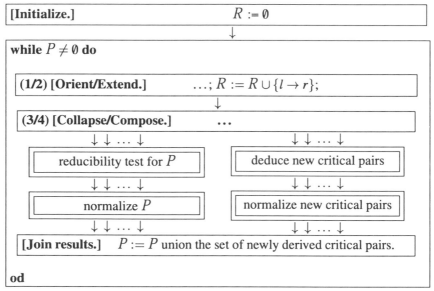

Figure 5.   Parallel completion in PaReDuX (overall structure).

to P and R in the steps of COMPLETE. Steps *orient* and *collapse* modify both P and R. Steps *extend* and *compose* read and modify R. Steps *deduce*, *simplify*, and *delete* modify P while requiring read-only access to R. Further, note that *simplification* and *deletion* of old equations is independent of *deduction, simplification,* and *deletion,* of new critical pairs. Therefore these two tasks can be computed in parallel. Figure 5 gives an outline of the resulting general parallel completion scheme realized in PaReDuX. Besides computing independent loops of COMPLETE in parallel, each of the loops depicted in doubly framed boxes is parallelized using the divide and conquer scheme described above. To keep the tasks as coarse-grained as possible the normalization and deletion procedures are called within the reducibility tests and critical pair computations, respectively. Thus the parallelization of the reduction tests and critical pair computations act as filters for the parallelization of the normalization and deletion procedures.

This kind of parallelization requires that all parallel threads have simultaneous access to R. The data structures for terms that support sharing R

without the need of copying rules are described in detail in (Bündgen et al., 1996b).

Modifications of R have a strong limiting influence on the parallelization of COMPLETE, because most of the operations depend on R. The problem caused by changing and deleting rules (steps *collapse* and *compose*) has become known as the *backward subsumption* (Slaney and Lusk, 1990) or *backward contraction* problem (Bonacina and Hsiang, 1995). In general, there are two ways to overcome these problems: We may permit reductions with outdated copies of R, which is correct but may possibly be inefficient, or we must synchronize accesses to R by locking. As our experiments indicate—cf. (Bündgen et al., 1996b, Sec. 7)—the time spent in steps *orient*, *extend*, *collapse*, and *compose* makes up less than 10% of the whole completion time— in most cases even less than 5%. Therefore it is not worth putting much effort into parallelizing this portion of the code.

All in all, the sequential ReDuX code had to be modified as follows:

1. Iterative loops had to be reorganized as divide and conquer style procedures;
2. Virtual S-thread-fork and Virtual S-thread-join instructions had to be inserted;
3. the term data structure had to be modified to allow parallel substitutions for rule sharing in reductions and critical pair computations.

It is known that the exact order in which equations are turned into rules is of the utmost importance for both term completion procedures and for Buchberger's algorithm. This is called the *completion strategy* and in our case it is encapsulated in the exact method after which the *best* of the equations is determined. Minute changes in this strategy can have huge effects (positive or negative) on the duration of completion. E. g., we must find the best of the equations (critical pairs) in P with respect to an ordering which is total on the equations. Simple quasi orderings (like those based on counting the symbols in each pair) are not sufficient: Changing the ordering of pairs in P which are equivalent with respect to the quasi-ordering (e. g., have the same number of symbols) may result in a different completion behavior. Our parallel completion procedures are designed to ensure that the priority queues containing P at each cycle exactly correspond to the queues in the respective cycle of the sequential completion.

Parallel Knuth-Bendix Completion

The scheme described above can be used in exactly the same way for the parallelization of the plain Knuth-Bendix completion procedure. Depending on our experiments, the parallelized portion of the code took 91 %–98 % of

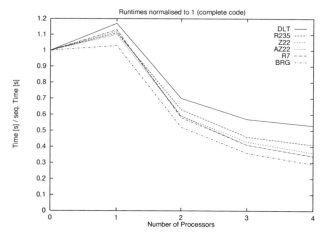

Timings for the sequential code ("0 processors") are normalized to 1.

Figure 6.   Normalized running times for parallel AC completion.

the total completion time. Our experiments indicate that for problems of a similar class the sequential bottleneck decreases with the size of the problem. For larger problems—i. e. one that take more than 100 seconds sequentially— we obtained speed-ups of 2.4–3.2 on four processors. Detailed results of our experiments with PaReDuX can be found in (Bündgen et al., 1996b, Sec. 7).

Parallel AC Completion
Although the differences between standard completion and AC completion seem moderate, their effect is tremendous from a complexity point of view. Searching for a most general AC-unifier of two terms results in general in a non-singleton set of substitutions. Thus more than one critical pair can be computed for each two rules with fixed superposition position. In addition, standard matches and unification can be computed in linear time. To test for AC-matchability or AC-unifiability however is NP-complete, and to compute a complete set of AC-unifiers is even of doubly exponential cost. This rise in complexity had to be taken into account when deciding on the parallelization grain size.

Parallelization of normalization tasks on the critical pair queue level turned out to be too coarse. We had to allow for normalizing the individual terms of an equation in parallel and we even allowed for fine grained parallelism, normalizing single terms in parallel.

For AC completion the sequential bottleneck was in general less than 2%. We obtained speed-ups of 2.4–3.5 on four processors. In Figure 6 the normalized runtimes for various examples are given. The time for the sequential code is normalized to 1 and we have used the slot "0 processors" for it. The numbers for 1, 2, 3, and 4 processors are the ones for the parallel code. A detailed description of the experiments that are summarized in Figure 6 can be found in (Bündgen et al., 1996b).

Parallel Unfailing Completion
In the unfailing completion procedure ordered rewriting is either done with respect to a term rewriting system R or a set of unorientable equations E. Thus in the orientation step either R or E is updated. The critical pairs between the new rule or equation on the one hand, and R or E on the other hand, can be computed and normalized independently. The computation of ordered reductions and critical pairs for ordered rewriting is more expensive than the corresponding operations for plain Knuth-Bendix completion because the application of equations in E involves a term comparison w. r. t. $\prec$, and, by symmetry, equations may be applied in both directions. The complexity is however much lower than the complexity of the corresponding AC-operations. Clearly, the minimal parallelization grain sizes for operations based on reductions by equations in E must be lower than the corresponding operations based on rewrite rules in R.

Unfailing completion is mainly intended to prove a single equation. Thus it may be an overkill to keep all critical pairs normalized. Keeping this consideration into account we implemented a second variant of the unfailing completion procedure that normalizes a critical pair only immediately after creation and just before selection in the orientation step. This eliminates the left column of tasks in Figure 5. Unless unfailing completion is used to complete a specification, both procedures spend less than 4 % of the total completion time in the sequential bottleneck, with slight advantages for the fully normalizing procedure. Surprisingly, none of the two procedures is uniformly better than the other. With both we obtained speed-ups of 3.0–3.8 on four processors. The detailed results of our experiments with PaReDuX are given in (Bündgen et al., 1996b, Sec. 7). Additional details on experiments with unfailing completion are given in (Maier et al., 1995).

4. PARALLEL SYMBOLIC COMPUTATION ON A HIERARCHICAL MULTIPROCESSOR

Hierarchical Multiprocessors are extremely important from a practical point of view. They keep the difficult network part of the communication relatively small and manageable while allowing standard multi-threading techniques on each node. In our example, it is much easier to parallelize and synchronize work between two 4-processor machines than between 8 uniprocessors.

In theory, one might just regard an HMP as a DMP with somewhat faster communication between nodes on the same cluster (the SMP components). Taking this view of our example, we might parallelize for 8 processors using PVM (Geist and Sunderam, 1991) and exploit a fast PVM implementation on the SMPs. This is *not* our approach because we wish to exploit the full power of an SMP with a multi-threaded operating system. Indeed, our results show (cf. Sec. 4.2) that with few exceptions our system is faster on one 4-processor machine than on two double-processor machines; this clearly points out the usefulness of an SMP component. Fig. 1 illustrates the overall architecture of *Distributed PaReDuX*.

Since *Virtual S-threads* (cf. Sec. 2.2) provides *virtual threads* which map a virtually unlimited amount of logical parallelism onto a small number of con-current kernel threads, we achieve fine-grained parallelization and synchro-nization, fast load balancing, and shared memory accesses by multi-threading on each SMP. We only use PVM on the network as a communication substrate for our DTS software (Bubeck et al., 1995) which realizes a much higher level of programming abstraction.

4.1. *The Master-Slave Approach to Unfailing Completion*

A master-slave unfailing completion procedure consists of one master per-forming an unfailing completion to find a proof for a goal, and a set of slaves employed by the master to search for rules, equations, and critical pairs, which can be fed into the master's proof process. Thus the slaves serve as 'oracles' providing valid lemmas to the master proof procedure.

Each slave is composed of an unfailing completion procedure that is stopped after a maximum number of cycles, and a procedure that subse-quently evaluates the results of the completion and passes the best results back to the master. Each slave uses a specific (not necessarily fair search) strategy different from the master strategy. When forked it obtains the fol-lowing arguments from the master: the maximum number of cycles the slave completion shall run, and the current critical pairs, equations, rules and the goal of the master. If a slave succeeds in proving the goal this is signaled to the master immediately. Otherwise the results obtained after the slave com-pletion procedure stopped are evaluated. Therefore, during the proof process, every slave counts the number of matches as well as successful *collapse* and *compose* operations for its system of rules and equations. This statistical in-formation is then used to compute a heuristic selecting the five best rules and equations which are sent back to the master.

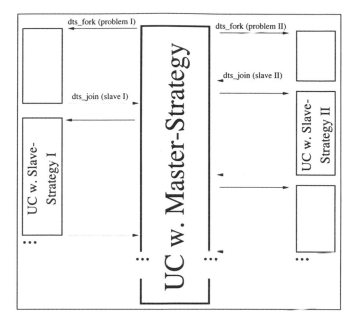

Figure 7.   The master-slave approach.

Thus, due to parallel search, super-linear speedups are possible. The master has to ensure that the slave setup is such that the master strategy remains fair during the search for the proof. Fig. 4.1 shows the principle of our master-slave approach.

The master process is an unfailing completion procedure that additionally can fork and join slaves over the network. Forking and joining slaves is done in the sequential part of the completion (e. g., at the beginning of the while loop of Fig. 5). During the initialization the master completion process forks a fixed number of slaves depending on the number of machines in the DTS environment. In the sequel the master checks once in every cycle whether any of its slaves can be joined (i. e., whether a slave has completed its work). If no, the master continues with its own completion task. Otherwise it either terminates with success if a slave could prove the goal, or it inserts the results of the slave into the master's queue of critical pairs with high priority. Subsequently a new slave is forked (with the same strategy as the slave just joined) and the master continues with its own completion work. Fig. 4.1 sketches the principle of our master-slave approach.

Note that the completion is fair if the master uses a fair strategy, and between joining and forking two slaves the master considers at least one critical pair it has deduced itself. Therefore the work of the slaves (number of cycles to run) should be scaled accordingly. Assuming the same time complexity for each completion cycle a slave should run more than $n \cdot m$ cycles where n is the number of slaves and m is the number of results returned by a slave.

Both the master and the slave can equally well run sequential or strategy compliant parallel completion procedures because the communication and synchronization between the master and the slaves is done in the sequential part of the completion procedure.

4.2. *Empirical Results*

The following problem specifications were used in an empirical evaluation of our concepts. They consist for all but Z_{22} of a set of equations, a reduction ordering, and an equation to be proved. The task for Z_{22} is to find a complete term rewriting system. The problems *glb1a*, *transa* and *p3b* describe proof tasks for lattice-ordered groups. These examples were taken from the TPTP library (Sutclife et al., 1994) (Version 1.2.0; Problems: GRP138-1.p, GRP162-1.p, GRP170-2.p); *glb1a* proves the greatest lower bound axiom, *transa* the transitivity axiom and *p3b* a general form of monotonicity from the given equational axiomatization. The *Lusk* examples are taken from (Lusk and Overbeek, 1985). *Lusk₃* proves that a ring with $x^2 = x$ is commutative. *Lusk₄* shows that the commutator $h(h(x,y),y) = e$ holds in a group with $x^3 = e$, and *Lusk₅* proves that $f(x, g(x), y) = y$ holds in a ternary algebra where the third axiom is omitted. The *Luka* examples are taken from (Tarski, 1956); cf. also (Lusk and McCune, 1990). The input equations are an equational axiomatization for propositional calculus by Frege. Lukasiewicz gave another set of axioms, which are the goals in our examples *Luka₂* and *Luka₃*. Finally, Z_{22} is a problem in combinatorial group theory taken from (Lusk and McCune, 1990). The task is to prove that the given finitely presented group with is the cyclic group of order 22.

To obtain information about the proof process that is independent of our parallel implementations, we use the synchronization point in step orient/unfail of parallel unfailing completion.

Strategy Compliant vs. Master-Slave Unfailing Completion
We have implemented and tested our parallel unfailing completion procedure on a network of Sun SPARC-stations 10.

The following machines were used in the experiments:

4-processor machines: 3 Sun SPARC-stations 10 two 90 MHz ROSS hyper-
    SPARC double processor modules;

2-processor machines: 3 Sun SPARC-stations 10, two 40 MHz Super-SPARC processors.

The operating system was Solaris 2.5 and our Virtual S-threads system was installed on Solaris threads. DTS used PVM version 3.3.10. All experiments were performed with 3,000,000 list cells (i. e., 24 MByte) of dynamic memory per workstation.

We have grouped these machines into different platforms and have performed the following experiments.

One master and one slave.   The strategy compliant experiments were run with *occnest* as selection function. The master-slave experiments that ran with one slave were performed on two different platforms: platform #1 consisted of two 4 processor machines, platform #2 of two 2-processor machines; *occnest* was the strategy of the master and *occnest (mod.)* the strategy of the slave.

The following result were obtained—cf. (Bündgen et al., 1996a, Table 1). $Lusk_3$, $Lusk_4$, Z_{22} and *transa* reveal very good speed-ups by strategy effects that can be further improved—in most of the cases—by work parallelism. $Lusk_4$ could not be solved within 20 minutes. using *occnest*, and the proof was found first by using a slave. The parallel master-slave speed-ups (w.r.t. the sequential implementation) for *transa*, $Lusk_3$, and Z_{22}, on platform #2 are 10.2, 14.5, and 13.6, on a total of 8 processors which is super-linear. With the exception of *transa* and $Lusk_4$, all problems are solved faster on one 4-processor machine than on two double-processors (after scaling the processor performance) which underscores the importance of the strategy compliant component.

One master, several slaves with different strategies.   In the following experiments, we used our three 4-processor machines, the one with 96 MByte of RAM as master, and the other two machines as slaves. The strategy compliant completions—sequential ones and parallel ones—were performed on the master machine only using the same strategy as for the master as in the master slave experiments. The strategy of the master was always *occnest* and *occnest (mod.)* was the strategy of one slave. The strategy of the other slave was as follows (cf. Sec. 2.4.2).

Experiment	#1	#2	#3
Strategy	cp2siz	cpnmsiz	cpmsiz

In all experiments a maximum of 500 critical pairs were given to a slave and 40 slave rounds were used.

Table I. General statistics for parallel and distributed unfailing completion. All times are in seconds; UC: unfailing completion; PUC: parallel unfailing completion; MS(s): Master-slave, each node sequential; MS(p): Master-slave, each node parallel. No time is given if no result could be obtained within 720 seconds. All times are inclusively the time for garbage collections.

		Strategy Compliant			Master-Slave				
		Time	Time	UC/	Time	Time	MS(s)/	PUC/	UC/
TRS	Exp.	UC	PUC	PUC	MS(s)	MS(p)	MS(p)	MS(p)	MS(p)
$glb1a$	#1	78.6	26.7	2.9	–	36.7	–	0.7	2.1
	#2	78.6	26.7	2.9	105	37.3	2.8	0.7	2.0
$transa$	#1	108	38.2	2.8	5.7	6.1	0.9	6.2	17.7
	#2	108	38.2	2.8	14.6	9.5	1.5	4.0	11.3
	#3	108	38.2	2.8	14.7	9.1	1.6	3.9	11.8
$p3b$	#3	179.2	60.1	2.9	192.8	66.9	2.8	0.8	2.6
$Lusk_3$	#1	974.4	271.9	3.5	24.2	10.3	2.3	26.3	94.6
$Lusk_3$	#2	974.4	271.9	3.5	276.0	106.0	2.6	2.5	9.0
$Lusk_3$	#3	974.4	271.9	3.5	255.2	106.7	2.6	2.5	9.0
$Lusk_4$	#1	41.9[†]	13.4[†]	3.1	42.2	18.4	2.2	0.7	2.3
$Lusk_4$	#2	41.9[†]	13.4[†]	3.1	42.2	18.0	2.3	0.7	2.3
$Lusk_4$	#3	41.9[†]	13.4[†]	3.1	42.8	18.6	2.2	0.7	2.3
$Luka_2$	#2	102.6	35.3	2.9	52.7	20.1	2.6	1.7	5.1
$Luka_2$	#3	102.6	35.3	2.9	52.7	21.3	2.4	1.6	4.8
$Luka_3$	#1	214.9	54.9	3.9	159.2	53.1	2.9	1.0	3.9
$Luka_3$	#2	214.9	54.9	3.9	159.0	53.7	2.9	1.0	3.9
$Luka_3$	#3	214.9	54.9	3.9	159.2	53.2	2.9	1.0	3.9
Z_{22}	#1	202.8	75.7	2.6	18.4	13.9	1.3	5.4	14.4
Z_{22}	#2	202.8	75.7	2.6	92.2	36.9	2.5	2.0	5.4
Z_{22}	#3	202.8	75.7	2.6	41.5	23.6	1.7	3.2	8.5

†: using *cpnmsiz* as selection function.

Z_{22}, $Lusk_2$, and *transa* reveal very good speed-ups by strategy effects that can be further improved—in most of the cases—by work parallelism. $Lusk_4$ could not be solved sequentially or on a 4-processor machine within 20 minutes using *occnest*. Using *cpnmsiz* it could be solved in 41.9 seconds sequentially and 13.4 seconds on a 4-processor machine and showed already in this case a good speed-up of 3.1. The speed-ups could not be improved in this case by a master slave-approach using the unsuccessful strategies for the master and the first slave.

On a total of 12 processors $Lusk_3$, *transa*, and Z_{22} showed super-linear speed-ups. In all these cases the strategy *cp2siz* was used for the second slave.

Discussion
The master-slave approach leads in a set of test cases to super-linear speed-ups w.r.t. the sequential implementation. Our system reaps strategy effects—which where first discovered and explored with the DISCOUNT system using the "teamwork method" (Avenhaus and Denzinger, 1993; Denzinger, 1993)—by a much simpler communication technique between the different proof processes. Moreover, it allows a simulation of different parallel completion techniques: E.g., we obtain a multi-threaded and strategy compliant theorem prover, if no slave is activated; we obtain a simple competitive theorem prover, if we join the slaves after an indefinite (very large) number of completion cycles; and finally, we obtain a restricted teamwork based theorem prover, where the supervisor can not be changed.

There are some examples where the slave contributions are not as good as expected. Of course, the setup overhead for slaves is not completely negligible, and moreover, slaves can slow down the master by misleading results. To avoid such effects, the setup of slaves and the evaluation of results have to be done very carefully and deserve further investigation. The communication between master and slave could be extended in such a way that good intermediate slave results can reach the master whenever they occur and not only after the termination of a slave. It should also be possible to distribute powerful rules and equations of the master system to all slaves. More elaborate master-slave communication techniques will reduce the number of slave setups and allow longer slave lifetimes, but go beyond the simple fork / join paradigm.

We have shown that our parallel master-slave unfailing completion procedure in its current form is a conceptionally simple but powerful approach towards the application of distributed and shared-memory based thread-systems for theorem proving. And it is a step forward towards the integration of shared memory multi-threaded components in existing distributed theorem provers.

4.3. *Comparison with Related Work*

Most of the completion systems employ search parallelism implicitly by maintaining several loosely synchronized completion loops on different subsets of critical pairs. All seem to be geared for, and implemented on, either SMPs or DMPs. Strategy compliant parallelization of term completion on an SMP has also been a major focus of our research, cf. Sec. 3. A similar design was applied to distributed Gröbner bases completion in (Attardi and Traverso, 1994). For a more detailed discussion about the common basis for parallel completion in term completion and Gröbner base completion we refer to (Amrhein et al., 1998). Shared memory search parallelism is exploited e. g., by the ROO prover (Lusk and McCune, 1990).

On the side of distributed search parallelism we can observe a wide range of solutions with simple competitive parallelism—e. g., (Schumann, 1995)— at one end of the scale and systems that support elaborate communication schemes between agents performing different strategies—e. g., the *Clause Diffusion* method used in Aquarius (Bonacina and Hsiang, 1995) and DIS-COUNT (Avenhaus and Denzinger, 1993). DISCOUNT even controls strategies by evaluating their success on a meta level using the *Teamwork* method (Avenhaus and Denzinger, 1993; Denzinger, 1993).

Our top-level scheme of search parallelism is inspired by the Teamwork concept and borrows strategies pioneered in that approach. However our master-slave approach lies in between the two extremes and can be parameterized to allow for more or less communication between different strategies. Due to its conceptual simplicity, the master-slave approach need not set up separate communication channels. It can be implemented with the same fork / join paradigm of multi-threaded programming employed in PaReDuX, but now supported over the network by our *Distributed Threads System* (DTS). This supports a seamless integration of both DMP and SMP parallelism, which is desirable on a hierarchical multiprocessor computer. It also allows to run the same code on more or fewer physical processors, and to use the code as a module within a larger parallel system.

5. CONCLUSION

We have designed and implemented parallel algorithms for term-rewriting, in particular Knuth-Bendix completion, AC completion, and unfailing completion. We have obtained good speed-up efficiency for work-parallel approaches on shared memory multiprocessors, and some large super-linear speed-ups for parallel search in unfailing completion on a network. Our elaborate parallel and distributed programming environments Virtual S-threads and DTS allow a high-level abstract programming technique based on the fork / join concept for all our work. Of course, the parallel speed-ups depend on the particular combination of hardware and software used.

Based on our experience, we can offer the following rough guidelines for parallelizing symbolic computations. Parallel Software is difficult to write and extremely difficult to debug. Therefore start from proven sequential code. Use a high-level but efficient and reliable programming environment. Note that passing a message is the parallel equivalent of a goto statement, but that the fork / join concept gives a parallel function call. Use shared memory multi-processors for better programming support and less communication overhead, and the network for scalability. Network parallelism demands coarse grain parallelization with little synchronization—so far it is ill suited for parallel Knuth-Bendix and AC completion. On the network, it is important to overlap communication and computation—this is why DTS supports multi-threading on each network node.

When designing a parallel symbolic computation algorithm—with the aim of speed-ups—first look for coarse grain search-parallelism. This parallelism can be implemented by forking the search threads in parallel and joining the first thread to finish. Multi-threaded code can be executed on conventional uniprocessors, and super-linear speed-ups due to strategy effects can thus be harvested first, independent of any physical parallelism. Next the work of parallel search can be spread over many processors, first on a shared memory machine, and—for scalability—over the network. This should result in further (sub-linear) speed-ups. If there are processors left idle or there is an imbalance in the grain size of the tasks, work parallelism and fine-grained parallelization can be added to fully utilize shared memory multiprocessors and to even out the runtimes of the parallel search routines. Often—for daily work rather than for large experiments—the multi-threaded program on a uniprocessor workstation or small shared memory multi-processor will be the most appropriate, and thanks to the super-linear portion of speed-ups it may be substantially faster than the sequential parent.

REFERENCES

Amrhein, B., R. Bündgen, and W. Küchlin: 1998, 'Parallel Completion Techniques'. In: M. Bronstein, J. Grabmeier, and V. Weispfenning (eds.): *Proc. 1995 Workshop on Symbolic Rewriting Techniques*. Monte Verità, Ascona, Switzerland. In print.

Anantharaman, D. and N. Andrianarievelo: 1990, 'Heuristical criteria in refutational theorem proving'. In: A. Miola (ed.): *Design and Implementation of Symbolic Computation Systems (DISCO '90)*, Vol. 429 of *Lecture Notes in Computer Science*. Capri, Italy, pp. 184–193.

Attardi, G. and C. Traverso: 1994, 'A Strategy-Accurate parallel Buchberger algorithm'. In: H. Hong (ed.): *First International Symposium on Parallel Symbolic Computation PASCO'94*. pp. 12–21. (Proc. PASCO'94, Linz, Austria, September 1994).

Avenhaus, J. and J. Denzinger: 1993, 'Distributing Equational Theorem Proving'. In: C. Kirchner (ed.): *Rewriting Techniques and Applications (LNCS 690)*. pp. 62–76. (Proc. RTA'93, Montreal, Canada, June 1993).

Bachmair, L., N. Dershowitz, and D. A. Plaisted: 1989, 'Completion Without Failure'. In: H. Aït-Kaci and M. Nivat (eds.): *Rewriting Techniques*, Vol. 2 of *Resolution of Equations in Algebraic Structures*. Academic Press, Chapt. 1.

Bonacina, M. P. and J. Hsiang: 1991, 'Completion procedures as Semidecision procedures'. In: S. Kaplan and M. Okada (eds.): *Conditional and Typed Rewriting Systems (LNCS 515)*. pp. 206–232. (Proc. CTRS'90, Montreal, Canada, June 1990).

Bonacina, M. P. and J. Hsiang: 1995, 'Distributed Deduction by Clause-Diffusion: Distributed Contraction and the Acquarius Prover'. *Journal of Symbolic Computation* **19**(1/2/3).

Bubeck, T., M. Hiller, W. Küchlin, and W. Rosenstiel: 1995, 'Distributed Symbolic Computation with DTS'. In: A. Ferreira and J. Rolim (eds.): *Parallel Algorithms for Irregularly Structured Problems, 2nd Intl. Workshop, IRREGULAR'95*, Vol. 980 of *Lecture Notes in Computer Science*. Lyon, France, pp. 231–248.

Buchberger, B., G. E. Collins, M. J. Encarnación, H. Hong, J. R. Johnson, W. Krandick, R. Loos, A. Mandache, A. Neubacher, and H. Vielhaber: 1993, 'SACLIB User's Guide'. Johannes Kepler Universität, 4020 Linz, Austria. Available via anonymous ftp at melmac.risc.uni-linz.ac.at in pub/saclib.

Bündgen, R.: 1993, 'Reduce the Redex $\longrightarrow$ ReDuX'. In: C. Kirchner (ed.): *Rewriting Techniques and Applications — 5th International Conference (RTA-93)*, Vol. 690 of *Lecture Notes in Computer Science*. Montreal, Canada, pp. 446–450.

Bündgen, R., M. Göbel, and W. Küchlin: 1994a, 'Experiments with Multi-Threaded Knuth-Bendix Completion'. Technical Report 94–05, Wilhelm-Schickard-Institut, Universität Tübingen, D-72076 Tübingen.

Bündgen, R., M. Göbel, and W. Küchlin: 1994b, 'A Fine-Grained Parallel Completion Procedure'. In: *Proc. Intl. Symposium on Symbolic and Algebraic Computation (ISSAC '94)*. Oxford.

Bündgen, R., M. Göbel, and W. Küchlin: 1994c, 'Multi-Threaded AC Term Rewriting'. In: H. Hong (ed.): *First International Symposium on Parallel Symbolic Computation PASCO'94*. pp. 84–93. (Proc. PASCO'94, Linz, Austria, September 1994).

Bündgen, R., M. Göbel, and W. Küchlin: 1995, 'Parallel ReDuX $\rightarrow$ PaReDuX'. In: J. Hsiang (ed.): *Rewriting Techniques and Applications (LNCS 914)*. pp. 408–413. (Proc. RTA'95, Kaiserslautern, Germany, April 1995).

Bündgen, R., M. Göbel, and W. Küchlin: 1996a, 'A Master-Slave Approach to Parallel Term Rewriting on a Hierachical Multiprocessor'. In: J. Calmet and C. Limongelli (eds.): *Design and Implementation of Symbolic Computation Systems — International Symposium DISCO '96*, Vol. 1128 of *Lecture Notes in Computer Science*. Karlsruhe, Germany, pp. 184–194.

Bündgen, R., M. Göbel, and W. Küchlin: 1996b, 'Strategy Compliant Multi-Threaded Term Completion'. *Journal of Symbolic Computation* **21**(4–6), 475–505.

Collins, G. E.: 1980, 'ALDES and SAC–2 Now Available'. *SIGSAM Bulletin* **12**(2), 19.

Cooper, E. C. and R. P. Draves: 1988, 'C Threads'. Technical Report CMU-CS-88-154, Computer Science Department, Carnegie Mellon University, Pittsburgh, PA 15213.

Denzinger, J.: 1993, 'Teamwork: Eine Methode zum Entwurf verteilter, wissensbasierter Theorembeweiser'. Ph.D. thesis, Universität Kaiserslautern, Postfach 3049, D-67663 Kaiserslautern.

Denzinger, J. and M. Fuchs: 1994, 'Goal oriented equational theorem proving using team work'. In: B. Nebel and L. Dreschler-Fischer (eds.): *KI-94: Advances in Artificial Intelligence — Proceedings of the 18th German Annual Conference on Artificial Intelligence*, Vol. 861 of *Lecture Notes in Artificial Intelligence*. Saarbrücken, Germany, pp. 343–354.

Denzinger, J. and J. Lind: 1996, 'TWlib—a Library for Distributed Search Applications'. In: *Proc. ICS'96-AI*. Kaohsiung, pp. 101–108.

Denzinger, J. and S. Schulz: 1996, 'Recording and Analyzing Knowledge-Based Distributed Deduction Processes'. *Journal of Symbolic Computation* **21**, 523–541.

Dershowitz, N. and J.-P. Jouannaud: 1990, 'Rewrite Systems'. In: J. van Leeuwen (ed.): *Formal Models and Semantics*, Vol. B of *Handbook of Theoretical Computer Science*. Amsterdam: Elsevier, chapter 6, pp. 243–320.

Fleischer, J., J. Grabmeier, F. W. Hehl, and W. Küchlin (eds.): 1994, 'Computer Algebra in Science and Engineering'. Bielefeld, Germany: Zentrum für Interdisziplinäre Forschung, World Scientific.

Fronhöfer, B. and G. Wrightson (eds.): 1990, 'Parallelization in Inference Systems', Vol. 590 of *Lecture Notes in Artificial Intelligence*. Dagstuhl Castle, Germany:, Springer-Verlag.

Fuchs, M.: 1995, 'Learning proof heuristics by adapting parameters'. Technical report, Fachbereich Informatik, Universiät Kaiserslautern, Postfach 3049, D-67663 Kaiserslautern.

Geist, G. A. and V. S. Sunderam: 1991, 'The PVM System: Supercomputer Level Concurrent Computation on a Heterogeneous Network of Workstations'. In: *Sixth Annual Distributed-Memory Computer Conference*. Portland, Oregon, pp. 258–261.

Halstead, Jr., R. H.: 1985, 'Multilisp: A Language for Concurrent Symbolic Computation'. *ACM Transactions on Programming Languages and Systems* **7**(4), 501–538.

Hsiang, J. and M. Rusinowitch: 1987, 'On Word Problems in Equational Theories'. In: T. Ottmann (ed.): *Automata, Languages and Programming (LNCS 267)*. pp. 54–71. (Proc. ICALP'87, Karlsruhe, Germany, July 1987).

Huet, G.: 1981, 'A Complete Proof of Correctness of the Knuth-Bendix Completion Algorithm'. *Journal of Computer and System Sciences* **23**, 11–21.

Ito, T. and R. Halstead, Jr. (eds.): 1990, 'US/Japan Workshop on Parallel Lisp', Vol. 441 of *Lecture Notes in Computer Science*. Sendai, Japan:, Springer-Verlag.

Knuth, D. E. and P. B. Bendix: 1970, 'Simple Word Problems in Universal Algebra'. In: J. Leech (ed.): *Computational Problems in Abstract Algebra*. (Proc. of a conference held in Oxford, England, 1967).

Küchlin, W.: 1982, 'A Theorem-Proving Approach to the Knuth-Bendix Completion Algorithm'. In: J. Calmet (ed.): *Computer Algebra (Proc. EUROCAM'82, LNCS 144)*.

Küchlin, W.: 1985, 'A Confluence Criterion Based on the Generalised Newman Lemma'. In: B. F. Caviness (ed.): *Eurocal'85 (LNCS 204)*. pp. 390–399. (Proc. Eurocal'85, Linz, Austria, April 1985).

Küchlin, W.: 1986, 'A Generalized Knuth-Bendix Algorithm'. Technical Report 86-01, Mathematics, Swiss Federal Institute of Technology (ETH), CH-8092 Zürich, Switzerland.

Küchlin, W.: 1990a, 'The S-Threads Environment for Parallel Symbolic Computation'. In: R. E. Zippel (ed.): *Second International Workshop on Computer Algebra and Parallelism*, Vol. 584 of *Lecture Notes in Computer Science*. Ithaca, USA, pp. 1–18.

Küchlin, W. W.: 1990b, 'PARSAC-2: A Parallel SAC-2 Based on Threads'. In: S. Sakata (ed.): *Applied Algebra, Algebraic Algorithms, and Error-Correcting Codes: 8th International Conference, AAECC-8*, Vol. 508 of *LNCS*. Tokyo, Japan, pp. 341–353.

Küchlin, W. W.: 1995, 'PARSAC-2: Parallel Computer Algebra on the Desk-Top'. in (Fleischer et al., 1994), pp. 24–43.

Küchlin, W. W. and N. J. Nevin: 1991, 'On Multi-Threaded List-Processing and Garbage Collection'. In: *Proc. Third IEEE Symp. on Parallel and Distributed Processing*. Dallas, TX, pp. 894–897.

Küchlin, W. W. and J. A. Ward: 1992, 'Experiments with Virtual C Threads'. In: *Proc. Fourth IEEE Symp. on Parallel and Distributed Processing*. Dallas, TX, pp. 50–55.

Lusk, E. L. and W. W. McCune: 1990, 'Experiments with ROO, A Parallel Automated Deduction System'. in (Fronhöfer and Wrightson, 1990), pp. 139–162.

Lusk, E. L. and R. A. Overbeek: 1985, 'Reasoning about Equality'. *Journal of Automated Reasoning* **1**, 209–228.

Maier, P., M. Göbel, and R. Bündgen: 1995, 'A Multi-Threaded Unfailing Completion'. Technical Report 95–06, Wilhelm-Schickard-Institut, Universität Tübingen, 72076 Tübingen, Germany.

Mohr, E., D. A. Kranz, and R. H. Halstead, Jr.: 1991, 'Lazy Task Creation: A Technique for Increasing the Granularity of Parallel Programs'. *IEEE Transactions on Parallel and Distributed Systems* **2**(3), 264–280.

Morisse, K. and G. Oevel: 1995, 'New Developments in MuPAD'. in (Fleischer et al., 1994).

Schreiner, W. and H. Hong: 1993, 'The Design of the PACLIB Kernel for Parallel Algebraic Computation'. In: J. Volkert (ed.): *Parallel Computation (LNCS 734)*. pp. 204–218. (Second International ACPC Conference, Gmunden, Austria, October 1993).

Schumann, J.: 1995, 'SiCoTHEO — Simple Competive parallel Theorem Provers based on SETHEO'. In: *Proc. PPAI'95, Montreal, Canada, 1995*.

Slaney, J. K. and E. L. Lusk: 1990, 'Parallelizing the Closure Computation in Automated Deduction'. In: M. E. Stickel (ed.): *10th International Conference on Automated Deduction, (CADE'90)*, Vol. 449 of *Lecture Notes in Compter Science*. Kaiserslautern, Germany, pp. 28–39.

Sutclife, G., C. Suttner, and T. Yemenis: 1994, 'The TPTP Problem Library'. In: A. Bundy (ed.): *12th International Conference on Automated Deduction (CADE'94)*. Nancy, France, pp. 252–266.

Tanenbaum, A. S.: 1992, *Modern Operating Systems*. Prentice Hall.

Tarski, A.: 1956, *Logic, Semantics, Metamathematics*. Oxford University Press.

Weiser, M., A. Demers, and C. Hauser: 1989, 'The Portable Common Runtime Approach to Interoperability'. In: *12th ACM SOSP*. pp. 114–122.

JOHANN SCHUMANN, ANDREAS WOLF, CHRISTIAN SUTTNER

CHAPTER 10

PARALLEL THEOREM PROVERS BASED ON SETHEO

1. INTRODUCTION

Since the development of high-performance parallel computers, a variety of attempts have been made to increase the power of automated theorem provers by exploitation of parallelism. The approaches are based on many different calculi and proof procedures and explore a large variety of different parallel models. Due to a lack of space, we cannot describe these models and systems in this chapter. Rather, we refer to extensive surveys in (Kurfeß, 1990; Suttner and Schumann, 1993).

Within the Automated Reasoning Group at the Munich University of Technology, a total of six parallel theorem provers has been designed, implemented, and evaluated in detail. They are all based on the same sequential theorem prover SETHEO which uses the Model Elimination calculus (Loveland, 1978). Model Elimination as a kind of Tableaux Calculi is discussed within this book in Chapter I.1.2. This common basis allows the description of the models of parallelism in a clear and uniform way and lets us draw comparisons between the different systems.

This article proceeds as follows: Firstly, we will describe a classification scheme for parallel theorem proving which, somewhat orthogonal to the standard AND-OR-parallelism, allows spanning of the space of possible models of parallel execution of logic, regardless of the underlying calculus. Then, we will describe the parallel systems which are based on SETHEO: PARTHEO (classical OR-parallelism) partitions the search space in a dynamic fashion, whereas SPTHEO (static partitioning with slackness) performs the partitioning before the parallel processes start. We also present three competitive systems: RCTHEO uses competition between different random traversals of the search space; p-SETHEO performs competition between different parameter settings of the sequential proof procedure, whereas SiCoTHEO explores parameter ranges in a competitive way. All these parallel provers (except for PARTHEO which requires Inmos transputers) run on networks of processors (i.e., workstations) with message passing as their means of communication.

For all provers, results of experiments are shown and assessed. Possible ways to use the remaining communication bandwidth for *cooperation* be-

W. Bibel, P. H. Schmitt (eds.), Automated Deduction. A basis for applications. Vol. II

tween the different proving processes are then discussed. The cooperative prover CPTHEO is presented and first results of simulations are presented.

We assume the reader is familiar with the basics of the Model Elimination calculus and goal-oriented top-down search procedures. A detailed discussion of Model Elimination can be found within this series in Chapter I.1.2.

2. CLASSIFICATION OF PARALLEL EXECUTION MODELS

Parallel execution models for automated theorem proving can be classified along two orthogonal issues: the primary issue is about how the search spaces explored by the parallel workers relate to each other. For this, two principles can be identified: *partitioning* and *competition*. Partitioning relies on partitioning the exploration of the search space between parallel workers. In contrast, competition relies on different approaches (one calculus or several calculi) for the same problem. This results in an individual search space for each worker which are independent of each other. Partitioning parallelization includes traditional AND/OR-parallelism (problem decomposition), which we will generalize to the operational concepts of completeness-based and correctness-based partitioning. The second issue is the degree of cooperativeness between the parallel workers. The distinguishing feature here is that information gathered during the processing of a computational task in the parallel system may or may not be shared with other tasks with the intention of saving work.

Table I shows the resulting classification matrix together with a classification of the systems discussed in this article. For the classification of other systems see e.g. (Kurfeß, 1990; Suttner and Schumann, 1993).

Table I. A matrix representation of the classification taxonomy.

		uncooperative	cooperative
Partitioning	completeness-based	PARTHEO, OR-SPTHEO	CPTHEO
	correctness-based	AND-SPTHEO	-
Competition	different calculi	-	-
	one calculus	RCTHEO, SiCoTHEO, p-SETHEO	-

2.1. *Partitioning Parallelization*

The idea of partitioning parallelization is to split the search space into parts which are searched in parallel. For *completeness-based partitioning*, independent parts of the search space are distributed among workers (for example OR-parallelism for AND/OR-trees, set-split parallelism for set-based computation). A solution found by an individual worker usually constitutes a solution to the overall problem. Dropping tasks which should be given to workers generally causes incompleteness of the proof search, and hence the name. The systems PARTHEO (Section 3.1) and OR-SPTHEO (Section 3.2) belong to this category.

For *correctness-based partitioning*, the tasks (representing parts of the search space) given to workers are interdependent, and an overall solution is built up from the corresponding partial solutions (e.g., AND-parallelism for AND/OR-trees). In general, dropping tasks which should be given to workers destroys the correctness of a deduction. AND-SPTHEO (Section 3.2) performs correctness-based partitioning.

2.2. *Competition Parallelization*

Competition parallelization is based on the attempt to solve the same problem using several different approaches. In particular, no partitioning of a common search space is performed. Of course, competition is relevant only if it is not possible to select the system with the best performance in advance as it is the case for automated theorem proving.

In competition parallelism, similarly as for OR-parallelism, the success of a single system is sufficient for a solution and allows the termination of the computation. Obviously, completeness of the overall computation is assured as long as at least one of the competing systems is deduction complete. Correctness is guaranteed if all competitors are correct.

The basic question for such an approach is how interesting competing systems are obtained. There are two principal choices: one option is to use different calculi based on the fact that each calculus has its particular merits and performs well in certain cases, while it may fail to do so in other cases. The other option is to use a single underlying calculus, and competition is achieved by using different search strategies for each competitor. This choice leads to the sub-distinction discussed below.

Competition using Different Calculi
Here, different algorithmic approaches are used simultaneously either to improve the solution quality or to obtain a solution earlier.

An uncooperative parallel system of this kind is quite easy to build, since several different systems are simply started in parallel. This has obvious performance potential for a well-chosen set of competitors when compared with any of the individual systems. For example, ILF (Dahn and Wolf, 1996) (see Chapter II.4.14, III.2.8) uses parallel competition between SETHEO, OTTER (McCune, 1990), and DISCOUNT (Denzinger, J. et al., 1997) (see also Chapter II.4.13).

Cooperative competition, on the other hand, poses many difficult questions regarding the parallel system design, e.g., the selection of the individual systems and the amount and type of information exchanged. These issues will be discussed in Section 5.

Competition using a Single Calculus

In this case, a set of competing systems is used, all of which are based on the same calculus. This is possible when the proof procedures are controlled by various parameters (for example, by choosing a particular search strategy or by issuing certain search bounds) or parameter values (for example, weight of a clause). Each distinct set of control parameters then leads to a different search behavior. For many systems, this gives rise to a large number of possible variants.

The speed-up obtainable with this scheme relies solely on the relation between the structure of the problem search space and the adequacy of the control parameters for achieving optimal exploration. For many types of uninformed search, variations of the control parameters cause significant changes in the run-time. Thus there is a significant potential for improvement over the average performance of the individual competitors. Our competitive systems RCTHEO, SiCoTHEO, and p-SETHEO will be described in Section 4.

3. PARALLELIZATION VIA SEARCH SPACE PARTITIONING

3.1. *Dynamic Partitioning: PARTHEO*

PARTHEO was the first parallel theorem prover which was developed on the basis of SETHEO. PARTHEO has been designed with the emphasis on theorem proving. In this field, the main problem is not the length of a proof but the large search space induced by the branching factor of the underlying calculus. Therefore, *dynamic partitioning* of the search space (OR-parallelism) is being used for PARTHEO. The system is designed for hardware relying on message passing and local memory. The target hardware is a network of Inmos transputers. This choice promises *scalable* implementations, i.e., the topology and size of the hardware may be varied easily.

Computational Model

The search space induced by the Model Elimination calculus together with a given proof procedure can be described as a tree of tableaux (i.e., a tree of trees). For a given formula F, the search tree is constructed as follows: the root is labeled with the empty tableau. For each node N labeled with an open tableau T, the children of N are nodes labeled with tableaux which can be constructed by applying one inference rule (start-, extension-, reduction-step) to T. Thus the construction (or traversal) of the OR-tree corresponds to a search for a proof of F.

During the proof process, the OR-search tree of a formula must be explored from the root to the leaves. OR-parallelism distributes the *branches* of the OR-search tree between the processors. Hence, each node of the OR-tree is comprised of an individual proof task. If a proof task is soluble, i.e., the corresponding tableau represents a proof, this event is reported to the user and usually leads to a successful finish. A task can fail if no expansion is possible (for example, no unifiable partner can be found or a resource bound is reached). Finally, a task may ramify into finitely many sub-tasks according to the different inference possibilities for the selected open leaf of the encoded tableau. This last case makes up the potential for parallelism. The respective proof task gives rise to a number of new tasks, which encode just the children in the OR-search tree. The processor cancels the old task, encodes the new tasks, and puts them into its local task memory, thus making them available to other processors.

Parallelism is achieved by communicating tasks over the network. In order to achieve a good load balance without too much communication overhead, PARTHEO employs a *task-stealing* model (Bose, S. et al., 1989). If a processor executes a task which fails, it tries to take a new one from its local task store. If it is empty, the processor asks one of its directly connected transputers for work. All processors must check from time to time for incoming task requests. Thus the idle processor will eventually get tasks from one of its neighbors with enough tasks. Thus tasks are only transmitted over the network if no work is available locally[1]. Proof tasks are encoded (8 byte × length of the branch in the OR-tree) to cut down the amount of communication. An efficient decoding scheme, the *partial restart* (Schumann and Letz, 1990), ensures low overhead for switching proof tasks.

The entire execution model can be summarized in the following C-like program.

---

[1] The proof process is initialized by setting all processors into the state of task request, and by simply giving the root task, which encodes the empty tableau, to an arbitrarily chosen processor.

```
on all nodes in parallel do {
  search: while true do {
      get_task T from local_task_store;
      do partial_restart;
      if unification succeeds then
        if T is closed then    // proof found
          report_success_to_host;
        else {
          generate_new_tasks T1,...,Tn from T;
          put T1,...,Tn into local_task_store }
      else ; }                 // fail
  request: if isempty(local_task_store) then
      request_tasks_from_neighbor;
  send: if request_for_tasks_received then {
      n = cardinality of local_task_store;
      if n <= 1 then send_back("no_tasks");
      else send_back_tasks(n/2); } }
```

Implementation

PARTHEO is designed to run on an arbitrary network of Inmos transputers.
We used a hardware system consisting of 16 + 1 T800's configured as a torus
or mesh (see Figure 1). One of the transputers is the *host* processor, which
is connected to a UNIX workstation. PARTHEO itself is implemented in 3L-
parallel C (3L Ltd., 1988). Synchronous message passing is used as the basic
communication mechanism. The access to global data structures is controlled
via semaphores inside one transputer.

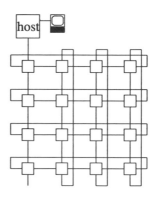

Figure 1. PARTHEO hardware configuration

Experimental Results

The performance of PARTHEO was evaluated[2] with a variety of examples
from the area of theorem proving and logic programming, as shown in Table II. Run-times of the sequential SETHEO and PARTHEO (same hardware, averaged over 5 runs) are compared. Problems without a TPTP number
are taken from the field of logic programming because they could be easily
modified to serve as good speed-up benchmarks. The *Queens* examples are
just standard Prolog programs for solving the queens problem. *Maximum* is
a quadratic program for computing the maximum of a list of integers, and
PropSat consists of a program for finding models of a propositional formula
(including one example formula). Finally, *FairyTale* is a plan generation problem, which consists in finding the happy end of a fairy tale.

Table II. Run-time comparison of SETHEO (T 800) and PARTHEO ($16 \times$
T 800) with speed-up s and efficiency $\mu = \frac{s}{16}$

Problem	SETHEO [s]	PARTHEO [s]	s	μ
GRP008-1	3413.0	1.8	1852.0	115.8
GRP012-1	64.0	2.2	29.7	1.9
GRP036-3	36490.0	5.9	6194.0	387.0
RNG006-1	575.3	114.2	5.0	0.3
RNG040-1	254.5	48.1	5.3	0.3
GRP012-2	1666.0	352.0	4.7	0.3
SET005-1	841.6	95.7	8.8	0.6
SET011-1	44.2	10.2	4.3	0.3
MSC006-1	4.5	1.8	2.4	0.2
SYN002-1.002:007	53.7	11.8	4.6	0.3
Queens8	16.9	1.8	9.2	0.6
Queens9	83.6	9.2	9.1	0.6
Queens10	396.0	38.0	10.4	0.6
Maximum	17.1	4.6	3.8	0.2
PropSat	48.8	7.3	6.7	0.4
FairyTale	24.0	1.8	13.7	0.9

Assessment

PARTHEO was the first running implementation of a parallel prover based
on SETHEO. It was designed for the transputer's message-passing paradigm.
As shown in the previous section, its performance was very encouraging. Although the figures strongly varied from example to example, a generally good
speed-up could be obtained. The scalability of PARTHEO is comparatively

---

[2] The transputer hardware (and thus PARTHEO) is not operational any more.
Hence, we can only present experimental data obtained in 1990-91.

good (up to about 100 processors). This could be established by modeling PARTHEO as an extended queuing model with subsequent simulation (Schumann and Jobmann, 1994).

However, the implementation effort for PARTHEO was very high. A minimal operating system had to be designed from scratch to perform the tasks necessary for PARTHEO's operation — for details see (Schumann, 1991). Therefore, PARTHEO is not portable and died due to the outdated hardware architecture.

3.2. *Static Partitioning: SPTHEO*

Computational Model

Static Partitioning with Slackness (SPS) (Suttner, 1995) is a method for parallelizing search-based systems. Traditional partitioning approaches for parallel search rely on a continuous distribution of search alternatives among processors ("dynamic partitioning"). The SPS-model instead proposes to start with a sequential search phase, in which tasks for parallel processing are generated. These tasks are then distributed and executed in parallel. No partitioning occurs during the parallel execution phase. The potential load imbalance can be controlled by an excess number of tasks (slackness) as well as an appropriate task generation. The SPS-model has several advantages over dynamic partitioning schemes: most importantly it assures that the amount of communication is strictly bounded and minimal. This results in the smallest possible dependence on communication latency, and makes efficient execution even on large workstation networks feasible. Furthermore, the availability of all tasks prior to their distribution allows optimization of the task set which is not possible otherwise. In addition to the classical OR-partitioning approach, AND-SPTHEO splits the generated tasks into literal groups which do not share variables. These literal groups become the new tasks.

Implementation

SPTHEO is a parallel implementation of the SETHEO system. It is implemented in C and PVM (Geist, A. et al., 1994), and runs on networks of workstations. The prover is based on the SPS model, and consists of three phases.

- In a first phase, an initial area of the search space is explored until enough tasks are generated. The number of generated tasks normally exceeds the number of processors by the *slackness* factor.
- In a second phase, the tasks are distributed.
- Finally, in a third phase, the tasks are executed on the individual processors.

For reasons of practical search completeness, each processor executes all its tasks concurrently (preemptive execution).

Experimental Results
All experiments were based on the TPTP v1.1.3 problem library (Sutcliffe, G. et al., 1994) containing 2571 usable problems[3]. Figure 2 compares the performance of SETHEO and SPTHEO. Given a run-time limit of 1000 seconds, SETHEO solves 858 problems. It can be expected that increasing the run-time limit only leads to a small increase in the number of additionally solved problems. An examination of the proof-finding performance of SPTHEO for different run-time limits is also given in Figure 2. The asymptotic upper bound on the number of proved problems is due to the run-time limit of 20 seconds per task.

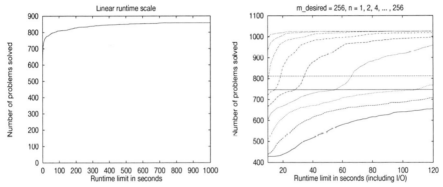

Figure 2.   Left: The number of problems solved by SETHEO as a function of the run-time limit. Right: The performance of SPTHEO for overall run-time limits between 10 and 120 seconds (20 seconds per task). Each curve denotes a particular number of processors ($n = 1, 2, 4, 8, 16, 32, 64, 128, 256$, from bottom). The horizontal lines show the SETHEO performance for run-time limits of 10 seconds (solid line) and 120 seconds (dotted line). During the static partitioning 256 sub-tasks have been generated.

Assessment
Runtime Assessment. The wall-clock run-time on some hardware platform can be estimated as follows: Let *task* be the chronologically first task that leads to a proof. The run-time of the parallel system then consists of the time

---

[3] From the 2652 TPTP problems 81 problems were excluded due to errors in the formulation or during the preprocessing of SETHEO. Thus 2571 problems were used.

t_{gen} for task generation, the time $t_{dist}(task)$ until $task$ is distributed, the run-time T_{task} for processing task $task$, the time $t_{sppdelay}$ for the processing delay of $task$ due to time sharing, and the time t_{term} until the success message from $task$ is received and processed by the master. t_{dist} can be estimated by measuring the average time it takes to distribute a task. $t_{sppdelay}$ depends on the number of tasks, the times when the processing of the tasks start, and the run-times of the tasks that are processed on the same processor as $task$. An upper bound on $t_{sppdelay}$ is obtained by assuming that all these tasks start at the same time as $task$. t_{term} can be obtained by appropriate communication time measurements. In case no proof is found within the run-time limit, the run-time can be computed as above, based on the task which terminates last.

The wall-clock time required for distributing a single task (including process startup) using the PVM message passing library under non-exclusive usage has been measured. The average time per task for up to 20 tasks is below 0.1 seconds, with maximal values of 0.7 seconds. The average value increases slightly (up to 0.13 seconds) as the number of tasks increases, with a maximal value of 2.1 seconds. The reason for this is that as the number of tasks increases, the probability increases that a processor with higher load is used.

Avoided Parallelization. A particular advantage of the SPS-model is the avoidance of parallel execution for very simple problems. A parallel execution phase is only initiated if at least the desired number of tasks (m_{des}) can be generated. For problems which exhibit a small degree of inherent parallelism, this does not arise before the problem is already solved during the generation phase.

All of the following evaluations are concerned with problems for which parallel processing occurs. For clarity, these problems are referred to as the *parallelized problems*. For $m_{des} = 16$ and 256 there remain 2339 and 2141 parallelized problems after task generation, respectively.

Number of Generated Tasks. Since the partitioning depends on the search space structure of an individual problem, it is not always possible to generate the desired number of tasks exactly. The SPTHEO task generation attempts to approximate the desired number, with a bias towards normally producing more tasks than specified. For $m_{des} = 16$, 22.1 tasks are generated in (geometric) average, and 267.0 tasks for $m_{des} = 256$.

Task Generation. The generation phase in SPTHEO operates sequentially, and increases the serial fraction of the computation. For the parallelized problems, the geometric averages for the run-time and the number of search steps are 0.06 seconds (86 steps) for $m_{des} = 16$, and 0.22 seconds (1346 steps) for $m_{des} = 256$.

Load Balance. Due to the static partitioning concept, load imbalances may arise whenever tasks terminate with no proof found (*failing tasks*), while the overall computation still continues. Idle processors can be avoided by attempting to avoid the generation of failing tasks, and by use of slackness.

In SPTHEO, a simple one step look-ahead is used to avoid the generation of tasks that would fail immediately. It reveals that on average 24-42% of the potential tasks are avoided for the most common iterative deepening levels for task generation.

For measuring the degree of load imbalance quantitatively, we define $LI(n)$ (for *Load Imbalance*) as

$$LI(n) = \frac{\Sigma_{i=1}^{n}(t_{pen} - T_i)}{(n-1) \times t_{pen}},$$

where T_i denotes the total run-time spent at processor i and t_{pen} denotes the total system run-time, and $n > 1$. The term $n - 1$ in the denominator represents the largest number of terms which can be different from zero, because there is at least one processor i with $T_i = t_{pen}$. $LI(n)$ is an absolute measure, not taking the best or worst possible balance that can be obtained for a particular set of tasks into account. It ranges from perfect balance ($LI(n) = 0$), which means that all processors finish working at the same time, to maximal imbalance ($LI(n) = 1$), where exactly one processor is busy during the execution.

Figure 3 shows the arithmetic average and the maximal load imbalance observed over all parallelized problems, as a function of the number of processors. It can be seen that a significant reduction of load imbalance is already obtained with $spp_{desired} = 4$ (desired slackness factor) for $m_{des} = 16$ and $spp_{desired} = 8$ for $m_{des} = 256$, where $spp_{desired} = m_{des}/n$.

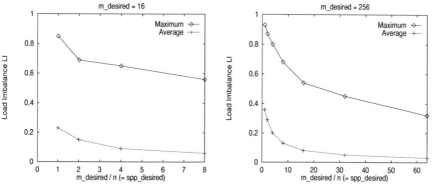

Figure 3. Maximal values observed and arithmetic average for the load imbalance as a function of n.

4. PARALLELIZATION VIA COMPETITION

Current trends lead away from special purpose, tightly coupled multicomputer systems like Intel Hypercube. Much more interesting and feasible seem to be clusters of workstations, connected by a local area network. Such a hardware configuration is readily available in many places. It features processing nodes with high processing power and comparatively large resources of local memory and disk space. The operating system (mostly UNIX) allows multitasking and multi-user operation. The underlying communication principle is message passing. Common data (for example, the formula to be proved) can be kept in file-systems which are shared between the processors (for example, by NFS). However, the bandwidth of the connection between the workstations is comparatively low and the latency for each communication is rather high.

Models of parallelism which are ideally suited for such networks of workstations must therefore obey the following requirements: (i) a small, limited amount of *necessary* communication between the processors for preserving completeness, and (ii) no dependency on short latencies. A parallel model which fulfills these requirements is *competition*: each processor tries to solve the entire problem, using different methods or parameters. As soon as one processor finds a solution, the entire system can be stopped. So, the only communication required are the start and stop messages.

Competitive parallel models have been studied in various approaches (cf. (Ertel, 1993) and (Suttner and Schumann, 1993) for competitive parallel theorem provers). In the following two sections, we describe three different parallel competitive models and their implementation: random competition with RCTHEO, competition by the exploration of parameter spaces with SiCoTHEO and strategy parallelism with p-SETHEO.

4.1. *Random Competition: RCTHEO*

Computational Model
RCTHEO is an acronym for Random Competition THEOrem prover. The name reflects the two basic ingredients of its execution model. Each competing prover process uses the same sequential SETHEO, but orders the choices for the Model Elimination extension steps during the search according to a random number sequence. Since each prover process uses a different sequence, the search spaces explored by the individual provers soon diverge due to the combinatorial explosion. Thus RCTHEO exploits a non-decompositional form of OR-parallelism. RCTHEO preserves the iterative deepening search strategy used by SETHEO.

Implementation
RCTHEO-II (Philipps, 1992) is based on SETHEO V2.63 and implemented in C using a RPC programming library (Corbin, 1991). It was evaluated on a network of 110 HP 9000/720 workstations.

Experimental Results
The performance of the RCTHEO model can be assessed quite easily in a sequential way, based on individual SETHEO runs. This, however, does not take into account the overhead due to startup, and termination. The overhead depends on the number of processors, but is mainly independent of the problem size. For RCTHEO-II, the overhead reported is in the range of 4 to 6 seconds (see (Philipps, 1992); variations are due to network traffic and processor utilization by other users). Table III shows results taken from (Ertel, 1993). They are based on individual SETHEO runs and exclude parallel overhead.

Table III. Speed-up s of randomized OR-branch selection with p processes with a fixed resource bound.

Problem	bound	$p = 25$	$p = 50$	$p = 100$	$p = 200$	$p = 400$
on-obvious	(-d 7)	40	58	72	87	-
10-queens	(-d 50)	13	25	50	98	195
s1	(-d 7)	13	25	63	240	435
lucas11	(-d 7)	13	40	108	141	-
ip1	(-i 19)	25	55	112	300	550

Assessment
The advantages of this scheme are that it is easy to simulate, easy to implement, and requires little communication. A theoretical speed-up prediction (compared to sequential random search) is possible if the run-time distribution of tasks is known. This, however, is usually not the case. For suitable applications good scalability is achieved.

4.2. *Parameter Competition: SiCoTHEO*

Computational Model

Given a sequential theorem proving algorithm[4] $A(P_1, \ldots, P_n)$ where the P_i are parameters which may influence the behavior of the system and its search (for example, completeness bounds or pruning methods). Then, a *homogeneous competitive theorem prover* running on n processors is defined as follows: on each processor p ($1 \leq p \leq n$), a copy of the sequential algorithm $A(P_1^p, \ldots, P_n^p)$ tries to prove the *entire* given formula. Some (or all) parameters P_i^p are set differently for each processor p. All processors start at the same time. As soon as one processor finds a solution, it is reported to the user and the other processors are stopped ("winner-takes-all strategy").

The efficiency of the resulting competitive system strongly depends on the influence of the parameter settings on the search behavior. The larger the difference, created by the values of P_i^p, the higher the probability that one processor finds a proof very quickly (if one exists, of course). Good scalability and efficiency can be obtained only if there are enough different values for a parameter, and if no good default estimation to set that parameter is known. Only then a large number of processors can be employed reasonably.

The prover variants introduced in this section compete on rather simple settings of parameters. Thus the system is called *Simple Competitive provers based on SETHEO* – SiCoTHEO (Schumann, 1996). SiCoTHEO-CBC competes via a combination of completeness bounds, and SiCoTHEO-DELTA via a combination of top-down and bottom-up processing. All SiCoTHEO versions explore *parameter ranges*.

Implementation

All versions of SiCoTHEO run on a network of UNIX workstations. The control of the proving processes, the setting of the parameters and the final assembly of the results is accomplished by the tool *pmake* (de Boor, 1989)[5], a parallel version of *make*. It exploits parallelism by exporting independent jobs to other processors. *Pmake* stops if all jobs are finished or an error occurs. In our "the winner takes all strategy" the system has to stop as soon as *one* job is finished. Therefore, we adapted SETHEO so that it returns "error" as soon as it found a proof. A critical issue in using pmake is its behavior with respect

---

[4] The definition of a parallel competitive system can easily be generalized to any search algorithm. An algorithm suitable for competition takes a problem as its input and tries to solve it. If a solution exists, the algorithm eventually must terminate with a message "solution found".

[5] This implementation of SiCoTHEO has been inspired by a prototypical implementation of RCTHEO.

to the load on the workstations: as soon as there is activity on workstations used by pmake, the current job will be aborted and restarted later. Therefore, the number of active processors and even the start-up times can vary strongly during a run of SiCoTHEO.

SiCoTHEO-CBC. The completeness bound, which is used for iterative deepening, determines the shape of the search space and therefore has a strong influence on the run-time the prover needs to find a proof. There exist many examples for which a proof cannot be found using iterative deepening over the depth of the proof tree within reasonable time. Whereas, iterative deepening over the number of inferences almost immediately reveals a proof, and vice versa[6]. In order to soften both extremes, SiCoTHEO explores a combination of the tableau depth bound d with the inference bound i_{max}. When iterating over depth d, the inference bound i_{max} is set according to $i_{max} = d^{\eta}$ where η is the mean length of the clauses. For our experiments, however, we have a slightly different approach by using a quadratic polynomial[7]: $i_{max} = \alpha d^2 + \beta d$

This polynomial approximates the structure of a tableau by a Taylor development with only the linear and quadratic terms. SiCoTHEO-CBC explores a set of parameters $\langle \alpha, \beta \rangle$ in parallel by assigning different values to each processor. For the experiments we selected $0.1 \leq \alpha \leq 1$ and $0 \leq \beta \leq 1$.

SiCoTHEO-DELTA. This competitive system affects the search mode of the prover. SETHEO normally performs a top-down search. The DELTA iterator (Schumann, 1994), on the other hand, generates small tableaux, represented as unit clauses in a bottom-up way during a preprocessing phase. These unit clauses are added to the original formula. Then, in the main proving phase, SETHEO works in its usual top-down search mode. The generated unit clauses now can be used to close open branches of the tableau much earlier, thus combining top-down with bottom-up processing. This decrease of the proof size can reduce the amount of necessary search dramatically. On the other hand, adding new clauses to the formula increases the search space. Thus, adding too many (or useless) clauses has a strong negative effect.

The DELTA preprocessor has various parameters to control its operation. Here, we focus on two parameters: the number of iteration levels l, and the maximal allowable term depth t_d. l determines how many iterations the preprocessor executes. In order to avoid an excessive generation of unit clauses, the maximal term depth (t_d) of any term in a generated unit clause can be restricted. For our experiments, we performed competition over l and t_d in

---

[6] This dramatic effect can be seen clearly in e.g. (Letz, R. et al., 1992), Table 3.

[7] $\alpha, \beta \in R_0^+$. For $\alpha = 0, \beta = 1$, we have inference-bounded search. $\alpha = \infty, \beta = \infty$ corresponds to depth-bounded search.

the range of $l \in \{1, 2, \ldots, 5\}$ and $t_d \in \{1, 2, \ldots, 5\}^8$. Furthermore, DELTA is configured in such a way that a maximum number of 100 unit clauses are generated to avoid excessively large formulae.

Experimental Results

All experiments with SiCoTHEO have been made with a selection of problems taken from the TPTP. All proof attempts (sequential and parallel) have been aborted after a maximal run-time of $T_{max} = 300s$ (on HP-9000/720 workstations); start-up and stop times are not considered in these experiments.

Table IV (the first group of rows) shows the mean speed-up values for the experiments with SiCoTHEO-CBC with various numbers of processors. The arithmetic mean $\bar{s}_a$, the geometric mean $\bar{s}_g$, and the harmonic mean $\bar{s}_h$ are shown for reference[9]. These figures can be interpreted more accurately when looking at the graphical representation of the ratio between T_{seq} and $T_{||}$, as shown in Figure 4. Each • represents a measurement with one formula. The dotted line corresponds to $s = 1$, the closely dotted line to $s = P$, where P is the number of processors. The area above the dotted line contains examples where the parallel system is *slower* than the sequential prover; dots below the closely dotted line represent experiments which yield a super-linear speed-up.

Table IV. SiCoTHEO: mean speed-up values for different numbers of processors P on 44 TPTP examples with $T_{seq} \geq 1s$.

		mean	$P = 4$	$P = 9$	$P = 25$	$P = 50$
SiCoTHEO-CBC	$\bar{s}_a$		61.21	77.30	98.85	101.99
SiCoTHEO-CBC	$\bar{s}_g$		5.92	12.38	18.18	19.25
SiCoTHEO-CBC	$\bar{s}_h$		2.12	3.37	4.34	4.41
			$P = 4$	$P = 9$	$P = 25$	
SiCoTHEO-DELTA	$\bar{s}_a$		18.39	63.84	76.50	–
SiCoTHEO-DELTA	$\bar{s}_g$		5.15	12.07	16.97	–
SiCoTHEO-DELTA	$\bar{s}_h$		1.43	2.92	3.71	–

---

[8] In order to overcome the negative effects of adding too many lemmas, we added a standard SETHEO to the competitors.

[9] In general, $\bar{s}_a$ yields values that are too optimistic, while $\bar{s}_h$ is decreased too much by single small values. Therefore, often $\bar{s}_g$ is considered as the best mean value for speed-up measurements.

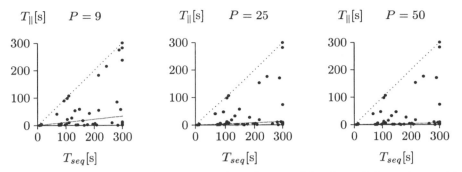

Figure 4. SiCoTHEO-CBC: Comparation of sequential and parallel run-times with different numbers of processors.

Figure 4 shows that even for few processors a large number of examples with super-linear speed-up exist. This encouraging fact is also reflected in Table IV which exhibits good average speed-up values for 4 and 9 processors. For our long-running examples and $P = 4$ or $P = 9$, $\bar{s}_g$ is even larger than the number of processors. This means that in most cases, a super-linear speed-up can be accomplished. Table IV furthermore shows that with an increasing number of processors, the speed-up values also increase. However, for larger numbers of processors (25 or 50), the efficiency $\eta = s/P$ decreases. This means that SiCoTHEO-CBC obtains its peak efficiency with about 15 processors and thus is only moderately scalable.

The experiments with SiCoTHEO-DELTA have been carried out with the same set of examples. In general, the speed-up figures of SiCoTHEO-DELTA (Table IV, lower section) show a similar behavior as those for SiCoTHEO-CBC.

Assessment

Although extremely primitive in its implementation, the SiCoTHEO system showed impressive speed-up results. Looking at examples with a run-time of more than 1 second and a comparatively low number of processors, a super-linear speed-up can be obtained in most cases. However, the scalability of SiCoTHEO is rather limited. For SiCoTHEO-CBC this is due to the fact that with larger numbers of processors the variations of α and β are too small to cause enough changes in run-time behavior. The coarse controlling parameters of DELTA limit the scalability of SiCoTHEO-DELTA. Speed-up and scalability could be increased substantially if one succeeded in producing a greater variety of preprocessed formulae. Furthermore, a combination of SiCoTHEO-CBC with SiCoTHEO-DELTA could be of great interest.

4.3.   Strategy Competition: p-SETHEO

Computational Model
Automated theorem provers use search strategies. Unfortunately, there is no single strategy which is uniformly successful on all problems. This motivates the *competitive* use of different strategies. In order to be successful with such an approach, the strategies must satisfy the following two conditions:

1. The function of the solved problems per unit time must be less than linear, i.e., with each new time interval fewer new problems are solved.

2. The strategies must be *complementary*, i.e., the set of problems solved in a given time must differ as much as possible.

Strategy parallelism. The selection of more than one search strategy in combination with techniques to partition the available resources (such as time and processors) with respect to the actual task defines a new parallelization method, which we call *strategy parallelism*. (Distributed) competitive agents try to solve the same problem, but use different methods. Often they traverse the same search space, but in a different order. Whenever an agent finds a solution, all other agents are stopped. With this method it is intended that the strategies traverse the search space in such a manner that, in practice, repeated consideration of the same parts can be avoided. In the pure form of strategy parallelism which is discussed here, there is no interaction between the competitive agents. This enables even completely different search paradigms (for example, resolution and model elimination) to be combined.

A combination of more than one strategy increases the chances of success. Limitations of resources, such as time or processors, enforce efficient use of these resources by partitioning the resources adequately among the involved strategies. This leads to an optimization problem.

The complexity of strategy allocation. In this article, we formulate the much easier *sequential* version of the problem, in which only one processor is available.[10] Given a set of (training) problems, a set of available strategies and a time limit, we want to determine an optimal distribution of time resources to each strategy, which solves a maximal number of problems from the training set.

GIVEN a set of *strategies* $S = \{s_1, \ldots, s_m\}$ with $s_i : F \to nat^+ \cup \{\infty\}$ (solution time) on a set of *problems* $F = \{f_1, \ldots, f_n\}$, and a time-limit t.

---

[10] The competitive strategies then need to be time-sliced on the processor. We do not consider the additional task of processor scheduling.

FIND nonnegative time resources $t_1, \ldots, t_m$ for each strategy such that $\sum_{i=1}^{m} t_i \leq t$, and $| \bigcup_{i=1}^{m} \{f : s_i(f) \leq t_i\} |$ is maximal.

Unfortunately, already the decision variant[11] of our problem is *strongly NP-complete*[12]. Therefore, in practice, the determination of an optimal solution for the full problem will not be possible, at least not on larger sets and with classical methods. One reasonable possibility is to use a gradient procedure (as it is currently done to select p-SETHEO's strategies).

Implementation
The implementation of p-SETHEO is an environment for the PVM based (Geist, A. et al., 1994) strategy parallel execution of theorem provers. p-SETHEO can be configured very easily by an ASCII file containing information about usable hosts, the maximum load allowed per processor, and the strategy allocation for the competitive strategies. Currently, all contained theorem provers are variants of SETHEO obtained by modifying the parameter settings. Because of the generic layout of the p-SETHEO controlling mechanism, new strategies and even new theorem provers can be integrated very easily. The implementation of p-SETHEO consists of approximately 1000 lines of code and is written in Perl, PVM, C, and shell tools.

The implementation of p-SETHEO works as follows:

1. Select a set of triples of strategies, computation times, and assigned processors according to the problem to be solved and the usable resources in terms of processors, processor load, and time.

2. Perform the preprocessing steps needed for all selected strategies using the available parallel resources[13] (reordering, equality treatment, lemma generation, etc.)

3. Start all prover strategies using the available parallel resources. The first prover that finds a proof stops the others.

---

[11] Determine, if there is a resource schedule which solves at least k problems for a given integer k.

[12] Provide a polynomial reduction of the strongly NP-complete *minimum cover problem* to the decision variant of our problem.

[13] The prover was told which processors it can use, and the maximal number of jobs allowed on each processor. The product of number of processors and allowed load gives the number of processes allowed to run simultaneously.

Experimental Results

CADE-14 competition configuration. In order to investigate the potential of strategy parallelism in practice, we have evaluated the method on different strategies of SETHEO. As training set F, we have taken the 420 problems in clausal form of the TPTP (V.2.0.0) that have been used for the theorem prover competition at CADE-14[14]. For a given (non-optimized) set of 20 strategies S, the solution times $s(f)$, for any $f \in F, s \in S$ were computed with a time limit T of 500 seconds per problem. The best single strategy solves 211 of the problems. The number of problems soluble with any of the 20 strategies is 266. After selecting a set of strategies and their corresponding execution times the resulting prover configuration solves 246 problems within the given time $T = 500$ seconds with time-slicing the strategies on a single processor.

A simple TPTP configuration. A further proof of the performance of the strategy parallel approach is the combination of four fixed strategies on four processors: iterative deepening on tableau depth, with and without fold-up, and the iterative deepening with the weighted depth bound, also with and without fold-up[15]. The experiments were performed on a test set of 230 tasks[16]. The results are shown in Table V[17].

Table V. p-SETHEO: Comparison with single strategies.

configuration	solutions	part (%)	time (s)	part (%)
-dr	151	66	106128	377
-dr with -foldup	159	69	91561	325
-wdr	122	53	128587	456
-wdr with -foldup	161	70	87321	310
p-SETHEO	230	100	28168	100

It can be seen that p-SETHEO solves on 4 processors within only 22...32% of time 43...89% more problems than the sequential provers. Even if we use

---

[14] A selection of problems which already have been solved by at least one existing ATP system, but at most by a small number of provers.

[15] For details on the parameters see (Letz, R. et al., 1994).

[16] The 230 tasks from TPTP, which can be solved by at least one of the four configurations within 1000 seconds but which are solved by at most two configurations within less than 20 seconds.

[17] The provers tried to solve each problem within at most 1000 seconds. p-SETHEO added the times of the best competitor.

parallel work instead of time, p-SETHEO solves significantly more problems with nearly the same costs.

Assessment.

In theorem proving, the system developer or advanced user can often tune a sequential system to a given problem by using his or her experience. This is not possible if the theorem prover is integrated into a larger interactive proof environment such as ILF (Dahn, B. I. et al., 1997). In this case, the configuration *must* be done automatically. Here, strategy parallelism is a good approach to fulfill these requirements.

Here, we do not discuss a number of issues of importance for strategy parallelism. These topics describe future research:

1. How can a set of strategies be obtained that solve as many problems as fast as possible and which are as *complementary* as possible?
2. Often, strategies are successful for a certain class of problems. For example, unit equality problems need different treatment than problems without equality. If such features can be identified, the selection of strategies can be made more specific and hence more successful.
3. The success of the selected strategies depends on the given training set. How do we obtain a training set which is representative for the considered domain of problems?
4. The number of sensible strategies which are successful and as complementary as possible seems to be bounded. This restricts the scalability of strategy parallelism. Can a systematic method be found for producing as many successful and complementary strategies as we want? Such a method will probably need to contain randomized elements.

5. COOPERATION

Sequential search procedures often employ large sets of heuristics and refinements of the underlying calculus to prevent unnecessary search, e.g. to avoid redundant search steps. The exchange of information between synchronously running sub-provers may cause synergetic effects by providing additional knowledge. This idea leads to cooperative theorem proving.

An automated theorem prover is *cooperative*, if and only if it offers *on request* and *immediately* all its data and results to other provers. *Immediately* means that the prover utilizes the available bandwidth of the communication channels and handles the request with the same priority as its own proof attempt. *On request* includes that other provers issue requests. Furthermore, only the requested data is transmitted.

Basically, there are two methods of exchanging data between theorem provers: on a *request* base (a prover asks for information) and on an *offer* base (a prover offers its results).

5.1. *A Classification of Cooperation in Theorem Proving*

We discuss a classification of the problems that are to be solved when constructing a cooperative parallel theorem prover. The aspects to be considered can be divided into two parts: first, *how* can cooperating provers work together, i.e., which topics concerning the *system architecture* have to be considered, and second, *in which way* they can work together, i.e. which *kinds of cooperation* can occur?

System Architecture
Homogeneous and Heterogeneous Systems. Most parallel provers ordinarily use the implementation of the inference machine of a sequential prover with minor changes. Their sub-provers are united by partitioning the search tree in a completeness or completeness based manner as previously described, or they can use different search strategies in their different instances. Due to the use of the same inference machine and an equal encoding of formulae and control parameters, the expected additional costs for the implementation of an information exchange in such *homogeneous systems* are relatively low.

Heterogeneous systems probably need more effort during implementation due to the necessary syntactical transformations and different semantics. In the interactive proof system ILF (Dahn, B. I. et al., 1997), the main part of implementation required for communication is syntactical transformation and adaptation of theories. The context needed for different systems will significantly increase the amount of information to be exchanged. Heterogeneous systems are justified only if the connection of the systems leads to a significant increase of the performance of the overall system. The performance can increased, e.g., by integrating a prover for equational problems into a system that is poor on equations. Another possibility is the integration of provers applying meta-mathematical knowledge.

Synchronous and asynchronous exchange of information. Using the *synchronous* communication mode, all partners are informed on the situation of their companion. Generally, a transmitted message will be confirmed. That means, at least one of the exchanging partners waits until the others are ready to receive or to send a message. The synchronous mode of information exchange has the advantage that all involved processes mutually know their status of information, i.e., if a message was written, it is read at the same predetermined time in the program scheme.

When communicating *asynchronously*, the sender cannot assume that the receiving prover got the message at the predetermined moment in the program cycle. But assuming the *message passing concept*, it is guaranteed that at least the temporal sequence of messages is preserved. Furthermore, it is guaranteed that all messages reach the receiver, if this process still exists. This method has the advantage that no time is wasted by waiting for a communication partner. It has the disadvantage that usually information may be not available at the moment it is needed.

Hierarchical structures of sub-provers. All sub-provers can be *on the same hierarchical level*. Because the processes do not need to take their place in a hierarchical order during the initial phase, the start of the whole proof system is very easy in this model. Thus, no communication is needed for that purpose. But using this model, processes that want to communicate must obtain knowledge about the way they can do so, if the operating system does not offer such support. The sub-provers can be arranged in a *hierarchical structure*. If a prover generates new sub-tasks, it will create the subordinated provers and transmit the tasks to them. The information exchange can be done easily in this model but the control of the globally used resources is complex.

The *combination* of different structures following both models is also possible. Nevertheless, each architecture needs facilities for information exchange, i.e., it should be possible to group processes and perform broadcasts to such groups.

Combination of goal oriented and saturating provers. It is one aim of using parallelization of theorem provers to decompose the search space for finding a proof, and to treat the parts of that search space at the same time in parallel. Using cooperative concepts it is possible to exchange information about solved subgoals to prevent redundancies, specifically, solving a subgoal more than once. Furthermore, it is possible to partition the search space in a *horizontal* manner, i.e. to combine *bottom-up* (generating, resolution like) with *top-down* (goal oriented, model elimination like) proof procedures. An example for such a combination is SiCoTHEO-DELTA (see Section 4.2).

Kinds of Cooperation

Different strategies of search. In many cases, the provers involved in a parallel proof system partition the search space. Unfortunately, it is possible that the same search strategy produces similar subgoals at the same time on similar proof tasks. That means a loss of chance that intermediate results solved by one prover could be interesting for another one: one prover having already solved that subgoal itself cannot use the external results. If different search strategies are used by the sub-provers it is possible that intermediate results generated by one prover can be re-used by one or more of the others for their further work. Therefore, it can be accepted that some of these provers work with *incomplete search strategies* if they solve tasks from their specific domain especially fast and efficiently. But it should be guaranteed that the whole system remains *fair* and *complete*. So sub-provers with specialized strategies can be used, for instance, to generate lemmas or to deal with subgoals belonging to special problem classes such as equality problems.

Cooperation and competition. The relation between cooperation and competition has been discussed in (Fronhöfer and Kurfeß, 1987). Often, competition is the basic concept of existing parallel provers (see Section 4). The reasons are the low cost of implementation and the small amount of inter-process communication. *Cooperation and competition are not necessarily contradictory.* If competitive provers exchange information, they lose time for their own work, but using the results of their competitors they may solve their tasks faster. An example for such a synergetic effect is the *Teamwork Method* (Denzinger, 1995). This concept includes the competition of some provers which exchange their intermediate results periodically. Cooperation *without* competition entails the risk that sub-provers with bad results on a special class of tasks can decrease the overall performance. This is possible if useless results increase the amount of information to be processed. Competitive concepts can eliminate such provers from the actual configuration of the system.

Exchange and optimization of configurations and control information. Speaking about cooperation in the context of parallel theorem provers, one has in mind the exchange of formulae or sets of formulae. But a further possibility is the exchange of all information that describes *how* an inference machine works such as, e.g., *control parameters*, *used heuristics*, *search strategies*, and *inference rates*. It is useful to terminate the sub-provers that are less successful in the previous time period and to start new provers with the control information similar to the successful ones. It is also imaginable that some running provers get the parameters to change their behavior in the intended sense. It is noteworthy that, in this model, information needs to be exchanged only infrequently. It has to be considered which criteria are suitable in determining the quality of the control configuration of a prover. A measure for that purpose can be, for instance, the depth of the proof structure relative to the other sub-provers or the number and quality of the generated lemmas. Using such criteria it should be considered that the resulting system must have a *fair* search strategy to save the completeness of the whole joined system. p-SETHEO (see 4.3) will be improved using a similar concept in the near future.

Exchange of intermediate results (lemmas). In the previous paragraph we have described the exchange of configurations. Now, we discuss exchange of proved intermediate results (proved formulae). Considering their structure, the formulae to be integrated will mostly be *unit clauses*. But it is also possible that *more complex formulae* have to be transmitted.

A transmission of all possible lemmas without filtering, even to a subset of the involved sub-provers, will cause an overloading of the network, and even worse, an overloading of the accepting sub-prover due to the search space explosion. Thus, the candidate lemmas must be evaluated to decide if they can be exchanged. In order to get a *measure of a lemma* one can use information on *the syntactical structure* of the lemmas (for example the high generality), and information on the *derivation of the lemmas* (for example the number of inferences). If provers can ask for goals that are important for their work, then the existence of such a question for a formula should be a criterion for a lemma. These questions should be filtered analogously to the lemmas to avoid network overload.

Exchange of failure information. In the paragraphs before, we discussed the exchange of information about successful events (successful configurations, lemmas). These informations can be considered to be *positive* knowledge. Furthermore, *negative* knowledge can be exchanged, e.g., the provers can communicate about what they cannot prove. Such messages should include the conditions under which a proof failed, i.e., the parameter settings and search bounds. This concept also demands a strong selection, as already explained considering lemma and request generation. It must be realized that not every *backtracking* step generates negative information. It should be possible to use analogous criteria as in the case of lemma generation.

5.2. *The Cooperative Prover CPTHEO*

Computational Model

The cooperative prover CPTHEO uses techniques of SPTHEO for the scheduling of tasks. In addition to SPTHEO, the evaluation of the search space of the sub-tasks takes place in components consisting of parallel provers itself. In the following, we describe only a single component. The lemma/request exchange within such a single component is enriched by a lemma exchange between the components, but the mechanisms to evaluate these lemmas are similar to those within the components.

Initially, one component consists of a lemma generating prover similar to DELTA (Schumann, 1994) and a top-down prover which generates requests of unit literals which can help to prove the sub-task the component has to solve. Both data streams are filtered to eliminate redundancies. Then, matches between lemmas and proof requests are searched. These matches are useful candidates to help proving the sub-task of the component. Small portions of the thus generated and ordered set of lemmas can be added to the given problem of the component. These newly created proof tasks form a queue. As many of those tasks in the queue as possible are given a certain user defined time to find the proof. After that time their execution is terminated.

The filters of the sub-task generators and the lemma and proof request generators consider the following: *identical formulae, subsumed formulae,* and *tests with models* or *model fragments* of the considered theory. Filters only delete multiply occurring and obsolete formulae. Referees rank the proof requests as well as lemmas by labeling them with measures to order them with respect to their expected relevance for the proof. The measures depend on: the *generality* of a literal, the derivation *cost* of a literal, the relatively *isolated position* of a subproof leading to a literal, the *multiple usability* of a literal in the already existing parts of the proof, and the *similarity* of the generated facts to the task to be proved.

Implementation

CPTHEO like all the other provers described in this article, uses the SETHEO abstract machine. In addition, it uses PVM (Geist, A. et al., 1994), C, Prolog, and Perl.

Evaluation

To determine the influence of the cooperation on the proof process, we present the results of experiments with a single top-down/bottom-up cooperating component with only one task in the generated task queue. Thus we eliminate the effects of partitioning from the underlying SPTHEO model. The new task was generated by adding the first lemmas to the problem (see the number in column 2 of Table VI). The results are compared with a sequential SETHEO using the same search strategy. Some typical tasks from the TPTP which can be solved cooperatively within 50 seconds are shown. The time limit of the sequential prover for finding a proof was 1000 seconds.

Assessment

Cooperation between sub-provers can achieve very high speed-ups. The main problem for cooperative theorem provers is the intelligent reduction of the produced data to the part which is relevant for the proof. That selection can fail, and no then improvement is reached. But using cooperation as a useful *additional* feature, we lose at least no performance. Methods and techniques for the information assessment and selection still need further research. The combination of cooperation with other parallelization paradigms like static partitioning in the full model of CPTHEO increases the scalability of cooperative approaches.

Table VI. Experimental results with CPTHEO. Some parts of the system are still inefficiently implemented in PROLOG, so we selected inferences (micro-steps) as the measure unit for the costs.

task	added lemmas	coop. proof micro-steps	seq. attempt micro-steps	sequential proof	speed-up
BOO003-1	3	640	13489	yes	21.1
CAT004-4	5	4069	73231	yes	18.0
COL061-1	3	611	23221	no	>38.0
COL066-2	3	177	2453	yes	13.9
FLD010-1	3	215	39908	no	>185.6
FLD031-5	3	133	2377	yes	17.9
GEO017-2	5	384	6554	yes	17.1
GEO026-2	3	259	121505	no	>469.1
HEN003-5	5	418	7341	yes	17.6
LCL080-1	10	660	13862	no	>21.0
LCL108-1	5	40	549	yes	13.7
NUM180-1	5	836	342886	no	>410.2
PUZ001-2	3	1452	14408	yes	9.9
RNG004-3	5	1764	241877	no	>137.1
ROB002-1	3	213	1408	yes	6.6
SYN310-1	3	247	56953	yes	230.6

6. CONCLUSIONS

In this article, we have presented six parallel theorem provers based on the sequential SETHEO system. Whereas PARTHEO and SPTHEO belong to the class of systems which partition the search space, RCTHEO, SiCoTHEO, and p-SETHEO perform competitive parallel computation. CPTHEO introduces cooperation into both paradigms. Each of the systems has been implemented and evaluated. Only with an extensive evaluation (sometimes combined with simulation), the advantages and shortcomings of each approach and system can be assessed.

In particular, when looking at applications (see Part III.2 in this series), a prover should reflect important characteristics of the proof tasks to be solved. Here, we can identify three major types: *hard problems* involve the exploration of huge search spaces. On the other hand, the proofs (i.e., number of inference steps in the proof) are often comparatively short. Classical theorem proving examples belong to this class. In particular in combination with interactive systems tasks are comparatively easy, but a *short answer-time* is of great importance. Often, only a few seconds of run-time are available. A third category comprises *logic-programming* type problems with comparatively little search, but long proofs.

As shown in the experiments, competitive provers seem to perform better in cases where short answer-times are expected. In many cases, one of the competitors is running a strategy (or parameters) which results in a fast proof. Complex proofs, on the other hand, are normally handled by partitioning systems in a better way. This observation also reflects the human approach to larger problems: first, the problem is broken down into several parts which (hopefully) can be handled more easily than the original problem. Here, additional cooperation can greatly enhance the prover's performance.

In general, we found that parallel search for a proof carries a vast potential. In many cases, even super-linear speed-up values can be obtained. Together with the reasonable to good scalability of our models (to about 50-300 processors) a tremendous increase in performance can be obtained.

Our experience with the parallel provers showed that those systems are very hard to use in practice. Most provers require elaborate knowledge about the underlying model and SETHEO. Furthermore, most parallel provers are outdated by new versions of the sequential prover or developments in hardware. Here, only a rigorous application of software engineering methods (modularization, reuse, version control) can tackle the first problem.

For practical applications, the start-up time of a parallel prover is a critical issue. This means that a theorem prover should work in parallel only if the anticipated run-time for the problem is considerably larger than its start-up time (usually up to several seconds). However, this problem can be handled by always starting a sequential prover several seconds before a parallel proof attempt.

The design of a parallel system must also reflect the available hardware architecture. Since our early developments we focus on multi-processing with message passing instead of shared memory. This decision allowed us to easily switch over to clusters of powerful workstations, connected by relatively fast networks. Although the communication bandwidth is limited and the latency is rather high, such hardware configurations are readily available at many locations. This enormously facilitates the application of parallel theorem provers. Modern developments of closely coupled multiprocessing within workstations (with shared memory) open up new possibilities to enhance our already powerful parallel models and will result in parallel high-performance theorem provers which we expect to be able to handle proof tasks arising from real-world applications.

Acknowledgments

This work is supported by the Deutsche Forschungsgemeinschaft within the Sonderforschungsbereich 342 subproject A5: Parallelization in Inference Systems (PARIS).

REFERENCES

3L Ltd.: 1988, *Parallel C – User Guide*. 3L Ltd., Livingston, Scotland.

Bose, S. et al.: 1989, 'Parthenon: A Parallel Theorem Prover for Non-Horn Clauses'. In: *Proc. LICS-4*. pp. 1–10, IEEE.

Corbin, J.: 1991, *The Art of Distributed Applications*. Springer–Verlag.

Dahn, B. I. and A. Wolf: 1996, *Natural Language Presentation and Combination of Automatically Generated Proofs*, Vol. 3 of *Applied Logic Series*, pp. 175–192. Kluwer Academic Publishers.

Dahn, B. I. et al.: 1997, 'Integration of Automated and Interactive Theorem Proving in ILF'. In: *Proc. CADE-14*, Vol. 1249 of *LNAI*. pp. 57–60, Springer–Verlag.

de Boor, A.: 1989, 'PMake – A Tutorial'. Berkeley Softworks, Berkeley, U.S.A.

Denzinger, J.: 1995, 'Knowledge-Based Distributed Search Using Teamwork'. In: *Proc. ICMAS-95*. pp. 81–88, AAAI-Press.

Denzinger, J. et al.: 1997, 'DISCOUNT. A Distributed and Learning Equational Prover'. *JAR* **18**(2), 189–198.

Ertel, W.: 1993, *Parallele Suche mit randomisiertem Wettbewerb in Inferenzsystemen (in German)*, Vol. 25 of *DISKI*. Infix-Verlag. Ph.D. thesis, Munich University of Technology.

Fronhöfer, B. and F. Kurfeß: 1987, 'Cooperative Competition: A Modest Proposal Concerning the Use of Multi-Processor Systems for Automated Reasoning'. Technical report, Munich University of Technology.

Geist, A. et al.: 1994, *PVM: Parallel Virtual Machine. A Users' Guide and Tutorial for Networked Parallel Computing*. MIT Press.

Kurfeß, F.: 1990, 'Parallelism in Logic – Its Potential for Performance and Program Development'. Ph.D. thesis, Munich University of Technology.

Letz, R. et al.: 1992, 'SETHEO: A High-Performance Theorem Prover'. *JAR* **8**(2), 183–212.

Letz, R. et al.: 1994, 'Controlled Integration of the Cut Rule into Connection Tableau Calculi'. *Journal Automated Reasoning (JAR)* **13**(3), 297–337.

Loveland, D. W.: 1978, *Automated Theorem Proving: a Logical Basis*. North–Holland.

McCune, W.: 1990, 'Otter 2.0'. In: *Proc. CADE-10*, Vol. 449 of *LNAI*. pp. 663–664, Springer–Verlag.

Philipps, J.: 1992, 'RCTHEO II, ein paralleler Theorembeweiser (in German)'. Fortgeschrittenenpraktikum, Munich University of Technology, Computer Science Department.

Schumann, J.: 1991, 'Efficient Theorem Provers based on an Abstract Machine'. Ph.d. thesis, Munich University of Technology.

Schumann, J.: 1994, 'DELTA – A Bottom-up Preprocessor for Top-Down Theorem Provers, System Abstract'. In: *Proc. CADE-12*, Vol. 814 of *LNAI*. pp. 774–777, Springer–Verlag.

Schumann, J.: 1996, 'SiCoTHEO: Simple Competitive Parallel Theorem Provers'. In: *Proc. CADE-13*, Vol. 1104 of *LNAI*. pp. 240–244, Springer-Verlag.

Schumann, J. and M. Jobmann: 1994, 'Analysing the Load Balancing Scheme of a Parallel System on Multiprocessors'. In: *Proc. PARLE 94*. pp. 819–822, Springer–Verlag.

Schumann, J. and R. Letz: 1990, 'PARTHEO: a High Performance Parallel Theorem Prover'. In: *Proc. CADE-10*, Vol. 449 of *LNAI*. pp. 40–56, Springer–Verlag.

Sutcliffe, G. et al.: 1994, 'The TPTP Problem Library'. In: *Proc. CADE-12*, Vol. 814 of *LNAI*. pp. 252–266, Springer–Verlag.

Suttner, C.: 1995, *Parallelization of Search-based Systems by Static Partitioning with Slackness*, Vol. 101 of *DISKI*. Infix-Verlag. Ph.D. thesis, Munich University of Technology.

Suttner, C. and J. Schumann: 1993, 'Parallel Automated Theorem Proving'. In: *Proc. PPAI-93*. pp. 209–257, Elsevier.

CHAPTER 11

MASSIVELY PARALLEL REASONING

1. INTRODUCTION

From its beginning, research in the field of connectionist systems[1] has always been concerned with the integration of symbolic and connectionist computation (McCulloch and Pitts, 1943). Although connectionist systems inherently enjoy many desirable properties like massive parallelism, robustness, graceful degradation, context–sensitivity as well as the ability to learn and adapt (see e.g. (Feldman and Ballard, 1982)), they are heavily criticized on the grounds that until now they cannot adequately represent and reason about structured objects and structure–sensitive processes (see e.g. (McCarthy, 1988; Fodor and Pylyshyn, 1988)).

Representing and reasoning about structured objects and structure–sensitive processes is the prime task for which (predicate) logic has been developed. Today logic is a well–understood, still very active research area and many powerful deductive systems for computing the logical consequence relation are known (see e.g. (Bibel, 1993; Bibel and Eder, 1993)). But many deductive systems are inadequate in the sense of (Bibel, 1988) since, roughly speaking, they occasionally solve simple problems much more slowly than more difficult ones. As shown in (Beringer and Hölldobler, 1993) for a deductive system to be adequate it must be massively parallel since otherwise it cannot adequately handle problems which are parallelizable. Besides this need for a massively parallel implementation, deductive systems are also heavily criticized for their brittleness; some noisy data may lead to a complete and sudden breakdown of the whole system. There are other problems as well such as the problem of learning and adapting a *good* heuristic to guide the search for a proof.

When comparing the respective strengths and weaknesses of connectionist and deductive systems they look like a perfect match — if one would only know how to amalgamate the two systems.

There is a straightforward relation between propositional logic and symmetric networks. E.g., Pinkas showed that finding a global minimum of a symmetric network corresponds precisely to finding a model for a propositional

---

[1] In this article we do not distinguish between *connectionist systems* and *artificial neural networks*. We prefer the term *connectionist system* as it expresses the fact that we are dealing with a computational model, where most of the knowledge is encoded in the connections between simple computing units.

W. Bibel, P. H. Schmitt (eds.), Automated Deduction. A basis for applications. Vol. II
© 1998 *Kluwer Academic Publishers. Printed in the Netherlands*

logic formula and vice versa (Pinkas, 1991). However, in a symmetric network one unit is updated and eventually flipped at a time as otherwise the network is not guaranteed to converge to a global minimum. This kind of sequential behavior is inadequate for certain propositional logic formulas and we will show in Section 3 how this problem can be solved.

The development of connectionist models which go beyond propositional logic has turned out to be extremely difficult, and no such connectionist model is known so far. Besides technical problems like the variable binding problem or the problem of representing potentially infinite structures, the difficulties are ultimately linked to the fact that predicate logic is undecidable. Typically, connectionist models aiming at going beyond propositional logic are structured connectionist networks using local coding (see e.g. (Lange and Dyer, 1989; Shastri and Ajjanagadde, 1993; Hölldobler, 1993)). To handle the undecidability of predicate logic, however, such networks have to be able to extend their topology during the inference process. Although recruitment learning (Feldman, 1982) can potentially cope with such problems, we are unaware of any application of recruitment learning or any other learning technique to learn the complex structures needed to represent the additional copies of the rules or clauses. Moreover, structured connectionist systems using local encodings are in many respects as brittle as conventional deductive processes.

To circumvent these problems, many researchers have proposed to build hybrid systems, where certain tasks like classification or rule extraction are performed by connectionist modules, whereas other tasks like deduction are performed by standard symbolic modules (see e.g. (Palm et al., 1991)). Although this is certainly a short term solution, it does not solve the real problems such as the brittleness of conventional logic–based systems. To paraphrase Smolensky, we need to *find ways of naturally instantiating the power of symbolic computation within fully connectionist systems* (Smolensky, 1988).

What requirements must a deductive system fulfill so that a massively parallel, connectionist implementation is feasible? What kind of deductive processes satisfy these requirements? What kind of connectionist model should be used to implement such deductive processes? These questions are investigated in Section 4. Guided by the respective answers, we then approach a massively parallel model generation process. We formally show that the meaning operator T_P associated with normal logic programs can be approximated arbitrarily well by feed–forward networks. For constructing such networks, we finally introduce a new finite representation of interpretations and a function $\widetilde{T}_P$ which simulates T_P on top of this representation. At the end we discuss our results and point out further research in Section 5.

2. Preliminaries

We assume the reader to be familiar with notions and notations concerning logic programs (see e.g. (Lloyd, 1987)). However, we briefly recall basic notions and notations concerning metrics following (Fitting, 1994) and connectionist models following (Feldman and Ballard, 1982).

2.1. *Metrics*

A *metric* or *distance function* on a space M is a mapping $d : M \times M \to \mathbf{R}$ such that $d(x,y) = 0$ iff $x = y$, $d(x,y) = d(y,x)$, and $d(x,y) \leq d(x,z) + d(z,y)$. Let (M,d) be a metric space and $S = s_1, s_2, \ldots, s_i \in M$, be a sequence on M. S *converges* if $\exists s \in M : \forall \varepsilon > 0 : \exists N : \forall n \geq N : d(s_n, s) \leq \varepsilon$. S is *Cauchy* if $\forall \varepsilon > 0 : \exists N : \forall n, m \geq N : d(s_n, s_m) \leq \varepsilon$. (M,d) is *complete* if every Cauchy sequence converges. A mapping $f : M \to M$ is a *contraction* on (M,d) if $\exists 0 < k < 1 : \forall x, y \in M : d(f(x), f(y)) \leq k \cdot d(x,y)$.

Consider a level mapping for a logic program P and M_P be the space of interpretations of P. An *associated* distance function d_P on M_P is defined as follows. Let v and w be two interpretations. If $v = w$ then $d_P(v,w) = 0$. Otherwise, $d_P(v,w) = \frac{1}{2^n}$, where v and w differ on some atom A of level n but agree on all atoms of lower level.

2.2. *Connectionist Models*

A *connectionist network* consists of a set of units and connections between these units. A *unit* U is characterized by its *potential* $p \in \mathbf{R}$, its *value* v, and its *input vector* $(i_1, \ldots, i_n)$. The units are connected via a set of directed and weighted connections. If there is a connection from unit U_j to unit U_k, then w_{kj} denotes the weight associated with this connection and $w_{kj} v_j$ denotes the input i_j received by U_k from U_j. The potential p_k and value v_k of the unit U_k are computed wrt to an *activation* and an *output* function respectively. An activation function $\phi : \mathbf{R} \to [0,1]$ is called *sigmoidal* if it is non–decreasing, $\lim_{\lambda \to \infty}(\phi(\lambda)) = 1$, and $\lim_{\lambda \to -\infty}(\phi(\lambda)) = 0$. A unit is said to be a *binary threshold* unit if its potential and output are computed as follows, where j iterates through the set of units which have a connection to U_k, $\theta_k \in \mathbf{R}$ is the *threshold* of U_k and t denotes linear time.

$$p_k(t) = \sum_j w_{kj} v_j(t-1) \qquad v_k(t) = \begin{cases} 1 & \text{if } p_k(t) > \theta_k \\ 0 & \text{if } p_k(t) < \theta_k \\ v_k(t-1) & \text{if } p_k(t) = \theta_k \end{cases}$$

The re–computation of a unit's potential and output is called an *update*. If the unit changes its output value, then this is called a *flip*. Units can be updated synchronously or asynchronously.

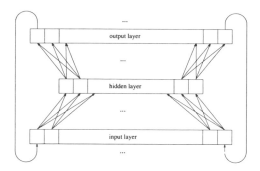

Figure 1.   Sketch of a 3–layered recurrent network.

A *layer* is a vector of units. An *n–layered feed–forward network* (*n*–FFN) F consists of an *input* layer, $n-2$ *hidden* layers, and an *output* layer, where $n \geq 2$. Each unit occurring in the i–th layer is connected to each unit occurring in the $i+1$–st layer, $1 \leq i < n$. All units are updated synchronously. Let r and s be the number of units occurring in the input and output layer, respectively. A multi–layered feed–forward network F computes a function $f_F : \{0,1\}^r \to \{0,1\}^s$ as follows. The input vector is presented to the input layer at time t_1 and propagated through the hidden layers to the output layer. At each time point all units update their potential and value. At time t_n the output vector is read off the output layer.

An *n–layered recurrent network* N consists of an *n*–FFN such that the number of units in the input and output layer are identical. Furthermore, each unit in the k–th position of the output layer is connected with weight 1 to the unit in the k–th position of the input layer, where $1 \leq k \leq n$ and n is the number of units in the output (or input) layer. Figure 1 shows a 3–layered recurrent network.

A *Hopfield (neural) network (HNN)* consists of symmetrically connected binary threshold units, i.e. $w_{ij} = w_{ji}$. We allow for so–called *higher–order connections*, i.e. connections that connect more than two units. We especially consider connections of order 3 with weights of the form w_{ijk}. In this case, the equation to compute the potential of a unit k evolves to

$$p_k(t) = \sum_{i<j} w_{ijk} v_i(t-1) v_j(t-1) + \sum_j w_{kj} v_j(t-1).$$

HNNs with 3–fold connections are suitable for representing propositional formulas in conjunctive normal form with at most three literals per clause (3–CNF). The 3–SAT problem then is the problem of determining the satisfiability of a propositional formula in 3–CNF and is known to be NP–complete.

The behavior of a HNN can be characterized by an energy function (Hopfield, 1982). HNNs with units $U_1, \ldots, U_n$ and higher–order connections of order 3 or less can be described by the energy function

$$(11.1) \quad E(v_1, \ldots, v_n) = - \sum_{i<j<k} w_{ijk} v_i v_j v_k - \sum_{i<j} w_{ij} v_i v_j + \sum_i \theta_i v_i,$$

In the standard Hopfield model, units are updated asynchronously, but sequentially, i.e. at each time step only one unit updates its output. The network has reached a *stable state*, if no unit has changed its output value during its last update. If a stable state is reached, the computation may be stopped as the network will not further change its state as long as there is no interference from "outside". If updating is done asynchronously, it can be shown that the HNN will eventually reach a stable state. Such stable states of an HNN correspond directly to the local minima of the energy function and vice versa.

3. PROPOSITIONAL LOGIC FORMULAE

HNNs are as expressive as propositional logic (Pinkas, 1991), so they can be used to solve satisfiability problems. Those problems consist of finding a model for a given propositional formula. Each propositional formula can be mapped uniquely to an energy function, whose global minima correspond to the models of the formula. These problems can be solved by incomplete algorithms like GSAT (Selman et al., 1992) and are easily mapped to HNN's or Boltzmann machines, the stochastic variant of HNNs (Beringer et al., 1994).

Connectionist networks like HNNs are often regarded as models for massively parallel computation. But this parallelism is sometimes rather limited, especially when considering symmetric networks. For instance, HNNs are updated sequentially to guarantee a continuous decrease of the network's energy level. On the other hand, the energy level can only rise if two units U_1 and U_2, which are updated in parallel, are adjacent, i.e. receive input from each other by a symmetric connection with a non–zero weight. Otherwise, flipping both units in parallel will result in an energy decrease equal to the sum of decreases achieved by only flipping U_1 or U_2 respectively (Aarts and Korst, 1989; Hertz et al., 1991). Thus, performing a simultaneous flip of non-adjacent units — which is called a *multi-flip* in the sequel — results in a faster decrease of energy. Consequently, a local minimum of the energy function corresponding to a stable state is reached faster. This is especially interesting for tasks where *all* stable states of a network provide a solution to the problem at hand such as for example, the maximal independent set problem for undirected graphs (Luby, 1986), the n–tower problem (Rojas, 1993), and associative memories. Allowing multi–flips in these algorithms means an enlargement of the "neighborhood" in the state space of the connectionist network, i.e. from a given state not only states differing in the activation of one unit are reachable, but also those that differ in the activations of any set of non–adjacent units. A similar approach enlarging the neighborhood was proven to be successful for the quadratic sum assignment problem (Boissin and Lutton, 1993).

Hence, it is interesting to be able to perform multi–flips, i.e. compute an independent set of the units of a HNN, which want to flip their value, and then flipping all units belonging to the independent set in parallel. If successful, such an approach could speed up computations while still guaranteeing termination. In the sequel, we will describe the multi–flip network, its correspondence to a given HNN, and look at some of the applications pointed out above together with a comparison of related approaches. To our knowledge, the parallelizability of GSAT–like algorithms has not been investigated so far, though it might be interesting because of a possible speed up and can be easily implemented with a connectionist network approach. On the other hand, Simulated Annealing is a theoretically complete, but inherently sequential algorithm, despite its strong connection with Boltzmann Machines (Aarts and Korst, 1989). By means of the approach presented in this section, we can exploit a limited form of parallelism for Simulated Annealing, while still preserving the theoretical convergence property. Section 3.2 gives the connectionist architecture for parallel versions of hill–climbing algorithms. Section 3.3 describes our experimental results obtained so far along with possible further interesting experiments which are discussed in Section 5.

3.1. Hopfield Networks and SAT Problems

From (Pinkas, 1991) we know that each propositional logic formula is related to an energy function, such that the models of the formula correspond precisely to the global minima of the energy function. For a 3–CNF formula F over the nullary predicate symbols (or variables for short) $X_1, \ldots, X_n$ the corresponding energy function has the form

$$(11.2)\; E(\vec{X}) = \sum_{i<j<k} c_{ijk}X_iX_jX_k + \sum_{i<j} c_{ij}X_iX_j + \sum_i c_iX_i + c.$$

Comparing (11.1) and (11.2) we obtain

$$w_{ijk} = -c_{ijk}, \qquad w_{ij} = -c_{ij}, \qquad \theta_i = c_i, \qquad v_i = X_i, \qquad c = 0$$

as parameters for a HNN associated with F such that the activation of the units in the HNN corresponds to the truth value assignments of the variables occurring in F.

Consequently, the execution of the update algorithm for Hopfield networks depicted in Figure 2 can be seen as a local hill–climbing procedure with a strong relation to GSAT–like algorithms. Whereas HNNs search for local minima of the energy function, GSAT–like algorithms only look for global ones.

```
procedure Hopfield(Units)
    for i := 1 to maxTries do
        A := initial(Units);
        for j:= 1 to maxFlips do
            if stable( A ) then return A
            else
                possFlips := hillClimb(Units);
                U := pick(possFlips);
                flip assignment of U in A ;
            end if;   end for;   end for;
        return ''no stable state found'';
    end Hopfield;
```

Figure 2. The algorithm executed by a HNN rewritten within the GenSAT frame (see Figure 3). *initial* produces a random output of 0 or 1 for each unit in the HNN. *stable* checks if any unit has changed its output during its last two updates. *hillClimb* determines the set of units, whose output would flip if updated in the current state of the network. Finally, *pick* picks those units that are actually allowed to flip. In the standard Hopfield model, this will be only one single unit in order to assure convergence of the network to an energy minimum. *maxTries* is the maximal number of attempts for finding a stable state. Within each attempt, units are flipped at most *maxFlips* times.

As already mentioned, the algorithm depicted in Figure 2 is sequential as simultaneous flips of two adjacent units may lead to an increase of the energy level. In the following section we will show how such possibly hazardous flips of adjacent units can be avoided within a truly parallel computation using so-called *multi–flip networks* (MFNs). Such MFNs may be constructed not only for any HNN and for any SAT–problem, they can also be used as truly parallel computational models for Simulated Annealing.

Before turning to the description of MFNs we like to draw the attention of the reader to the general description of local search algorithms for SAT. These methods can be formalized by means of a very general algorithmic frame called GenSAT. The input to such an algorithm is always a propositional formula ϕ in conjunctive normal form (CNF). We restrict ourselves to formulas in 3–CNF. Roughly speaking, the algorithm measures the quality of a given truth assignment for the variables in ϕ by counting the number of clauses left unsatisfied by the assignment. This number can be seen as the *energy* of the system. The general method can be described within the GenSAT-frame first introduced in (Gent and Walsh, 1993) as given in Figure 3.

Different variants of this algorithm differ mainly in the function *hillClimb* building the set possflips. In our work, we use an indifferent GenSAT–variant called I_2 SAT (Hölldobler et al., 1994) for comparison that prefers any energy–reducing flips over those, that increase or do not change the energy, i.e. it is indifferent to the extent of energy decrease if any can be achieved. Because

```
procedure GenSAT( φ )
    for  i  := 1 to maxTries do
      A  := initial( φ );
        for  j:= 1 to maxFlips do
          if  A  satisfies  φ  then return  A
          else
            possFlips := hillClimb( φ,A );
            V  := pick(possFlips);
            flip assignment of  V  in  A ;
          end if;   end for;   end for;
      return ''no satisfying assignment found'';
    end GenSAT;
```

Figure 3.   The GenSAT frame

of its indifference to the extent of improvement caused by a variable flip, it is better suited for parallelization as it needs only local information. Empirical results give evidence that I_2 SAT performs as well as GSAT (Hölldobler et al., 1994). The algorithm used for the MFN mainly relies on this principle. We further like to emphasize the fact that the basic GSAT algorithm for solving SAT–problems is less successful than probabilistic variants. The most successful GenSAT variants are always those using *random walk* (Selman et al., 1994): With a given probability, the function *hillClimb* consists of all variables that occur in a currently unsatisfied clause independently of the change of energy caused by their flip. Taking this observation into account we will design extended MFNs that also mirror the clausal structure of the propositional formula at hand which is normally lost during the transformation via energy functions. This enables the MFNs to also perform parallel *walk* steps. Furthermore, this method can be seen as a general possibility to enable symmetric networks to search for a global energy minimum. Usually, Boltzmann machines using Simulated Annealing are used for this task, but in a comparison of Simulated Annealing and GSAT–like search methods we learned that GSAT with walk always outperforms the simulated annealing approach (Beringer et al., 1994). We expect that the use of walk instead of simulated annealing might also result in better performance of procedures for solving for problems like Constraint Satisfaction or Traveling Salesman, where the constraints can also be mirrored by the structure of the MFN.

Obviously, many steps of the algorithm are parallelizable: *initial* can be executed for all variables in parallel and in *hillClimb*, all variables can compute the energy resulting from a potential flip simultaneously. Our approach to parallelize these algorithms goes still further: The idea is to enlarge the possible neighborhood of each variable assignment: In the sequential case, the neighbors of an assignment, i.e. the assignments reachable in one step of the algorithm, are those assignments differing in only one truth value. A larger neighborhood then means that we allow the simultaneous flip of more than one variable, a *multi-flip*. This parallelization is implemented by the MFN.

3.2. *Multi-Flip Networks*

A multi-flip network (MFN) is a recurrent network with a layered structure. A basic multi–flip network corresponds to an arbitrary HNN, whose multi–flips it has to compute (and whose energy function it has to minimize). Its extended version represents a propositional formula in CNF preserving information about the clause structure. By this extension we also achieve that only global minima of the energy are stable states of the network.

In order to allow multiple flips at the same time, we have to change the update–algorithm depicted in Figure 2 in two ways: (i) The function *pick* has to be redefined so that it selects an independent subset of the set `possFlips`. (ii) All units left in this independent set have to flip their activation simultaneously.

The network must be designed so that the new function *pick* can be simulated, which is done in a two–step computation. Informally, in the first step each HNN–unit announces whether it wants to flip; this will be called a *flip–request* in the sequel. It then waits for permission to flip. This is determined in the second step by an ensemble of winner–take–all (WTA-) subnetworks, which, for competing adjacent units, selects those that will be allowed to flip.

The Components
The structure of an extended MFN is completely determined by the given propositional formula: It consists of four types of units, namely the H-, C-, S-, and M–units. If the C–units are omitted, we obtain a basic MFN. The H–units have two different states corresponding to the two–step computation described above and are used to represent the corresponding HNN. In other words, for each variable X_i occurring in a given SAT–problem there is an H–unit H_i. In its first state each H–unit determines if it wants to flip and, if so, announces a flip request. In the second state, the flips selected by the WTA subnetwork are performed.

All other units have only a single state. The S- (for *select*), and M–units (for *maximum*) form the WTA–subnetwork, where the S–units select priorities for each flip and the M–units determine the unit with highest flip priority among adjacent units. For each variable X_i occurring in a given SAT–problem there is a corresponding cluster consisting of the units H_i, S_i and M_i. Each C–unit C_i represents a clause ϕ_i of the given formula and determines the units participating in a walk step. These units can be omitted, if the walk option is not used or Simulated Annealing is to be implemented instead of GSAT. The activation and output functions of the different units are given in Table I.

The H–units switch to their second state immediately after they have produced an output in the first state. They remain in the second state for three

Table I. Activation and output functions of the different units in an MFN. All units are assumed to generate an output only once in the time step directly after they have received an input, otherwise they remain passive. The thresholds $\theta_h(i)$ of H–units and the weights on connections between H–units are determined by the energy function of the HNN or propositional formula, resp., as shown in Section 3.1. rnd(0,1) yields a random number taken from $[0,1]$. prob$(p) = 1$ if $p \leq$ rnd$(0,1)$ and 0 otherwise. pos$(x) = 0$ if $x \leq 0$ and 1 otherwise. sgn is the signum function. The two states of H–units are considered in the first and last row, respectively.

Unit	activation
$H_i(1)$	$p_{H_i}(t+1) = p_{H_i}(t)$
C_i	$p_{C_i}(t+1) = \sum_j wc_{ij} v_{H_j}(t)$
S_i	$p_{S_i}(t+1) = v_{H_i}(t) + \text{prob}(p) \cdot \sum_{\{j \mid wc_{ji} \neq 0\}} v_{C_j}(t)$
M_i	$p_{M_i}(t+1) = \max_j \{v_{S_j}(t) \mid w_{ij} \neq 0\} - v_{S_i}(t)$
$H_i(2)$	$p_{H_i}(t+1) = v_{M_i}(t) \cdot (2(p_{H_i}(t-3) - \theta_h(i)) - 1)$

Unit	output
$H_i(1)$	$v_{H_i}(t+1) = \text{pos}((1 - 2v_{H_i}(t)) \cdot (p_{H_i}(t) - \theta_h(i))$
C_i	$v_{C_i}(t+1) = \text{pos}(p_{C_i}(t+1) - \theta_c(i))$
S_i	$v_{S_i}(t+1) = \text{rnd}(0,1) \cdot \text{pos}(p_{S_i}(t+1))$
M_i	$v_{M_i}(t+1) = \text{sgn}(p_{M_i}(t+1))$
$H_i(2)$	$v_{H_i}(t+1) = \text{pos}(p_{H_i}(t+1) - \theta_h(i))$

time steps (until they receive new input from the M–units). After processing this input in the fourth time step, they switch again back to the first state. The complete connection structure is as follows. The weights wc on connections between H–units and C–units and the thresholds $\theta_c(i)$ of C–units are determined as follows:

$$wc_{ij} = \begin{cases} 1 & \neg X_j \in \phi_i \\ -1 & X_j \in \phi_i \\ 0 & \text{else} \end{cases} \qquad \theta_c(i) := |\{j \mid \neg X_j \in \phi_i\}| - 0.5$$

Consequently, unit C_i is activated if the corresponding clause ϕ_i in the given formula is left unsatisfied by the variable assignment represented by the current activation values of the H units. There are connections of weight 1 from each unit H_i to the corresponding S_i unit, and connections of weight -1 from the latter to the corresponding M_i unit. If $w_{ij} \neq 0$, then the unit S_i is connected to unit M_j with weight 1. If $wc_{ji} \neq 0$, then the unit C_j is also connected to the unit S_i with weight 1. Figure 4 depicts a small example.

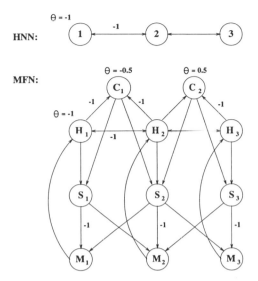

Figure 4.   The HNN and corresponding extended MFN for the for-
mula $\phi = (X_1 \lor X_2) \land (\neg X_2 \lor X_3)$ and the corresponding energy function
$E(X_1,X_2,X_3) = X_1X_2 - X_2X_3 - X_1$. In the MFN, the HNN is represented by the
H –units. Each unit i of the HNN and each unit H_i , S_i , and M_i , resp., of the MFN
corresponds to the variable X_i . The units C_1 and C_2 represent the first and the second
clause of the formula ϕ . Weights are 1 and thresholds are 0 if not specified otherwise.

The Multi–Flip Algorithm

The network is initialized by the procedure *initial* as follows: $p_{H_i}(0)$ is set to
-1 or 1 with probability 0.5 each, $v_{H_i}(0)$ is set to $\text{pos}(p_{H_i}(0))$, $v_{C_i}(0)$ and
$v_{S_i}(0)$ are set to 0 , and v_{M_i} is set to 1 . The functions *hillClimb* and *pick* are
defined as follows, where H and C denote the set of H –units and C –units
in the MFN, respectively.

```
function hillClimb(Units)              function pick(possFlips)
    for all U ∈ H ∪ C in parallel do       for each i ∈ possFlips in parallel do
        update( U );                           update( Mᵢ );
    for each Sᵢ in parallel do             return {i | mᵢ(t) = −1}
        update( Sᵢ );                      end pick
    return {i | sᵢ(t) > 0}
end hillClimb
```

Walk steps are realized by the S –units as follows. The C –units are activated if the current assignment does not satisfy the corresponding clause. But
their output is used by the corresponding S –units only with *walk probability*
p . The realization of the walk steps can be done in several ways. In our experiments, we used *global walk*: With walk probability p all variables occurring

in an unsatisfied clause request a flip. This variant is the direct parallelization of walk as it is used in GSAT–variants (Selman et al., 1994).

The function *hillClimb* then returns the variables (i.e. indices of all units) that request a flip. The set of indices returned by *pick* is then independent and identical to the indices of units H_i that actually will flip their activation value in the following update step. The algorithm now only has to stop if a global energy minimum (of the HNN represented by the H–units) is reached, viz. if for four time steps no unit of the MFN flips. Depending on the walk probability p local energy minima may not result in stable states if a walk step is performed, as in a local but non–global minimum always some clauses remain unsatisfied.

We assume that each unit produces an output only one time step after it has received an input. Consequently, there is no need to control the input to the H–units: In their first state, all M–units are passive, hence the H–units receive input only via their symmetric connections. In the second state the H–units are passive and produce no output, such that they only receive input from the M–units. But it is still necessary to control the output of the H–units. In state 1 the output is directed only to S- and C–units, whereas in state 2 the output is only propagated along the symmetric connections to all adjacent H–units.

In the worst case, e.g. when a fully connected HNN is represented, only one unit will participate in a multi–flip — the one with the absolutely maximal priority of all candidate units. One multi–flip can be implemented such that it takes three times as long as a single flip in the sequential algorithm. (For reasons of simplicity and clarity the algorithm presented herein needs four steps per multi-flip). As a consequence we obtain the following result.

THEOREM 1. *The Multi–Flip–Algorithm for 3–SAT has the same time complexity as the standard local GenSAT algorithm.*

3.3. *Experiments*

In order to compare the performance of MFNs with that of GSAT–like algorithms, we used the same random 3–CNF–formulas also commonly used for GSAT–experiments. We generated formulas with 100, 200, and 300 variables, each with a fixed ratio of clauses/variables of 4.3, which is approximately the crossover point or hardest region of random formulas (Crawford and Auton, 1993). We tested I_2 SAT, an indifferent variant which prefers all improving flips equally with and without the walk option (the walk probability set to the optimal value 0.5; (Beringer et al., 1994; Hölldobler et al., 1994)).

Regarding the results for the parallel algorithm we found a rather stable ratio of flips per multi–flip. This ratio obviously depends on the number of variables in the formula and the walk probability. The higher the walk probability, the more units request a flip per step, this results in more dependencies between requested flips and, therefore, in the end less flips are performed. As was pointed out in (Spears, 1993), in a random 3–CNF formula with a ratio clauses/variables = 4.3, each variable shares clauses with approximately 25.8 other variables. These variables will be called "depending variables". So a successful flip request automatically blocks 25.8 other depending variables on average. On the other hand, this number is independent from the number of variables in the formula. Consequently, when the number of variables is higher, the connection structure in the network is less dense, which results in a higher number of simultaneous flips. Comparing the corresponding quotient *(number of depending variables)/(total number of variables)* with the number of flips per multiflip in our experiments, we find the following

Observation: In the case of pure hill–climbing (without random walk), the parallelism of the MFN–algorithm is nearly optimal.

For example, we found 3.77 (optimum: 3.87) flips per multiflip for MFN without walk applied to formulas with 100 variables, and 10.86 (optimum: 11.63) flips per multiflip for MFN without walk applied to formulas with 300 variables. For a more extensive description of our experiments we refer the reader to (Strohmaier, 1997b) and (Strohmaier, 1997a).

4. NORMAL LOGIC PROGRAMS

In this section we propose to step beyond propositional logic. What requirements must a first–order deductive system fulfill such that a massively parallel, connectionist implementation is feasible? Since in present–day connectionist systems the structure of a system cannot be extended easily and quickly enough, the computation should be based on a fixed structure which very likely depends on the form of the given formula. Consequently, the potentially infinite number of objects which are dynamically created during a deductive process are not locally coded in the connectionist implementation but are instead distributed and subsymbolically represented by real–valued activation vectors. Furthermore, each step in a computation should be completely local so that a unit can compute its output solely based on its input and need not communicate with other units. The function performed by each unit should be simple so that it can be computed in essentially one step. The computation should be massively parallel, albeit respecting the results from complexity theory. Additionally, the system should exhibit sufficient expressive power, i.e., it should be beyond the expressive power of propositional logic. Last but not least the system should be adequate (Bibel, 1988).

Having identified the requirements, the next question to be addressed concerns the kind of deductive processes which satisfy the aforementioned requirements and are thus prime candidates for a massively parallel, connection-ist implementation. Analytical deductive processes like e.g. resolution applied to goal clauses do not meet the locality constraint as the subgoals occurring in a goal clause are not independent. Consequently, instantiations of a variable in the search for a solution of one subgoal must be communicated and agreed upon as soon as this variable occurs in more than one subgoal. On the other hand, synthetic deductive processes like e.g. model generation based on the meaning function T_P of a logic program P do meet the locality constraint. The structure of such a system depends only on the form of the given formula and starting from the empty set in each step a set of completely independent literals is generated.

What kind of connectionist model should be used to implement a model–generating process? We know from (Hornik et al., 1989) that multi–layered feed–forward networks with an input layer, a so–called hidden layer of units with sigmoidal activation function and an output layer can approximate any (Borel–measurable) function arbitrarily well. If we consider a model–generat-ing approach towards reasoning,[2] then, for a given program P functions like T_P can be recursively applied to generate a model in the limit (see e.g. (Lloyd, 1987)).[3] This observation suggests the following connectionist model.

We may use a multi–layered feed–forward network to approximate the function T_P for some logic program P. We may then turn this network into a recurrent one by propagating its outputs back to its inputs (see Figure 1). We may finally initialize the network by a certain external activation pattern and let it run until it reaches a stable state, i.e. a fixed point. Unique stable states are eventually reached for acceptable programs (Fitting, 1994). In such cases the activation pattern of the output layer represents the (approximated) least model of the program P and, thus, the logical consequences of P.[4] If this idea works out, then we may indeed have a fully connectionist model dealing with structured objects and structure–sensitive processes.

In Section 4.2 we will prove that for each acceptable program P there ac-tually exists a 3–FFN that approximates T_P. The remaining question then is: how to obtain such a network from a given program? In Section 4.1 we present such a construction for propositional programs. For first–order programs how-ever, the proof shown in Section 4.2 does not give an idea of how to map them

---

[2] It is an interesting question on its own whether reasoning should be done via model–generation or not (see e.g. (Johnson-Laird and Byrne, 1991)).

[3] Let I be an interpretation and P a normal program. $T_P(I)$ is the set of all atoms A where $A \leftarrow A_1, \ldots, A_m, \neg A_{m+1} \ldots, \neg A_n$ is a ground instance of a clause in P, $\forall 1 \leq i \leq m : A_i \in I$ and $\forall m < i \leq n : A_i \notin I$. An interpretation I is a model of P iff $T_P(I) \subseteq I$. Furthermore, I is a model of $comp(P)$ iff $I = T_P(I)$, where $comp(P)$ denotes the completion of P (see (Clark, 1978)).

[4] One should observe that the recurrent networks envisioned here are similar to the kind of networks used in (Jordan, 1986; Elman, 1989; Pollack, 1990).

to appropriate networks. In Section 4.3, we propose a certain construction and discuss associated problems. One of these problems is that the application of T_P to an interpretation may result in an infinite interpretation. Therefore, in Section 4.4 we introduce a new so–called *rational* representation for interpretations that is always finite and maintains the advantageous properties of T_P.

4.1. *Propositional Programs*

The very idea presented in the previous section works fine for the propositional case as shown in (Hölldobler and Kalinke, 1994). In fact, T_P and its fixed point — and not only approximations thereof — can be computed in this case. More precisely, it can be proven that for each program P there exists a 3–FFN of binary threshold units which computes T_P. The proof is constructive as follows. For each predicate symbol occurring in P a unit in the input and output layer is assigned. For each clause of the form $A \leftarrow L_1,\ldots,L_n$ there is a unit in the hidden layer, which becomes active as soon as the conjunction $L_1 \wedge \ldots \wedge L_n$ is satisfied by the activation pattern of the input layer. This activation is propagated to the unit in the output layer representing A. The size of this network is bounded by $O(n \times m)$, where n is the number of clauses in P and m is the number of predicate symbols occurring in P. The application of T_P to an interpretation I is performed in two time steps. As the sequential time to compute $T_P(I)$ is bounded by $O(n \times m)$ (assuming that no literal occurs more than once in the conditions of a clause), the parallel computational model is optimal.[5] Furthermore applying the results in (Fitting, 1994) we learn that for acceptable programs there exists a 3–FFN such that each computation starting with an arbitrary initial input (i.e. interpretation) converges and yields the unique fixed point of T_P.

The time needed by the network to settle down in the unique stable state is equal to the time needed by a sequential machine to compute the least fixed point of T_P in the worst case. In particular, let P be a definite program containing n clauses. The time needed by the network to settle down in the unique stable state is $3n$ in the worst case and, thus, the time is linear w.r.t. the number of clauses occurring in the program. This comes as no surprise as it follows from (Jones and Laaser, 1977) that satisfiability of propositional Horn formulas is P–complete and, thus, it is unlikely to be in the class NC (e.g. (Karp and Ramachandran, 1990)).

In (Hölldobler and Kalinke, 1994) it is also shown that for each recurrent 3–layered network of binary threshold units which always settles down in a unique stable state there is a program P such that T_P admits a unique fixed point.

---

[5] A parallel computational model requiring $p(n)$ processors and $t(n)$ time to solve a problem of size n is *optimal* if $p(n) \times t(n) = O(T(n))$, where $T(n)$ is the sequential time to solve this problem (see e.g. (Karp and Ramachandran, 1990)).

4.2. Acceptable Programs

The question is now whether the very idea presented in the previous sections does also work for first–order logic programs. The approach for propositional programs shown in Section 4.1 can easily (and naively) be extended to first–order programs. To compute the function T_P for programs P we can use a similar construction as before, where each atom in the Herbrand base B_P is represented by a unit in the input and output layer. Hence, for each normal program P there exists a 3–FFN which computes T_P. But since B_P may contain denumerable many elements, the input and output layer may contain infinitely many units to represent all elements of B_P. In this section we will show how to overcome this problem by accepting to approximate the function T_P arbitrarily well. Using a theorem by Hornik et al. (Hornik et al., 1990), we show that a feed–forward network with at least one hidden layer can approximate the function T_P for an acceptable program P to any desired degree of accuracy.

Mapping Interpretations to Real Numbers

In the following we consider logic programs P with at least one non–nullary function symbol. Let B_P be a corresponding Herbrand–Base. We can assign a number $n \in \mathbf{N}$ to each atom $A \in B_P$ by using a *level mapping* $| | : B_P \to \mathbf{N}$. Such a level mapping is said to be *bijective* if $\forall A, A' \in B_P : A \neq A' \Rightarrow |A| \neq |A'|$ and $\forall n \in \mathbf{N} : \exists A \in B_P : |A| = |n|$.

In (Fitting, 1994) it is shown that for an acceptable logic program P there exists a level mapping such that T_P is a contraction on the complete metric space $(2^{B_P}, d_P)$, where d_P denotes the associated distance function defined in Section 2. We will prove an even stronger result, viz. there even exists a *bijective* level mapping.

PROPOSITION 2. *Let P be an acceptable logic program. There exists a bijective level mapping for P such that T_P is a contraction on the complete metric space $(2^{B_P}, d_P)$.*

PROOF (Sketch): By (Lloyd, 1987) we know that for each acceptable program P we can find a level mapping $| |$ such that the function T_P is a contraction on $(2^{B_P}, d_P)$ where the distance between two interpretations I_1 and I_2, $d_P(I_1, I_2)$ (viz. $d_P(I_1, I_2)$) is defined by $1/2^n$, where n denotes the level of the atom A with smallest level such that $I_1(A) \neq I_2(A)$. Since the number of literals in the bodies of the clauses is finite, we can show that in the case we find such a level mapping we can find such a *bijective* level mapping too.   □

A bijective level mapping $| | : B_P \to \mathbf{N}$ for a program P can be extended to a mapping $R : 2^{B_P} \to \mathbf{R}$ as follows. R assigns to each element $I \in 2^{B_P}$ a real number $r \in [0, 1)$ such that on the i–th position behind the point in r

there is a 1 if for the atom A with $|A| = i$ we find $A \in I$, otherwise there is a 0. One should observe that R is injective and, consequently, there exists a mapping R^{-1}.

Mapping T_P to a Real–valued Function f_P

Using the mapping R interpretations can be mapped to real numbers. Thus we can define a function f_P on real numbers such that applying f_P to a real number r is equivalent to applying T_P to the corresponding interpretation $I = R^{-1}(r)$ (given such an I exists in 2^{B_P}).

The domain $D_f \subseteq \mathbf{R}$ of f_P is defined by the set $\{r \in \mathbf{R} \mid \exists I \in 2^{B_P} : R^{-1}(I) = r\}$ [6] and the mapping itself by

$$f_P : D_f \to D_f : f_P(x) \mapsto R(T_P(R^{-1}(x))).$$

Now we want to prove that the function f_P is continuous. For a real–valued function $f : D \to D$, where $D \subseteq \mathbf{R}$, the continuity property is defined as follows: f is *continuous at a point* $x_0 \in D$ if and only if for every $\varepsilon > 0$ there exists a $\delta > 0$ such that $|f(x) - f(x_0)| < \varepsilon$ for all x such that $|x - x_0| < \delta$. The function f itself is said to be *continuous* if it is continuous at all points.

Exploiting the property of acceptable logic programs that the meaning function T_P is a contraction on the corresponding complete metric space $(2^{B_P}, d_P)$, we are able show that the real–valued function f_P for these programs is continuous.

PROPOSITION 3. *If P is acceptable then f_P is continuous.*

PROOF (Sketch): Since for an acceptable program P the function T_P is a contraction, we find an $\varepsilon > 0$ such that for the Euclidean distance between any real numbers r and r', where $r' = f(r)$ and therefore $r = R(I)$ and $r' = R(I') = R(T_P(I))$ $(I, I' \in B_P)$ yields: $d_e(r, r') \leq \varepsilon$. Since the digits in r and r' correspond to atoms from B_P and are arranged according to the iteration step in which these atoms occur in an interpretation during the iteration process of T_P, r and r' differ at the earliest at the n–th position behind the point ($n \in \mathbf{N}$) but agree in the first $n-1$ positions. Since T_P is a contraction the function f_P does not change the value of numbers at the first $n-1$ positions and hence the distance between r and r' decreases or remains the same.
□

Using lattice theory it is shown in e.g. (Lloyd, 1987) that for definite programs[7] the meaning function T_P is monotonic, that is $\forall I, I' \in B_P : I \subseteq I' \Rightarrow$

---

[6] Note that since our domain is restricted to the binary numbers in $[0, 1)$, there occurs the problem that the number $0.\bar{1}$ equals the number 1.0. This can be avoided by using the number system with basis 3 , factorization and forming equivalence classes.

[7] Note that there exist definite programs that are not acceptable.

$T_P(I) \subseteq T_P(I')$. In this case it is straightforward to show that the corresponding function f_P is continuous as well.

Approximating the Meaning Function T_P

Let C^n denote the set of continuous functions from $\mathbf{R}^n$ to $\mathbf{R}$ $(n \in \mathbf{N})$, and $\Sigma^n(\phi)$ the class of output functions $f : \mathbf{R}^n \to \mathbf{R}, (n \in \mathbf{N})$ for 3–FFN with sigmoidal activation function ϕ in the hidden layer and the identity function as activation function in the output layer. The following theorem states the approximation capability of the elements of $\Sigma^n(\phi)$.

THEOREM 4.*(Hornik et al., 1990) For every sigmoidal activation function ϕ and every $n \in \mathbf{N}$, $\Sigma^n(\phi)$ is uniformly dense on compacta in C^n.*

In other words, 3–FFN can approximate any continuous function uniformly on any compact set, regardless of the sigmoidal activation function (continuous or not) of the hidden units and regardless of the dimension of the input space.

The density concept describes closeness of one class of functions to another. If the class $\Sigma^n(\phi)$ is uniformly dense on compacta in the class C^n, then for any element in C^n we can find an element in $\Sigma^n(\phi)$ that is arbitrarily close to it and consequently can approximate this element to any desired degree of accuracy. For the complete definitions see (Hornik et al., 1990).

COROLLARY 5. *Let P be an acceptable program. For the function f_P there exists a 3–FFN with sigmoidal activation function in the hidden layer and identity function as activation functions in the output layer, that is capable of approximating f_P to any desired degree of accuracy.*

In other words, the meaning function T_P for an acceptable logic program P can be approximated by a 3–FFN. The proof follows directly by the application of Proposition 3 and Theorem 4. If, however we want to compute and respectively approximate the *semantics* of a logic program we have to compute the *iteration* of the meaning function T_P. This can be realized by extending the network to a 3–layer recurrent neural network by connecting the output units of the network with the corresponding input units. Which units actually correspond to each other depends (w.r.t. a distributed representation) on the actual used representation for the atoms that constitute the interpretations.

4.3. *Problems*

Corollary 5 does not give any hint on how to construct a network for approximating T_P given a program P. For propositional programs, we showed how to construct suitable networks by representing each element of the underlying Herbrand–Base by one unit. Since infinite networks are infeasible in practice, this way of representation does not work for normal programs. But, there is another point of view on these networks: the units of the input layer represent the body literals of the considered clauses, and the units of the output layer represent the corresponding head literals. The hidden units still represent the program clauses. From this point of view, the same construction maps normal programs to finite networks. If we use such a network for computing T_P, then the potentially infinite number of atoms which are dynamically created during a deductive process have to be distributed over the resulting network and subsymbolically represented by real–valued activation vectors. Hence, the following question remains: What does this subsymbolic representation look like?

Two problems have to be solved by a suitable subsymbolic representation of atoms:

- The *first problem* is that, in general, T_P results in infinite interpretations. Hence, if each element of a considered interpretation is represented for its own, then an infinite number of subsymbolic representations must be processed. This problem can be solved at the level of symbolic representation and computation, respectively. To this end, in Section 4.4 we introduce a new so–called *rational* representation for interpretations. The meaning function T_P can be simulated on top of this representation. This simulation always results in finite expressions and maintains the properties of T_P which motivated its use.
- The *second problem* has already been mentioned: If a certain element of some domain occurs during computation, there must be a corresponding connectionist representation of this element. But it cannot be determined in advance what part of an infinite domain has to be considered. Hence, an infinite number of connectionist representations is needed, even if only a finite number of them will actually occur during computation. This is still an open problem and will be discussed in Section 5.

4.4. *Rational Models*

In this section, a new so-called *rational* representation for interpretations is developed, which is always finite even if the interpretation satisfies infinitely many ground atoms. Moreover, it is shown that the application of T_P can be simulated[8] on top of this representation.

As already mentioned in the previous section the application of T_P to an interpretation may result in an interpretation which satisfies infinitely many ground atoms. Hence the usual representation of interpretations as the set of satisfied ground atoms is infeasible in practice. E.g., infinitely many ground atoms are satisfied if the head of clause contains a variable which does not occur in some *positive* body literal of the clause.

---

[8] More precisely, we iterate a function which is homomorphic to T_P.

For this reason, approaches to model generation which are based on a representation of interpretations by sets of ground atoms like (Manthey and Bry, 1988) are only to be applied to correspondingly restricted classes of programs.

In the following section we introduce rational interpretations. We adapt the concept of most general unifiers to our representation. Moreover, in order to avoid nested representations, we introduce a simple way of disunification, which is enabled by our way of representation. Then, we give a surprisingly simple definition of a function $\widetilde{T}_P$, which simulates T_P. In contrast to T_P, the finite iteration of $\widetilde{T}_P$ always results in finite representations. Iterating $\widetilde{T}_P$ appears more simple and elegant than related approaches, and results in significantly shorter expressions. This is due to the representation we introduce, which is carefully adapted to the processed computation. Finally, the relation to other approaches is discussed.

Rational Models

Let A be an atom and G be a function from atoms to sets of ground atoms such that $G(A)$ is the set of all ground instances of A. For example,

$$G(p(X)) = \{p(a)\} \cup \{pf^i a \mid i > 0\}$$

if a is the only constant and f the only other function symbol in the underlying alphabet. For some infinite interpretations, there are equivalent finite expressions of the form $G(A_1) \cup \ldots \cup G(A_n)$ where $A_1 \ldots A_n$ are atoms and $n \in \mathbf{N}$. Accordingly, such interpretations can be finitely represented by the corresponding set $\{A_1 \ldots A_n\}$ (see eg. (Falaschi et al., 1988; Fermueller and Leitsch, 1996)).

However, some interpretations resulting from the application of T_P cannot be represented in this way as the following example demonstrates. Let $P = \{p(X,Y) \leftarrow \neg q(X,Y)\}$. Then $T_P(q(X,X)) = G(p(X,Y)) \setminus G(p(X,X))$. In other words, $T_P(q(X,X))$ consists of all instances of $p(X,Y)$ *except* the instances of $p(X,X)$. If the underlying alphabet contains a function symbol with arity ≥ 0, then there is no finite join $G(A_1) \cup \ldots \cup G(A_n)$ which is equal to $T_P(q(X,X))$.

We use a more sophisticated representation incorporating the general idea underlying constructive negation (Chan, 1988) wrt analytic calculi, i.e., we represent interpretations by sets of atoms, where each of these atoms is restricted by a set of substitutions representing exceptions. A *restricted atom* $A\Sigma$ consists of an (possibly non–ground) atom A and a finite set of substitutions Σ. A *rational interpretation* is a set of restricted atoms.

A restricted atom $A\Sigma$ *represents* all ground instances of the atom A *except* all ground instances of any $A\sigma$ where $\sigma \in \Sigma$.[9] The intended meaning of ratio-

---

[9]   The use of non–ground substitutions for representing sets of substitutions is similar to the use of substitution trees in (Ohlbach, 1990).

nal interpretations is formalized by extending the definition of G as follows. If $A\Sigma$ is a restricted atom and $\tilde{I}$ a rational interpretation, then

$$G(\tilde{I}) = \bigcup_{A\Sigma \in \tilde{I}} G(A\Sigma) \text{ where } G(A\Sigma) = G(A) \setminus \bigcup_{\sigma \in \Sigma} G(A\sigma).$$

Coming back to the previous example we find that

$$T_P(q(X,X)) = G(p(X,Y)\{\{Y \setminus X\}\}).$$

According to the mapping of rational interpretations to (usual) interpretations given by G, we define the notion of rational models as follows. A rational interpretation $\tilde{I}$ is a *rational model* of a completed program P iff $G(\tilde{I})$ is a model of the completion of P.

In the sequel, let P be a normal program, I an interpretation, $\tilde{I}$ a rational interpretation, A an atom, R a restricted atom, σ and ϕ substitutions, and Σ a finite set of substitutions, where $A, R, \sigma, \phi, \Sigma$ may be indexed.

There are interpretations which cannot be represented by a rational interpretation such as, for example, $\{p(a), p(ffa), p(ffffa), \ldots\}$. But the crucial point is that the application of T_P is strict as to whether the corresponding interpretations can be finitely represented in this way:

THEOREM 6. $\forall I : [\exists \tilde{I}_1 : G(\tilde{I}_1) = I \Rightarrow \exists \tilde{I}_2 : G(\tilde{I}_2) = T_P(I)]$.

The following section is a constructive proof of this result and we will see that Theorem 6 is an immediate consequence of Theorem 14. Since every finite interpretation can be represented by a rational interpretation, this Theorem 6 admits the following corollary stating that every interpretation generated during the iteration of T_P can be finitely represented by a rational interpretation.

COROLLARY 7. $\forall P : \forall n \in \mathbf{N} : \exists \tilde{I} : G(\tilde{I}) = T_P{}^n(\emptyset)$.

Generating Rational Models
In this section we define a function $\widetilde{T}_P$ processing rational interpretations such that the iteration of T_P is simulated by a corresponding iteration of $\widetilde{T}_P$. More precisely, in the following we define the function $\widetilde{T}_P$ to map rational interpretations to rational interpretations such that $\forall \tilde{I}, P : G(\widetilde{T}_P(\tilde{I})) = T_P(G(\tilde{I}))$. This definition of $\widetilde{T}_P$ will be quite similar to the definition of T_P, but so–called restricted substitutions (see below) are used to represent sets of substitutions. Corresponding to the concept of restricted substitutions, most general rational unifiers, general rational disunifiers and a function glb are introduced; the function glb is used for representing the intersection of infinite sets of substitutions.

Restricted substitutions   Restricted substitutions and their application to atoms are defined as follows. A *restricted substitution* $\sigma\Sigma$ consists of a substitution σ and a finite set of substitutions Σ with $A(\sigma\Sigma)=(A\sigma)\Sigma$. Furthermore, $\sigma_1(\sigma\Sigma)=(\sigma_1\sigma)\Sigma$. One should observe that by this definition a restricted substitution is only applied to an atom and maps this atom to a restricted one.

In analogy to restricted atoms, a restricted substitution $\sigma\Sigma$ *represents* the set of ground substitutions $G(\sigma\Sigma)$ where the definition of G is extended to restricted substitutions $\sigma\Sigma$ as follows: $G(\sigma)$ is the set of all ground instances of σ, and

$$G(\sigma\Sigma) = G(\sigma)\setminus\bigcup_{\sigma_2\in\Sigma} G(\sigma\sigma_2).$$

In the sequel, let θ (which might be indexed) be a rational substitution.

Most general rational unifiers   We now extend the notion of most general unifiers wrt restricted atoms. The intended meaning of unifying restricted atoms $A_1\Sigma_1,\ldots,A_n\Sigma_n$ is to compute a representation of the set $\bigcap_{i=1}^n G(A_i\Sigma_i)$. This is in analogy to the unification of atoms as follows: For atoms $A_1,\ldots,A_n$, let

$$\bigcap_{i=1}^n G(A_i) = \begin{cases} G(A_1 mgu(A_1,\ldots,A_n)), & \text{iff } mgu(A_1,\ldots,A_n) \text{ exists} \\ \emptyset, & \text{otherwise,} \end{cases}$$

where $mgu(A_1,\ldots,A_n)$ denotes the most general unifier of $A_1,\ldots,A_n$.

According to the concept of rational substitutions, *most general rational unifiers* (*mgru*) of, say, $A_1\Sigma_1$ and $A_2\Sigma_2$, map the atoms A_1 (and A_2, resp.), to a representation of $G(A_1\Sigma_1)\cap G(A_2\Sigma_2)$, namely $A_1 mgru(A_1\Sigma_1,A_2\Sigma_2)$. To this end, $mgru(A_1\Sigma_1,A_2\Sigma_2)$ has to restrict A_1 (and A_2, resp.) to those common instances of A_1 and A_2 which are not an instance of some exception represented in Σ_1 or Σ_2, respectively. Most general rational unifiers (*mgru*) are defined as follows: For all $A_1\Sigma_1,\ldots,A_n\Sigma_n$, the most general rational unifier (*mgru*) of $A_1\Sigma_1,\ldots,A_n\Sigma_n$ exists iff $A_1,\ldots,A_n$ are unifiable. Then,

$$mgru(A_1\Sigma_1,\ldots,A_n\Sigma_n) = mgu(A_1,\ldots,A_n)\bigcup_{1\leq i\leq n}\bigcup_{\sigma\in\Sigma_i}\phi_i^\sigma,$$

where

$$\phi_i^\sigma = \{mgu(A_1 mgu(A_1,\ldots,A_n),A_i\sigma)\,|\,A_1,\ldots,A_n,A_i\sigma \text{ are unifiable}\}.$$

PROPOSITION 8.

$$\bigcap_{i=1}^n G(A_i\Sigma_i) \neq \emptyset \Rightarrow \bigcap_{i=1}^n G(A_i\Sigma_i) = G(A_1 mgru(A_1\Sigma_1,\ldots,A_n\Sigma_n))$$

PROOF (Sketch): Iff $\bigcap_{i=1}^n G(A_i\Sigma_i) \neq \emptyset$, then $A_1,\ldots,A_n$ are unifiable, and $G(A_1 mgu(A_1,\ldots,A_n))$ is the set of common instances of $A_1,\ldots,A_n$. For each A_i (where $1 \le i \le n$) and each $\sigma \in \Sigma_i$, the common instances of $A_i\sigma$ and A_1 mgu $(A_1,\ldots,A_n)$ are just represented by their most general unifier mgu $(A_1$ mgu $(A_1,\ldots,A_n),A_i\sigma)$. Hence, G $(A_1$ mgu $(A_1\Sigma_1,A_2\Sigma_2))$ is the set of common instances of $A_1,\ldots,A_n$ which are not an instance of some exception $A_i\sigma$ where $\sigma \in \Sigma_i$.   □

Concerning $T_P(I)$, we are interested in the ground instances of body atoms A which are satisfied by I. Now, if $I = G(R)$, where R is a restricted atom, then according to Proposition 8, $mgru$ $(A0,R)$ represents just these substitutions. This result is easily extended wrt some $\tilde{I}$ instead of R:

COROLLARY 9. $A\sigma \in G(\tilde{I}) - \exists R \in \tilde{I}: \sigma \in G(mgru(A0,R))$.

Similar to the unification of restricted atoms, for all $\theta_1,\ldots,\theta_k$ a unique restricted substitution glb $(\theta_1,\ldots,\theta_k)$ representing $\bigcap_{i=1}^{k} G(\theta_k)$ can be computed by using the function $mgru$. To this end, the considered restricted substitutions are appropriately mapped to restricted atoms, and then the $mgru$ of these atoms is computed. The function glb is defined as follows: For all θ_1, glb $(\theta_1)=\theta_1$; if p is an arbitrary predicate symbol, and $X_1,\ldots,X_n$ are the variables occurring in some $\theta_1,\ldots,\theta_k$, where $k \ge 2$, then

$$glb(\theta_1,\ldots,\theta_k)=G(mgru(p(X_1,\ldots,X_n)\theta_1,\ldots,p(X_1,\ldots,X_n)\theta_k))$$

PROPOSITION 10. $G(glb(\theta_1,\ldots,\theta_k)) = \bigcap_{i=1}^{k} G(\theta_k)$.

PROOF (Sketch): For all θ the definition of G yields that $G(\theta)$ is the set of all substitutions which map each atom A to an instance of $G(A\theta)$. Hence, according to Proposition 8, for each atom A the set $G(A$ $mgru$ $(A\theta_1,\ldots,A\theta_k))$ is the set of all $A\sigma$ where $\sigma \in \bigcap_{i=1}^{k} G(\theta_k)$. Since each variable occurring in some $\theta_1,\ldots,\theta_k$ does also occur in $p(X_1,\ldots,X_n)$, the substitutions mapping $p(X_1,\ldots,X_n)$ to an element of $G(p(X_1,\ldots,X_n)$ glb $(\theta_1,\ldots,\theta_k))$ are just the substitutions mapping all atoms A to an element of $G(A$ glb $(\theta_1,\ldots,\theta_k))$. □

General rational disunifiers As a next step, we define *general rational disunifiers (grdu)* of restricted atoms. Wrt. the simulation of T_P by $\tilde{T}_P$, consider an atom A corresponding to a negative body literal, and a restricted atom R from the considered rational interpretation. Then, we are interested in the substitutions σ where $A\sigma \notin G(R)$ (and $A\sigma$ is ground). In general, we cannot represent all these σ by a single rational substitution. But there is a finite set of rational substitutions (namely general rational disunifiers), such that each such σ is represented by at least one of them: For all $A_1\Sigma_1$, $A_2\Sigma_2$, the set of all $grdu$ $(A_1\Sigma_1,A_2\Sigma_2)$ is

$$\{0\Sigma_1 \cup \{mgu(A_1,A_2)\}\} \cup \{mgru(A_1\Sigma_1,A_2\sigma0) \mid \sigma \in \Sigma_2\}.$$

PROPOSITION 11. $G\,(A\Sigma) \setminus G\,(R) \;=\; G\,(\{A\theta \mid \theta \text{ is a } grdu\,(A\Sigma, R)\})$.

PROOF (Sketch): Applied to A_1, $\emptyset\Sigma_1 \cup \{\, mgu\,(A_1, A_2)\}$ restricts $A_1\Sigma_1$ to *all* of its instances *except* those[10] which are also instances of A_2. In the same way, $mgru\,(A_1\Sigma_1, A_2\sigma\emptyset)$ restricts $A_1\Sigma_1$ to those of its instances which are instances of A_2, but, nevertheless, not an element of $G\,(A_2\Sigma_2)$, since they are an instance of some exception $A_2\sigma$, too.   □

In other words, $\emptyset\Sigma_1 \cup \{\, mgu\,(A_1, A_2)\}$ and each $mgru\,(A_1\Sigma_1, A_2\sigma)$, where $\sigma \in \Sigma_2$, are rational disunifiers of $A_1\Sigma_1$ and $A_2\Sigma_2$. Proposition 11 immediately yields that the $grdu$'s of some $A\emptyset$ and R represent the substitutions σ where $A\sigma \notin G\,(R)$:

COROLLARY 12.

σ is ground $\wedge\, A\sigma \notin G(R) \quad - \quad \exists grdu(A\emptyset, R) : \sigma \in G(grdu(A\emptyset, R))$.

So far, a representation θ of substitutions σ of an atom A can be computed where $A\sigma \notin G\,(R)$ for some restricted atom R. This is easily extended to rational interpretations: Let $\tilde{I} = \{R_1, \ldots, R_k\}$ and each θ_i ($1 \le i \le k$) represent a set of substitutions σ where $A\sigma \notin G\,(R_i)$. Then the intersection $\bigcap_{1 \le i \le k} G\,(\theta_i)$ of these sets is a set of substitutions σ where $A\sigma \notin G\,(\tilde{I})$. According to Proposition 10, this intersection is represented by $glb\,(\theta_1, \ldots, \theta_k)$:

COROLLARY 13. *Let* $\tilde{I} = \{R_1, \ldots, R_k\}$. σ *is ground and* $A\sigma \notin G\,(\tilde{I})$ *iff*

$$\exists \theta_1, \ldots, \theta_k : [\sigma \in G(glb(\theta_1, \ldots, \theta_k)) \wedge \forall 1 \le j \le k : \theta_j \text{ is a } grdu(A\emptyset, R_j)].$$

The function $\widetilde{T}_P$ By using the functions $mgru$, $grdu$, and glb a function $\widetilde{T}_P$ can be defined as follows: Let P be a normal logic program, and $\{R_1, \ldots, R_k\}$ a rational interpretation. Then

$$\begin{aligned}
\widetilde{T}_P(\{R_1, \ldots, R_k\}) = \{A\theta \mid\ & A \leftarrow A_1, \ldots, A_m, \neg A_{m+1}, \ldots, \neg A_n \in P, \\
& \theta = glb(\theta_1, \ldots, \theta_n), \\
& \forall 1 \le i \le m : [\exists 1 \le j \le k : \theta_i = mgru(A_i\emptyset, R_j)], \\
& \forall m < i \le n : [\exists \delta_1, \ldots, \delta_k : [\theta_i = glb(\delta_1, \ldots, \delta_k) \wedge \\
& \qquad \forall 1 \le j \le k : \delta_j \text{ is a } grdu(A_i\emptyset, R_j)]]\}
\end{aligned}$$

THEOREM 14. *Consider a rational interpretation* $\tilde{I}$ *and a normal logic program* P. *Then* $\widetilde{T}_P\,(\tilde{I})$ *is a rational interpretation such that* $G\,(\widetilde{T}_P\,(\tilde{I})) = T_P(\,G\,(\tilde{I}))$.

---

[10]   Note that, if $mgu(A_1, A_2)$ does not exist, then $\{mgu(A_1, A_2)\} = \emptyset$.

PROOF (Sketch): Let $\tilde{I}$ be the rational interpretation $\{R_1, \ldots, R_k\}$. Applied to restricted atoms, $mgru$ and $grdu$ do always result in (finite) rational substitutions. Applied to a finite number of restricted substitutions, glb always results in one restricted substitution. Hence, in the above definition of $\tilde{T}_P$, θ is a restricted substitution, and $A\theta$ is a restricted atom. There is only a finite number of such θ's, since, for all A and R, there is only one $mgru(A\theta, R)$ and only a finite number of glb's of $A\theta$ and R. Hence, for all rational substitutions $\tilde{I}$, $\tilde{T}_P(\tilde{I})$ is a (finite) rational interpretation.

According to Corollary 9 and Corollary 13, each set $G(A_i\theta_i)$ consists of ground instances of the body literals, which occur ($i \leq m$) or do not occur ($i > m$) in the interpretation $G(\tilde{I})$. The intersection $\bigcap_{1 \leq i \leq k} G(\theta_i)$ (represented by $glb(\theta_1, \ldots, \theta_n)$) of such sets yields substitutions σ where $A\sigma \in T_P(G(\tilde{I}))$. Hence, each element of $G(\tilde{T}_P(\tilde{I}))$ is an element of $T_P(G(\tilde{I}))$. Analogously, for each $A' \in T_P(G(\tilde{I}))$ there is an $A\theta$ according to the above definition such that $A' \in G(A\theta)$ and, hence, $A' \in G(\tilde{T}_P(\tilde{I}))$. □

The following corollary follows immediately from Proposition 14, and formalizes our idea of simulating an iterated application of T_P by iterating $\tilde{T}_P$:

COROLLARY 15. $\forall P, n \in \mathbf{N}: T_P{}^n(\emptyset) = G(\tilde{T}_P{}^n(\emptyset))$

Related Work
Most of the recent approaches to model generation use some kind of iterated hyperresolution (Robinson, 1965) or corresponding hyper tableaux (e.g. I.1.3). In particular, SATCHMO (Manthey and Bry, 1988) uses positive unit hyperresolution, and splitting of the resulting positive ground clauses. Like (Fujita et al., 1992; Bry and Yahya, 96; Inoue et al., 1992) SATCHMO can only be applied to range restricted[11] programs. By this restriction, the problem tackled in this section is avoided: During the generation of models of range restricted programs, head atoms and negative body literals are grounded by the variable bindings to positive body literals. Therefore, only finite sets of ground atoms are generated. On the other hand, this means that no generalizing inference can be modeled, which is quite unsatisfying wrt applications in the field of Artificial Intelligence.

The use of (possibly not ground) atoms instead of ground atoms (see eg. (Falaschi et al., 1988; Fermueller and Leitsch, 1996)) enables for generating models of certain programs which are not range restricted, but, as we demonstrated in Section 4.4.1, not for all normal programs. In a way quite similar to our approach, the problem of representing infinite model candidates is solved by Caferra, Peltier and Zabel (e.g. (Caferra and Peltier, 1995b; Peltier,

---

[11] A program P is called range restricted iff each variable occurring in a clause C of P occurs in a negative literal (wrt log. programs these are the positive body literals) of C.

1997)) wrt monotonic reasoning. They use *constrained atoms* instead of re-stricted atoms. A constrained atom is an atom together with a conjunction of simple constraints. The same representation is used by Turi in (Turi, 1991), which is the work most similar to ours: A function T_P^{CS} is computed in corre-spondence to T_P and $\widetilde{T}_P$, but operating on sets of constrained atoms. Con-strained atoms are structured in a more simple way than restricted atoms, but the computation of direct consequences in accordance to the nonmono-tonic operator T_P is more complex than by using restricted atoms. This is because restricted atoms reflect the generation performed by T_P: The sub-stitutions resulting from disunification wrt negative body literals mean the disjunction of the simple constraints expressed by their elements. For rep-resentations by conjunctions of simple constraints, the results of unification and disunification must be normalized. For example, consider the program $P = \{p(X,Y,Z) \leftarrow \neg q(X,Y,Z), \neg r(X,Y,Z)\}$ and the rational/constraint[12] interpretation

$$\tilde{I} = \{q(fUb, fVb, fWb), r(faU, faV, faW)\}.$$

Then,

$$\widetilde{T}_P(\tilde{I}) = \{p(X,Y,Z) \quad \{\{X \setminus fUb, Y \setminus fVb, Z \setminus fWb\}, \\ \{X \setminus faU, Y \setminus faV, Z \setminus faW\}\},$$

while $T_P^{CS}(\tilde{I}) =$

$$p(X,Y,Z)\{X \neq fUb, X \neq fbU\}, p(faU,Y,Z)\{U \neq b, Y \neq faV\}, \\ p(faU, faV, Z)\{U \neq b, Z \neq faW\}, p(fUb, Y, Z)\{Y \neq fVb, U \neq a\}, \\ p(fab, Y, Z)\{Y \neq fVb, Y \neq aV\}, p(fab, faV, Z)\{V \neq b, z \neq faW\}, \\ p(fUb, fVb, Z)\{z \neq fWb, U \neq a\}, p(fab, fVb, Z)\{z \neq fWb, V \neq a\}, \\ p(fab, fab, Z)\{Z \neq fWb, Z \neq faW\} \}$$

In general, iterating T_P^{CS} seems to be more complex and to yield much more extensive representations than iterating $\widetilde{T}_P$. Actually, Turi's work does not focus on a computation of T_P^{CS}, but on extending S-Semantics (Falaschi et al., 1993) to normal programs. While Turi shows the relation of his resulting CS–Semantics to completion semantics, we just adopt the latter via Defini-tion 7 (which would be a proposition, otherwise). Since rational models can trivially be transformed to CS-models, the relations to other semantics given in (Turi, 1991) are easily transfered to our approach. Similarly, (Caferra and Peltier, 1995a) might yield relations to decision procedures. Finally, our work is closely related to constructive negation (Chan, 1988), which is the corre-sponding backward reasoning method. In contrast to SLDNF-resolution (Clark,

---

[12] Interpreting this $\tilde{I}$ by means of a set of constraint atoms results in the same classic interpretation $G(\tilde{I})$.

1978), the use of constructive negation avoids floundering, which is, in backward reasoning, the problem corresponding to the computation of infinite interpretations by model generation.

5. DISCUSSION

In Section 3, we have shown how a feed–forward network equivalent to a symmetric network can be used to parallelize GSAT–like algorithms for solving propositional SAT–problems. With the MFN we can simulate a symmetric network performing multiple flips, thus overcoming its sequential updating schema without sacrificing the convergence property. We have pointed out its applications for SAT–problems and have shown how the MFN can parallelize local search algorithms. We have shown experimentally that MFNs indeed lead to a faster convergence in terms of algorithmic steps. We even found that the speed–up achieved matches the theoretically optimal value.

Nevertheless, we have to take into account that this method is not a "maximal" solution to the problem of multiple flips, as we cannot guarantee that always a maximal set of possible flips will be selected. But this clearly cannot be expected as this problem is equivalent to the problem of finding a maximal independent set in an undirected graph, which requires polylogarithmic time. This problem itself can be mapped to an HNN in such a way that all stable states of the HNN or, equivalently, all local minima of the corresponding energy function represent maximal independent sets of vertices in the graph (Luby, 1986). So MFN without clause units represents a kind of heuristics to update units in parallel.

Further advantages of MFN are the following: The updating heuristics for the H–Units, especially the choice of the initial configuration, can be influenced in a way such that the search for a satisfying assignment needs only linear time in the number of variables for Horn formulas. Moreover, a learning algorithm can be applied which augments the energy of local minima encountered during search, leading to a "smoother" energy landscape (Strohmaier, 1997a).

Furthermore, MFNs can be used to represent so–called penalty logic (Pinkas, 1992). This logic consist of formulas or clauses associated with a given penalty that influences clause weights. By weighting some clauses more than others, constraints of different "importance" can be modelled. Pinkas also used HNNs for representing validity problems in predicate logic. Consequently, these problems and their algorithmic solution can also be represented in MFNs. In (Strohmaier, 1997a) we also show that MFNs capture probabilistic approaches such as the Boltzmann machine (Aarts and Korst, 1989).

In Section 4, we turned our attention to normal programs. We proved that for all acceptable programs P there exists a connectionist network which approximates T_P arbitrarily well. By our proof, we established a strong relationship between logic programming and connectionist models of computation. Furthermore, we presented a new representation of interpretations of normal logic programs, and adapted the meaning function T_P to this representation. For all normal programs P the resulting function $\widetilde{T}_P$ simulates T_P and always results in finite representations of tractable length. This enables a simple and elegant (still symbolic) generation of models of normal logic programs; semantics and properties concerning fixed points are just adopted from T_P. Like T_P, $\widetilde{T}_P$ generates sets of independent[13] (restricted) atoms. Hence, generated rational interpretations can be distributed and locally processed within a fixed massively parallel architecture. A massively parallel implementation is also supported by the simplicity of $\widetilde{T}_P$, which is constructed by unification and the union of sets. Hence, $\widetilde{T}_P$ fulfills the general conditions for inference methods we outlined in the introduction, and avoids the problems we discussed in the previous section.

As a next step, a suitable subsymbolic representation of rational interpretations and a corresponding subsymbolic computation of *mgu*'s and the union of sets is needed. This is still an open problem. In (Kalinke, 1997) we analyse the necessary properties of such a representation and discuss existing approaches.

Even without a suitable subsymbolic representation and computation, $\widetilde{T}_P$ might be used within a *micro level hybrid architecture* which overcomes the brittleness problem: $\widetilde{T}_P$ is computed by a distributed symbolical system as sketched in Section 4.3. A connectionist network of the same structure controls this computation, i.e., it might switch off some of the units for some steps of computation. This might yield incorrect results, but restricts the number of processed objects, and, hence, enables to proceed computation within given limits in time and space. The connectionist network then might be trained to force the symbolical system to "reason about the right thing".

References

Aarts, E. and J. Korst: 1989, *Simulated Annealing and Boltzmann Machines*. John Wiley & Sons.

Beringer, A., G. Aschemann, H. Hoos, M. Metzger, and A. Weiß: 1994, 'GSAT versus Simulated Annealing'. In: A. G. Cohn (ed.): *Proceedings of the European Conference on Artificial Intelligence*. pp. 130–134.

Beringer, A. and S. Hölldobler: 1993, 'On the Adequateness of the Connection Method'. In: *Proceedings of the AAAI National Conference on Artificial Intelligence*. pp. 9–14.

---

[13] There are no variable bindings ranging over multiple elements of a rational interpretation.

Bibel, W.: 1988, 'Advanced Topics in Automated Deduction'. In: R. Nossum (ed.): *Fundamentals of Artificial Intelligence II*. Springer, LNCS *345*, pp. 41–59.

Bibel, W.: 1993, *Deduction*. Academic Press.

Bibel, W. and E. Eder: 1993, 'Methods and Calculi for Automated Deduction'. In: *Handbook of Logic in Artificial Intelligence and Logic Programming*, Vol. 1. Oxford University Press, pp. 67–182.

Boissin, N. and J.-L. Lutton: 1993, 'A Parallel Simulated Annealing Algorithm'. *Parallel Computing* **19**, 859–872.

Bry, F. and A. Yahya: 96, 'Minimal Model Generation with Positive Unit Hyper-Resolution Tableaux'. Technical Report PMS-FB-1996-1, LMU Munic.

Caferra, R. and N. Peltier: 1995a, 'Decision Procedures Using Model Building Techniques'. In: *Computer Science Logic*. pp. 130–144.

Caferra, R. and N. Peltier: 1995b, 'Model Building and Interactive Theory Discovery'. In: *TABLEAUX 95*, Vol. 918 of *LNAI*. pp. 154–168.

Chan, D.: 1988, 'Constructive Negation Based on the Completed Database'. In: *Proceedings of the International Conference on Logic Programming*. pp. 111–125.

Clark, K. L.: 1978, 'Negation as Failure'. In: H. Gallaire and J. Minker (eds.): *Logic and Databases*. New York, NY: Plenum, pp. 293–322.

Crawford, J. M. and L. D. Auton: 1993, 'Experimental Results on the Crossover Point in Satisfiability Problems'. In: *Proceedings of the AAAI National Conference on Artificial Intelligence*. pp. 21–27.

Elman, J. L.: 1989, 'Structured Representations and Connectionist Models'. In: *Proceedings of the Annual Conference of the Cognitive Science Society*. pp. 17–25.

Falaschi, M., G. Levi, M. Martelli, and C. Palamidessi: 1988, 'A New Declarative Semantics for Logic Languages'. In: R. A. Kowalski and K. R. Bowen (eds.): *Proceedings of the International Conference and Symposium on Logic Programming*. pp. 993–1005.

Falaschi, M., G. Levi, M. Martelli, and C. Palamidessi: 1993, 'A Model–Theoretic Reconstruction of the Operational Semantics of Logic Programs'. *Information and Computation* **102**(1), 86–113.

Feldman, J. A.: 1982, 'Dynamic Connections in Neural Networks'. *Biological Cybernetics* **46**, 27–39.

Feldman, J. A. and D. H. Ballard: 1982, 'Connectionist Models and Their Properties'. *Cognitive Science* **6**(3), 205–254.

Fermueller, C. and A. Leitsch: 1996, 'Hyperresolution and Automated Model Building'. *Journal of Logic and Computation* **6**(2), 173–203.

Fitting, M.: 1994, 'Metric Methods – Three Examples and a Theorem'. *Journal of Logic Programming* **21**(3), 113–127.

Fodor, J. A. and Z. W. Pylyshyn: 1988, 'Connectionism and Cognitive Architecture: A Critical Analysis'. In: Pinker and Mehler (eds.): *Connections and Symbols*. MIT Press, pp. 3–71.

Fujita, M., R. Hasegawa, M. Koshimura, and H. Fujita: 1992, 'Model Generation Theorem Provers on a Parallel Inference Machine'. In: *Proceedings of the International Conference on Fifth Generation Computer Systems*.

Gent, I. P. and T. Walsh: 1993, 'Towards an Understanding of Hill–Climbing Procedures for SAT'. In: *Proceedings of the AAAI National Conference on Artificial Intelligence*. pp. 28–33.

Hertz, J., A. Krogh, and R. G. Palmer: 1991, *Introduction to the Theory of Neural Computation*. Addison-Wesley Publishing Company.

Hölldobler, S.: 1993, 'Automated Inferencing and Connectionist Models'. Technical Report AIDA–93–06, Intellektik, Informatik, TH Darmstadt. (Postdoctoral Thesis).

Hölldobler, S., H. Hoos, A. Strohmaier, and A. Weiß: 1994, 'The GSAT/SA-Familiy – Relating Greedy Satisifability Testing to Simulated Annealing'. Technical Report AIDA-94-17, TH Darmstadt.

Hölldobler, S. and Y. Kalinke: 1994, 'Towards a Massively Parallel Computational Model for Logic Programming'. In: *Proceedings of the ECAI94 Workshop on Combining Symbolic and Connectionist Processing*. pp. 68–77.

Hopfield, J. J.: 1982, 'Neural Networks and Physical Systems with Emergent Collective Computational Abilities'. In: *Proceedings of the National Academy of Sciences USA*. pp. 2554 – 2558.

Hornik, K., M. Stinchcombe, and H. White: 1989, 'Multilayer Feedforward Networks are Universal Approximators'. *Neural Networks* **2**, 359–366.

Hornik, K., M. Stinchcombe, and H. White: 1990, 'Universal Approximation of an Unknown Mapping and its Derivative Using Multilayer Feedforward Networks'. *Neural Networks* **3**, 551–560.

Inoue, K., M. Koshimura, and R. Hasegawa: 1992, 'Embedding Negation as Failure into a Model Generation Theorem Prover'. In: D. Kapur (ed.): *Proceedings of the Conference on Automated Deduction*. pp. 400–415.

Johnson-Laird, P. N. and R. M. J. Byrne: 1991, *Deduction*. Hove and London (UK): Lawrence Erlbaum Associates.

Jones, N. D. and W. T. Laaser: 1977, 'Complete Problems for Deterministic Sequential Time'. *Journal of Theoretical Computer Science* **3**, 105–117.

Jordan, M. I.: 1986, 'Attractor Dynamics and Parallelism in a Connectionist Sequential Machine'. In: *Proceedings of the Annual Conference of the Cognitive Science Society*. pp. 531–546.

Kalinke, Y.: 1997, 'Using Connectionist Term Representation for First–Order Deduction – A Critical View'. In: F. Maire, R. Hayward, and J. Diederich (eds.): *Connectionist Systems for Knowledge Representation Deduction*. Queensland University of Technology. CADE–14 Workshop, Townsville, Australia.

Karp, R. M. and V. Ramachandran: 1990, 'Parallel Algorithms for Shared-Memory Machines'. In: J. van Leeuwen (ed.): *Handbook of Theoretical Computer Science*. Elsevier Science Publishers B.V., Chapt. 17, pp. 869–941.

Lange, T. E. and M. G. Dyer: 1989, 'High-Level Inferencing in a Connectionist Network'. *Connection Science* **1**, 181 – 217.

Lloyd, J. W.: 1987, *Foundations of Logic Programming*. Springer.

Luby, M.: 1986, 'A Simple Parallel Algorithm for the Maximal Independent Set Problem'. **15**(4), 1036–1053.

Manthey, R. and F. Bry: 1988, 'SATCHMO: A Theorem Prover Implemented in Prolog'. In: E. Lusk and R. Overbeek (eds.): *Proceedings of the Conference on Automated Deduction*, Vol. 310 of *Lecture Notes in Computer Science*. pp. 415–434.

McCarthy, J.: 1988, 'Epistemological Challenges for Connectionism'. *Behavioural and Brain Sciences* **11**, 44. Commentary to (Smolensky, 1988).

McCulloch, W. S. and W. Pitts: 1943, 'A Logical Calculus and the Ideas Immanent in Nervous Activity'. *Bulletin of Mathematical Biophysics* **5**, 115–133.

Ohlbach, H. J.: 1990, 'Abstraction Tree Indexing for Terms'. In: L. C. Aiello (ed.): *Proceedings of the European Conference on Artificial Intelligence*. pp. 479–484.

Palm, G., U. Rückert, and A. Ultsch: 1991, 'Wissensverarbeitung in Neuronaler Architektur'. In: W. Brauer and D. Hernandez (eds.): *Verteilte Künstliche Intelligenz und kooperatives Arbeiten*. Springer, Informatik-Fachberichte *291*.

Peltier, N.: 1997, 'Simplifying and Generalizing Formulae in Tableaux. Pruning the Search Space and Building Models'. In: *TABLEAUX 97*, Vol. 1227 of *LNAI*. pp. 313–327.

Pinkas, G.: 1991, 'Symmetric Neural Networks and Logic Satisfiability'. *Neural Computation* **3**.

Pinkas, G.: 1992, 'Logical Inference in Symmetric Connectionist Networks'. Ph.D. thesis, Washington University, Sever Institute of Technology, St. Loius, Missouri.

Pollack, J. B.: 1990, 'Recursive distributed representations'. *Artificial Intelligence* **46**, 77–105.

Robinson, J. A.: 1965, 'Automatic Deduction with Hyper–Resolution.'. *Int. J. Computational Mathematics* **1**, 227–234.

Rojas, R.: 1993, *Theorie der Neuronalen Netze*. Springer.

Selman, B., H. A. Kautz, and B. Cohen: 1994, 'Noise Strategies for Improving Local Search'. In: *Proceedings of the AAAI National Conference on Artificial Intelligence*, Vol. 1. pp. 337–343.

Selman, B., H. Levesque, and D. Mitchell: 1992, 'A New Method for Solving Hard Satisfiability Problems'. In: *Proceedings of the AAAI National Conference on Artificial Intelligence*. pp. 440–446.

Shastri, L. and V. Ajjanagadde: 1993, 'From Associations to Systematic Reasoning: A Connectionist Representation of Rules, Variables and Dynamic Bindings using Temporal Synchrony'. *Behavioural and Brain Sciences* **16**(3), 417–494.

Smolensky, P.: 1988, 'On the Proper Treatment of Connectionism'. *Behavioral and Brain Sciences* **11**, 1–74.

Spears, W. M.: 1993, 'Simulated Annealing for Hard Satisfiability Problems'. Technical report, Naval Research Laboratory, Washington D.C.

Strohmaier, A.: 1997a, *Logisches Schließen mit Massiv Parallelen Methoden*, Vol. 169 of *DISKI*. Sankt Augustin, Germany: infix Verlag.

Strohmaier, A.: 1997b, 'Multi–flip Networks: Parallelizing GenSAT'. In: G. Brewka, C. Habel, and B. Nebel (eds.): *KI–97: Advances in Artificial Intelligence*. pp. 349–360.

Turi, D.: 1991, 'Extending S–Models to Logic Programs with Negation'. In: *Proceedings of the International Logic Programming Symposium*. pp. 397–411.

Part 4

Comparison and Cooperation of Theorem Provers

Editor: Jürgen Avenhaus

JÜRGEN AVENHAUS

INTRODUCTION

This part contains three chapters and is devoted to the question of how the performance of theorem provers can be enhanced. The problems discussed are fundamental to automated deduction in general and are not related to special implemented systems nor to special proof tasks.

The first chapter by Baaz, Egly, and Leitsch discusses approaches to shortening the proof length. The authors discuss two techniques: they first show that the transformation of the input into a normal form—e.g., the process of Skolemization—is of great importance. They show that an appropriate transformation may result in a non-elementary reduction of the length of the shortest proof compared to an uncritical transformation. Secondly, they study extension techniques. In its simplest form this is done by extending the signature by introducing new names for complicated formulas. (See Tseitin (1968) as an early reference for using this technique.) Again, it is shown that this technique may result in non-elementary shortening of the minimal proof length. The deeper reason behind this shortening of the proof length is that cuts are simulated. (This can be compared to introducing lemmas in mathematics.) The results presented in the chapter are proved by giving examples where the phenomena appear. The techniques described are not yet extensively tested in existing provers (see (Egly and Rath, 1996)). But the results show that enormous savings in finding a proof are possible, and hence they open the door for future research on how the techniques presented can be used constructively in implemented proof systems.

The second chapter by Denzinger and Fuchs starts from the observation that automated theorem proving is to a large extent a search problem, where the most successful provers employ a sophisticated heuristic to guide the search for a proof. So the question arises as to which kinds of heuristics one may use and how one can encode the structure of the search (or of the proof searched for) into a search heuristic. Also it would be interesting to develop learning techniques to improve a given heuristic. The authors show that for problems from a special domain one can indeed give hints as to which heuristics with which parameter settings are promising. They also describe a technique dealing with how to learn from a "training problem" and how to tune a given heuristic for similar problems. This tuning is based on the idea to find a proof for the problem at hand that is derivationally similar to the proof of the training problem. (See Carbonell and Veloso (1988) for similar ideas.)

W. Bibel, P. H. Schmitt (eds.), Automated Deduction. A basis for applications. Vol. II
© 1998 Kluwer Academic Publishers. Printed in the Netherlands

The third chapter by Dahn and Denzinger studies the problem of how to couple different provers in order to cooperate in finding a proof for a given problem. The chapter starts with an analysis of known cooperation techniques, e.g., in the field of artificial intelligence. Then the authors describe the TEAMWORK architecture for coupling homogeneous provers, i.e., provers using the same logic and the same proof calculus, but different search heuristics. This architecture is intended to support the concepts of competition, cooperation, and evaluation of generated knowledge for pruning the current database. Next the authors present the TECHS architecture for coupling heterogeneous provers. This, of course, is by far more complicated than coupling homogeneous provers. Finally, techniques are presented for building an interactive prover in which the user plays the same role as the provers, i.e., the role of an expert. In this role the user can determine the proof plan, create appropriate sub-tasks, and submit these proof tasks to other experts in the net.

The three chapters in this part have many connections to other chapters. First of all, they develop fairly general concepts and techniques that can be incorporated into many existing provers.

More specifically, chapter I.1.4 by Egly studies in more detail the role of the cut rule in analytic tableaux. Cuts can be used to shorten proofs, and they offer canonical representations for various refinements of the calculus. This is strongly related to the first chapter of this part.

The technique to equip a prover with a learning component, as described here in the second chapter, is complemented by the techniques described in chapter II.2.8. If one has knowledge on the structure of the proof being searched for, then it makes sense to create and store proof patterns and, for a given problem, to retrieve a suitable proof pattern and to carry out the induced proof. For example, in inductive theorem proving the induction steps often lead to first-order proofs with a structure relatively well known from experience in the application domain. Since for first-order theorem proving in general no such knowledge on the structure of the proof is available, it makes sense to learn proof heuristics. So here the idea of reusing proofs (or proof patterns) is replaced by reusing (or adapting) heuristics for finding a proof.

The third chapter of this part is related to chapter II.3.10 of Schumann, Suttner, and Wolf. They study parallel theorem provers based on the sequential prover SETHEO. Hence they are interested in coupling homogeneous theorem provers. The simplest idea is to start several copies of SETHEO in parallel, each copy with a different search heuristic or a different permutation of the axioms as input, and then to wait for the first copy to report success. This idea, though successful in many examples, neglects the aspects of cooperation and pruning the search space by evaluating sub-results. So these concepts are incorporated to some extent into other parallel theorem provers on top of

SETHEO. We also refer the reader to chapter II.3.9 on parallel term rewriting. There results on combining cooperative and parallel rewriting are reported.

We now add some remarks on related research done outside the Schwerpunkt Deduktion on which these books report.

There are relatively few theoretic results on estimating the proof length for a given problem. So the efficiency of a prover is usually measured with experiments. Note that the work needed for finding a proof is not only determined by the proof length. The amount of work needed is also heavily depending on the branching degree of the search space. Since it is impossible to traverse a large part of the search space, the importance of a good search heuristic becomes evident. (Experiments have shown that for non-trivial problems only small fractions of one percent of the work done is really needed for the proof found.) This shows that realistic bounds for the work needed to find a proof are hard to obtain. Nevertheless, some results on the efficiency of theorem proving strategies are known. We refer the reader to Plaisted and Zhu (1997) for a very good overview.

The importance of good search heuristics has been noticed early. For example the power of the well-known proof system OTTER (McCune, 1994) is partly due to the sophisticated search heuristics used by the autonomous mode of this system. Also the improvement of the WALDMEISTER system (Buch et al., 1996), winner of the CADE-14 competition in the unit-equality category, over the version that started at the CADE-13 competition one year earlier, is to some extent based on the fact that better search heuristics were used. Nevertheless, there are almost no papers studying systematically the quality of search heuristics. For the aspect of improving the efficiency of theorem provers with machine-learning techniques one may distinguish several directions of research: for knowledge-based learning we refer the reader to the dissertation of Fuchs (1997). Remarkable speed-ups gained by techniques presented in this dissertation are reported on in several conference papers. For learning based on neural networks and analogy-based learning we refer the reader to Schulz et al. (1997) and Melis and Whittle (1997), respectively. For learning in connection with proof patterns see chapter II.2.8.

Cooperative theorem proving means to couple independent sequential provers to work on a common problem. In general each prover runs on its own processor and the processors form a net without shared memory. This is in contrast to parallel theorem proving where the work of one prover is parallelized and distributed on several processors with shared memory. For parallel theorem proving we refer the reader to part II.3 of this book. Cooperative theorem proving started with the DARES system (Conry et al., 1990). It was inspired by ideas from multi-agent systems, so each prover had only parts of the common knowledge in its memory and communication was demand-driven. In Avenhaus and Denzinger (1993) the TEAMWORK architecture was developed. It was intended to couple homogeneous provers. Each prover had all the common knowledge in its memory and communication was success-driven and restricted to designated communication phases. Today, the main research interest focuses on coupling heterogeneous provers, as described in the third chapter of this part. Provers as different as SPASS (completion based), SETHEO (tableau based), and DISCOUNT (completion based, using only unit-equality clauses as input) have been coupled successfully and re-markable synergetic effects have been obtained. We also mention the clause diffusion method of Hsiang and Bonacina (Bonacina and Hsiang, 1995). Here the provers mainly work on their own data (called settlers), but also have access to the data of the other provers. In the beginning the concept was close to parallel theorem proving, but was realized on a net of processors without shared data. Now (Bonacina, 1997) the concept incorporates more elements of cooperative theorem proving.

REFERENCES

Avenhaus, J. and J. Denzinger: 1993, 'Distributing Equational Theorem Proving'. In: *Proc. 5th Conference on Rewriting Techniques and Applications (RTA-93)*. pp. 62–76.

Bonacina, M.: 1997, 'The Clause-diffusion Theorem Prover Peers-mcd (System Description)'. In: *Proc. 14th Conference on Automated Deduction (CADE-14)*. pp. 53–56.

Bonacina, M. and J. Hsiang: 1995, 'The Clause-diffusion Methodology for Distributed Theorem Proving'. *Fundamenta Informaticae* **24**, 177–207.

Buch, A., T. Hillenbrand, and R. Fettig: 1996, 'Waldmeister: High Performance Equational Theorem Proving'. In: *Proc. DISCO-96*.

Carbonell, J. and M. Veloso: 1988, 'Integrating Derivational Analogy into a General Problem Solving Architecture'. In: *Proc. DARPA Workshop on Case-based Reasoning*.

Conry, S., D. MacIntosh, and R. Meyer: 1990, 'DARES: A Distributed Automated Reasoning System'. In: *Proc. AAAI-90*. pp. 78–85.

Egly, U. and T. Rath: 1996, 'On the Practical Value of Different Definitional Translations to Normal Form'. In: *Proc. 13th Conference on Automated Deduction (CADE-13)*. pp. 403–417.

Fuchs, M.: 1997, *Learning Search Heuristics for Automated Deduction*, Ph.D. Dissertation. Hamburg: Verlag Dr. Kovač. ISBN 3-86064-623-0.

McCune, W.: 1994, 'OTTER 3.0 Reference Manual and Guide'. Technical Report ANL-94/6, Argonne National Laboratory.

Melis, E. and J. Whittle: 1997, 'Analogy as a Control Strategy in Theorem Proving'. In: D. D. Dankel II (ed.): *Proc. 10th Florida AI Research Symposium (FLAIRS-97)*. pp. 367–371. ISBN 0-9620-1739-6.

Plaisted, D. and Y. Zhu: 1997, *The Efficiency of Theorem Proving Strategies*. Vieweg.

Schulz, S., A. Küchler, and C. Goller: 1997, 'Some Experiments on the Applicability of Folding Architecture Networks to Guide Theorem Proving'. In: D. D. Dankel II (ed.): *Proc. 10th Florida AI Research Symposium (FLAIRS-97)*. pp. 377–381. ISBN 0-9620-1739-6.

Tseitin, G.: 1968, 'On the Complexity of Derivation in Propositional Calculus'. In: A. Slisenko (ed.): *Studies in Constructive Mathematics and Mathematical Logic, Part II*.

M. BAAZ, U. EGLY AND A. LEITSCH

CHAPTER 12

EXTENSION METHODS IN AUTOMATED DEDUCTION

1. INTRODUCTION

Calculi of first-order logic were developed for many different purposes, e.g., for (i) *reconstructing* mathematical proofs in a formal framework or (ii) for *automated proof search*. The old and "traditional" calculi, like Hilbert-type and Gentzen-type calculi, belong to the first category. Either they serve as a framework to a theory of provability, where not actual proofs but only their existence is of relevance, or they are used as instruments for proof transformations (e.g., cut-elimination in the sequent calculus). Specific rules like cut and modus ponens serve as proof building tools, which allow to combine lemmata to more complex proofs. The substitution rule (or elimination rule for quantifiers) is mostly formulated as a *unary* rule which can be applied independently of others. As the discipline of automated theorem proving evolved in the early sixties, the cut-rule and the unrestricted substitution rule were clearly defective features in a discipline of proof search. The first attempts to use Herbrand's theorem directly and to reduce a first-order formula to a propositional one failed because of tremendous complexity. The paper of J. A. Robinson on the resolution principle (Robinson, 1965) then brought the decisive breakthrough: resolution was the first first-order calculus with a binary and minimal substitution principle — the well-known unification principle. Moreover, it works on quantifier-free conjunctive normal forms (clausal forms) which are *logic-free* (clauses can be represented as sequents of atoms). Resolution uses an atomic cut-rule in combination with the unification principle which allows for *most general* substitutions only. Thus, the key feature of resolution is *minimality*. Due to this minimality there are only finitely many deductions within a fixed depth, a property we call *local finiteness*. The price paid for this minimality is a loss of structure and an increase of proof complexity. The high proof complexity of computational calculi (like resolution, tableau calculi and connection calculi) forms a serious barrier to proof search for more complex theorems. The question arises, whether it is possible to combine the strong structural potential of the traditional logic calculi with the economical search features of computational calculi. To this aim, it is necessary to give up the strict minimality of the computational calculus without, at the same time, allowing the creation of arbitrary structures, signatures or

W. Bibel, P. H. Schmitt (eds.), Automated Deduction. A basis for applications. Vol. II
© 1998 *Kluwer Academic Publishers. Printed in the Netherlands*

lemmata. It is the purpose of this chapter to give a survey of different methods
for introducing and preserving structure in automated deduction. The inference methods we present are *extension methods* in the sense that either the
syntax is extended by predicate or function symbols or formulae are introduced which do not fulfill the (strict) subformula property. But all the rules
we present here are computational in the sense that they are locally finite.

A well-known extension method in automated deduction is *skolemization*, where the elimination of (existential) quantifiers is combined with an
introduction rule for new function symbols. Traditionally, skolemization is
considered as a rather trivial preprocessing step in the transformation to normal form. However, we show in Section 3 that different forms of skolemization may lead to extremely different situations concerning minimal proof
length and proof search. In fact, minimizing the range of quantifiers prior
to skolemization may lead to a *nonelementary* speed-up of proof complexity. Thus, the particular form of extension and handling of the quantifiers
is crucial to preservation or creation of *logical structure*. A result of the
same strength holds for structural transformations into conjunctive normal
form (CNF) versus nonstructural ones: indeed, structural CNF-transformation
(which requires the introduction of new predicate and function symbols) gives
a nonelementary speed-up over the naive methods. It should be emphasized
that it is not only the length of the obtained CNF which is of importance; the
real power of structural CNF-transformation lies in the encoding of the full
structure of the original formula.

Besides of preserving structure in normalization, extension methods can
be applied to *create structure*. In Section 4, we present a method, called *F-
extension*, of introducing new function symbols in clause logic. F-extension
is merely based on quantifier distribution and reskolemization; it is simple
and locally finite, but may produce a nonelementary speed-up of proof complexity. The proof-theoretical principle behind this effect is the potential of
the calculus to simulate the full cut rule. Thus, F-extension combines the advantage of computational calculi (normal forms, unification, local finiteness)
with that of full calculi (short proofs by use of strong lemmata). F-extension
is a very flexible principle, e.g., it can be used to formalize order induction.
From a proof theoretical point of view, F-extension is a technique of cut-
introduction. By the introduction of formulae which are not subformulae of
the theorem to be proved, the principle of cut-introduction is problematic to
proof search. In the conclusion, we point to this general feature which is common to all extension calculi. A blind search for cuts and extensions clearly is
very inefficient and would make the inference systems completely unpracti-
cal. We mention some results from proof theory concerning cut-syntax and
proof complexity which might help to distinguish effective and ineffective

cuts by their syntactical structure.

A major drawback of most automated theorem provers is a lack of potential to reuse knowledge, i.e., to work with lemmata. We believe that the extension techniques presented here define a theoretical basis for the development of stronger and more intelligent computational calculi.

2. EXTENSION: SOME MOTIVATING EXAMPLES

In searching for mathematical proofs we always try to find lemmata (of theorems proved in advance) and to look for analogous situations. The key technique in this procedure is *problem reduction*, i.e., we transform the original problem into one or more new problems; these problems should be already solved or at least be easier to handle. Traditional computational calculi like resolution, the connection method, the tableau calculus and many others only provide quite weak tools for problem reduction and use of lemmata. This leads to the rather rough practice to test hundreds and even thousands of theorem proving problems separately — without reuse of knowledge or simple reduction steps. The following subsections show that there are simple (extension) rules in clause logic which may help to change this situation.

2.1. *Problem Reduction*

Let A, B be two formulae of the following form.

$$
\begin{aligned}
L &= (\forall x)(\forall y)P(x,y) \wedge (\forall x)(\exists y)Q(x,y) \\
A &= L \to (\forall x)(\exists y)(P(x,y) \wedge Q(x,y)) \\
B &= L \to (\exists x)(\exists y)(P(f(x),y) \wedge Q(f(x),y))
\end{aligned}
$$

Obviously A and B represent simple theorems with equal antecedents but different consequents. If we transform the negated formulae into CNF then we get two sets of clauses C_A, C_B and two (corresponding and quite similar) resolution proofs shown in Figure 1.

If we manage to reduce the problem C_B to C_A, we can use the proof of C_A for showing C_B, i.e., A serves as lemma in the proof of B. We achieve this goal by reducing C_B to C_A via *extension*. The following steps define a rule based on $\exists$-introduction and reskolemization which is called *functional extension*:

$$
\frac{(\forall x)(\forall y)\,(\neg P(f(x),y) \vee \neg Q(f(x),y))}{(\exists z)(\forall y)\,(\neg P(z,y) \vee \neg Q(z,y))} \quad \exists\text{-introduction}
$$

$$
\frac{}{(\forall y)\,(\neg P(c,y) \vee \neg Q(c,y))} \quad \text{skolemization}
$$

$$C_A$$

C_1: $P(x,y)$
C_2: $Q(x,g(x))$
C_3: $\neg P(c,y) \vee \neg Q(c,y)$

$$C_B$$

C_1: $P(x,y)$
C_2: $Q(x,g(x))$
C_3': $\neg P(f(x),y) \vee \neg Q(f(x),y)$

refutation of C_A refutation of C_B

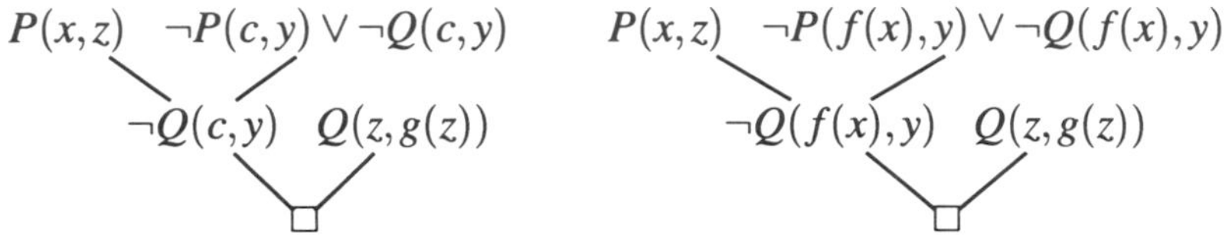

$P(x,z)$ $\neg P(c,y) \vee \neg Q(c,y)$ $P(x,z)$ $\neg P(f(x),y) \vee \neg Q(f(x),y)$

$\neg Q(c,y)$ $Q(z,g(z))$ $\neg Q(f(x),y)$ $Q(z,g(z))$

Figure 1.   Resolution proofs for C_A and C_B.

Observe that the resulting clause C_3 cannot be derived from C_B via the underlying resolution calculus. Indeed, c is a new constant symbol replacing the term $f(x)$; clearly such a replacement is impossible in every term-minimal calculus based on unification. Note that C_3 is not an instance of $\neg P(f(x),y) \vee \neg Q(f(x),y)$!

2.2. *Splitting*

Let C be a clause set consisting of the following clauses:

$$C_1 = P(x,y) \vee Q(x,y) \qquad C_2 = \neg P(x,y) \vee R(x,z)$$
$$C_3 = \neg Q(x,y) \vee R(x,z) \qquad C_4 = \neg R(x,f(x)) \vee \neg R(x,g(x))$$

Since we do not have a clause which decomposes into variable-disjoint sub-clauses, splitting (via the law of distributivity) is impossible. In fact, none of the clauses can be transformed into a "real" disjunction under preservation of logical equivalence (keep in mind that clauses represent universally closed disjunctions). But let us consider the following lemma L on C_1:

$$(\forall x)(\forall y)\,(P(x,y) \vee Q(x,y)) \to (\forall x)(\exists y)\,P(x,y) \vee (\exists x)(\forall y)\,Q(x,y).$$

Since L is valid and C_1 is given, the consequence of L can be obtained by modus ponens. Skolemization then yields the clause

$$C_1' = P(x,h(x)) \vee Q(a,y).$$

The clause C_1' is decomposable into variable-disjoint subclauses. Hence, we can split C_1' and obtain two subproblems of the form

$$C_1 = C \cup \{P(x,h(x))\} \qquad C_2 = C \cup \{Q(a,y)\}.$$

Both clause sets are unsatisfiable and can be refuted by the following resolution proofs:

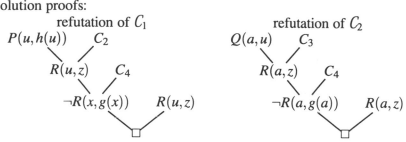

2.3. *Strong Factoring*

The factoring rule in resolution is not only necessary for the completeness
of the deduction principle, but also leads to shorter and simpler proofs. But
sometimes, factoring is impossible due to a symbol clash or due to the occur
check in unification. The following example shows that the use of extension
can lead to an essential change of the term structure enabling a *strong factoring* step and reducing the problem to a simpler one.

Let $C = \{P(x,x) \vee P(x,g(x)), \neg P(y,z)\}$. In every resolution refutation,
two new clauses (including $\Box$) have to be generated and three ground instances are required in order to obtain an unsatisfiable set of ground clauses.
Moreover, it is easy to see that every unsatisfiable set of ground instances
of C consists of at least three clauses. Again, we use a lemma introducing
existential quantifiers:

$$(\forall x)(P(x,x) \vee P(x,g(x))) \rightarrow (\forall x)(\exists y)P(x,y).$$

Using this lemma together with modus ponens and skolemization, we obtain
the new clause $P(x,h(x))$ for a new function symbol h. Clearly, the new set of
clauses

$$C' = \{P(x,x) \vee P(x,g(x)), P(x,h(x)), \neg P(y,z)\}$$

has a resolution refutation in only one step. Only two ground instances suffice
to obtain an unsatisfiable set of ground clauses.

The effect of factoring in the last example is only marginal (a single resolution step has a similar effect). But, like the extension techniques shown
in the first two subsections, it may also lead to strong proof compressions
and decrease of proof complexity. The complexity of this phenomena will be
discussed in Section 4.

3. EXTENSION AND NORMALIZATION

In automated deduction, using simple inference rules on syntactically simple formulae helps to reduce the number of possible inferences and thus to prune the search space. In most computational calculi, the formulae are transformed to quantifier-free form and even to conjunctive normal form. In refutational theorem proving, the typical situation is the following one: take the first-order sentence A (the theorem to be proved), transform $\neg A$ to a normal form $\gamma(\neg A)$, and finally apply a calculus on γ-forms to refute $\gamma(\neg A)$. In order to be useful, such a normalizing transformation must (at least!) be correct and terminating on every first-order formula. However we do not need strong correctness (preserving logical equivalence) but merely sat-equivalence ($\sim_{sat}$).

3.1. *Skolemization*

A simple and very important normalization is the elimination of *strong* quantifiers called *skolemization*.

DEFINITION 1. *Let A be a formula in predicate logic. If $(\forall x)$ occurs positively (negatively) in A then (the occurrence of) $(\forall x)$ is called a strong (weak) quantifier in A. Similarly, if $(\exists x)$ occurs positively (negatively) in A then (the occurrence of) $(\exists x)$ is called a weak (strong) quantifier in A.*

In a prenex form, all universal quantifiers are strong and the existential quantifiers are weak. In case of refutational theorem proving and negation normal forms, the strong quantifiers are exactly the existential ones. We first define structural skolemization as an operator on closed first-order formulae.

DEFINITION 2. *Let A be a closed first-order formula. The structural skolemization of A, denoted by $sk(A)$, is defined as follows. If A does not contain strong quantifiers, we define $sk(A) = A$.*
Suppose now that A contains strong quantifiers and (Qy) is the first strong quantifier occurring in A (in an ordering compatible with the subformula relation). If (Qy) is not in the scope of weak quantifiers then

$$sk(A) = sk(A_{-(Qy)}\{y \leftarrow c\})$$

where $A_{-(Qy)}$ is the formula A after omission of (Qy) and c is a constant symbol not occurring in A. If (Qy) is in the scope of the weak quantifiers $(Qx_1)\ldots(Qx_n)$ (appearing in this order) then

$$sk(A) = sk(A_{-(Qy)}\{y \leftarrow f(x_1,\ldots,x_n)\})$$

where f is a function symbol not occurring in A.

Iterating *sk* until no more strong quantifiers are present yields the skolemized form of *A*. In the definition above strong quantifiers are eliminated "at the spot", i.e., the formula is not subjected to any other transformation. By performing additional transformations prior to the application of *sk*, we obtain other forms of skolemization which may differ in the functional structure of the resulting formula.

DEFINITION 3.   *Let A be a closed first-order formula and A' be a prenex form of A; then sk(A') is called a* prenex skolemization *of A. If A' is an antiprenex form of A (i.e., the range of quantifiers is minimized by quantifier shifting rules) then sk(A') is called an* antiprenex skolemization *of A.*

We will see below in this section that the particular form of skolemization actually matters; it may strongly influence proof complexity and proof search — although the different normal forms are of about the same length.

According to Definition 2, skolemization eliminates $\forall$-quantifiers in a prenex form reducing the formula to a purely existential one. However, in refutational theorem proving, the $\exists$-quantifiers are those actually eliminated. In using a sequent notation, this problem is easily solved (refutational skolemization takes place on the left-hand side).

DEFINITION 4.   *Let $A_1, \ldots, A_n, B_1, \ldots, B_m$ be formulae in first-order predicate logic (for $n, m \geq 0$). Then the expression*

$$A_1, \ldots, A_n \vdash B_1, \ldots, B_m$$

is called a sequent.
    Let S be the sequent $A_1, \ldots, A_n \vdash B_1, \ldots, B_m$ and let $(A'_1 \wedge \ldots \wedge A'_n) \rightarrow (B'_1 \vee \ldots \vee B'_m)$ be the structural skolemization of $(A_1 \wedge \ldots \wedge A_n) \rightarrow (B_1 \vee \ldots \vee B_m)$ (if $m = 0$ then the consequent of the implication is set to falsum, if $n = 0$ then the antecedent is set to verum). The sequent

$$A'_1, \ldots, A'_n \vdash B'_1, \ldots, B'_m$$

is called the (structural) skolemization *of S. If $A''_1, \ldots, A''_n, B''_1, \ldots, B''_m$ are prenex forms of $A_1, \ldots, A_n, B_1, \ldots, B_m$ then the structural skolemization of*

$$A''_1, \ldots, A''_n \vdash B''_1, \ldots, B''_m$$

is called a prenex skolemization *of S.*

EXAMPLE 1. *Let $A = (\forall x)(\exists y)P(x,y)$. Then $sk(A) = (\exists y)P(c,y)$ for a constant symbol c. But $sk(A \vdash) = (\forall x)P(x, f(x)) \vdash$.*

The example above shows that the "traditional" refutational skolemization of A corresponds to a skolemization of $A \vdash$. The following example serves as an illustration for structural, prenex and antiprenex skolemization. Following the tradition of refutational skolemization, we consider a sequent of the type $A \vdash$.

EXAMPLE 2. *Let $A = (\forall x)(\forall y)(\exists z)(P(x,y) \vee P(y,z)) \wedge (\exists u)\neg P(u,u)$. The structural skolemization of $A \vdash$ is*

$$(\forall x)(\forall y)(P(x,y) \vee P(y,f(x,y))) \wedge \neg P(c,c) \vdash .$$

There are of course different forms of prenex skolemizations for A due to the different possibilities of quantifier shifting. In producing the quantifier prefix

$$(\forall x)(\forall y)(\exists z)(\exists u)$$

obtained by simply working from left to right, we get the sequent

$$(\forall x)(\forall y)[(P(x,y) \vee P(y,f(x,y))) \wedge \neg P(g(x,y),g(x,y))] \vdash$$

as prenex skolemization.

Finally, we construct an antiprenex skolemization of $A \vdash$. Due to the *logical structure of A, there is only one antiprenex form*

$$(\forall y)[(\forall x)P(x,y) \vee (\exists z)P(y,z)] \wedge (\exists u)\neg P(u,u).$$

Note that, in constructing antiprenex forms, we may also commute quantifiers of the same type. The skolemized form of the corresponding sequent is

$$(\forall y)[(\forall x)P(x,y) \vee P(y,h(y))] \wedge \neg P(c,c) \vdash .$$

In the last form, only a one-place function symbol and a constant symbol is introduced, while both structural and prenex skolemization contain binary Skolem symbols.

Still the question remains whether the particular form of skolemization really matters. If we only focus on the requirements of correctness and completeness, every form does the job. Even under the point of view of length, there is not much difference among the different skolemizations (in all cases, the increase of the number of symbol occurrences is at most quadratic). However, the situation changes if we focus on *proof complexity*. In fact, if $B_1 \vdash$

and $B_2 \vdash$ are two different skolemizations of $A \vdash$, the length of a shortest proof of $B_1 \vdash$ (or the length of a shortest refutation of B_1) may strongly differ from that of $B_2 \vdash$. This phenomenon is almost independent of the specific inference system and can be expressed in terms of Herbrand complexity.

DEFINITION 5. *Let S be the valid sequent* $A_1, \ldots, A_n \vdash B_1, \ldots, B_m$ *containing weak quantifiers only (i.e., $sk(S) = S$) such that each quantifier binds a different variable, and let* $A'_1, \ldots A'_n, B'_1, \ldots, B'_m$ *be the sequent formulae without quantifiers. Furthermore, let* $A'_{j,1}, \ldots, A'_{j,r_j}$ *be substitution instances of the* A'_j *and, similarly, for* $B'_{j,1}, \ldots, B'_{j,s_j}$ *and* B'_j. *A valid sequent of the form*

$$A'_{1,1}, \ldots, A'_{1,r_1}, \ldots, A'_{n,1}, \ldots, A'_{n,r_n} \vdash B'_{1,1}, \ldots, B'_{1,s_1}, \ldots B'_{m,1}, \ldots, B'_{m,s_m}$$

is called a Herbrand sequent *of S. Let λ denote the number of formula occurrences in a sequent. Then the number*

$$HC(S) = \min\{\lambda(S') \mid S' \text{ is a Herbrand sequent of } S\}$$

is called the Herbrand complexity *of S.*

By Herbrand's theorem, every valid sequent defines a corresponding Herbrand sequent; thus Herbrand complexity is well-defined. Still, it can be defined in several ways — depending on our choice of a complexity measure for sequents (an alternative would be to define λ as the number of symbol occurrences in a sequent). The following theorem on the complexity of skolemization shows that it is largely "robust" under different choices of Herbrand complexity. The following theorem is proved in (Baaz and Leitsch, 1994b) (as Theorem 4.1).

THEOREM 1. *There exists a sequence of sequents* $(S_n)_{n \in \mathbf{N}}$ *and a constant* $c \in \mathbf{N}$ *such that*

1. $HC(S'_n) \leq 2^{2^{2^{cn}}}$ for the structural skolemizations S'_n of S_n.
2. There exist prenex skolemizations S''_n of S_n such that

$$HC(S''_n) \geq \frac{1}{2}\mathsf{s}(n)$$

where s is defined by $\mathsf{s}(0) = 1$ and $\mathsf{s}(n+1) = 2^{\mathsf{s}(n)}$.

As the function s is nonelementary, a direct consequence of Theorem 1 is the nonexistence of an elementary function f such that $HC(S''_n) \leq f(HC(S'_n))$ for $n \in \mathbf{N}$; this result also holds if we define Herbrand complexity via the number of symbol occurrences (instead of the number of formula occurrences) in a sequent. The proof of Theorem 1 is based on a worst case sequence for cut-elimination as constructed by R. Statman (Statman, 1979); it is shown that prenex skolemization may destroy the "logical information" stored in the cut formulae of short proofs. In (Baaz and Leitsch, 1994b), it is also shown that minimizing the range of quantifiers (via shifting rules) never leads to an increase of Herbrand complexity; on the contrary, if a corresponding sequence of prenex sequents is given in advance, antiprenex skolemization may give a nonelementary speed-up over the structural one. Thus we see that different forms of skolemization — defining different forms of *functional extension*! — lead to very different proofs. As a consequence of this result, we should consider skolemization as an integral part of the inference process and not as a mere "preprocessing".

3.2. *CNF-Transformations*

As already mentioned, normalizing formulae is crucial to calculi in computational logic. There are two main approaches for normalization, namely "traditional" (nonstructural) transformations described in most textbooks on automated deduction, and structural transformations. We first recall nonstructural transformations into normal form. We do not concentrate on a concrete transformation but on all transformations satisfying significant properties. The main disadvantage of all nonstructural transformations is the loss of the formula's structure, mainly due to applications of distributivity laws in order to get clauses. Structural transformations avoid the disruption of the input formula by introducing *labels* (or names) for its subformulae. Hence, any subformula (occurrence) of the input formula has a unique abbreviation resulting in a preservation of the input formula's structure. We will introduce two variants of definitional transformations and compare nonstructural transformations with definitional transformations and both variants of the definitional transformations. We conclude this section with a discussion of practical merits of definitional transformations.

We start with a definition of nonstructural transformations. Recall that $\forall F$ denotes the universal closure of F.

DEFINITION 6. *A nonstructural transformation into clause form is any mapping $\gamma(\cdot)$ from first-order sentences into sets of clauses that fulfills the following conditions:*

1. If $L_1, \ldots, L_n$ and $M_1, \ldots, M_m$ are literals then

$$\gamma(\forall(L_1 \wedge \ldots \wedge L_n \to (M_1 \vee \ldots \vee M_m))) = \{L_1^d, \ldots, L_n^d, M_1, \ldots, M_m\},$$

where L^d denotes a literal which is dual to L.

2. $\gamma(A \wedge B) = \gamma(A) \cup \gamma(B)$ *for all sentences A and B.*

3. *If P is an atom and F is any formula then, for all $C \in \gamma(\forall (P \vee \neg P \vee F))$, $\{P, \neg P\} \subset C$. Observe that $\gamma(\forall (P \vee \neg P \vee F))$ may be empty, i.e., tautological clauses are removed from the clause set.*

$\gamma(F)$ *is called a* nonstructural clause form *of F.*

Observe that there is no assumption about how strong quantifiers are removed. Indeed, there are several skolemization algorithms which can be used for this purpose. We will come back to this point below.

A well-known drawback of nonstructural transformations, even for propositional formulae, is the exponential explosion of the length of the resulting clause set in terms of the length of the input formula. But even if the increase is not so drastic, the structure of the input formula is completely lost. This is of less importance, if we expect either Yes or No as the theorem prover's answer, but becomes very important in an interactive setting or if we want to get a readable proof of the *input formula*, which can be used for further tasks.

If we need a preservation of the formula's structure, then we have two possibilities, namely to use a calculus based on a rich syntax where formulae can have depth greater than 2, or to use a calculus based on clause form together with a structural transformation. We will discuss the second approach because it is better suited for proof search.

In the context of automated deduction, Tseitin (Tseitin, 1968) was the first who defined a structural transformation for propositional logic. Variants of this transformations were used to translate derivations between calculi with different connectives and in p-simulation proofs (Reckhow, 1976). First-order generalizations were provided independently by Eder (Eder, 1984; Eder, 1992) and Plaisted and Greenbaum (Plaisted and Greenbaum, 1986).

DEFINITION 7. *Let F be a first-order formula. Then $\Sigma(F)$ denotes the set of all occurrences of subformulae of F. Moreover, $\Sigma^+(F)$ denotes the set of all occurrences of subformulae with positive polarity in F and $\Sigma^-(F)$ denotes the set of all occurrences of subformulae with negative polarity in F.*

DEFINITION 8. *Let G be an occurrence of a first-order formula and let $\vec{x} = x_1, \ldots, x_k$ be the free variables of G (without duplicates). The atom $L_G(\vec{x})$ is called an abbreviation (or* label*) for G.*

DEFINITION 9. *Let F be a closed first-order formula. For each $G \in \Sigma(F)$ with free variables $\vec{x} = x_1, \ldots, x_k$, a label is introduced. Let $\vec{y} = y_1, \ldots, y_l$ be the free variables of H, $\vec{z} = z_1, \ldots, z_m$ be the free variables of I, where $\{\vec{y}\} \subseteq$*

G is atomic	$C_G^+ : \forall \vec{x}(\neg G \vee L_G(\vec{x}))$ $C_G^- : \forall \vec{x}(\neg L_G(\vec{x}) \vee G)$
$G = \neg M$	$C_G^+ : \forall \vec{x}(L_M(\vec{x}) \vee L_G(\vec{x}))$ $C_G^- : \forall \vec{x}(\neg L_G(\vec{x}) \vee \neg L_M(\vec{x}))$
$G = H \wedge I$	$C_G^+ : \forall \vec{x}(\neg L_H(\vec{y}) \vee \neg L_I(\vec{z}) \vee L_G(\vec{x}))$ $C_G^- : \forall \vec{x}(\neg L_G(\vec{x}) \vee L_H(\vec{y})) \wedge \forall \vec{x}(\neg L_G(\vec{x}) \vee L_I(\vec{z}))$
$G = H \vee I$	$C_G^+ : \forall \vec{x}(\neg L_H(\vec{y}) \vee L_G(\vec{x})) \wedge \forall \vec{x}(\neg L_I(\vec{z}) \vee L_G(\vec{x}))$ $C_G^- : \forall \vec{x}(\neg L_G(\vec{x}) \vee L_H(\vec{y}) \vee L_I(\vec{z}))$
$G = H \rightarrow I$	$C_G^+ : \forall \vec{x}(L_H(\vec{y}) \vee L_G(\vec{x})) \wedge \forall \vec{x}(\neg L_I(\vec{z}) \vee L_G(\vec{x}))$ $C_G^- : \forall \vec{x}(\neg L_G(\vec{x}) \vee \neg L_H(\vec{y}) \vee L_I(\vec{z})))$
$G = H \leftrightarrow I$	$C_G^+ : \forall \vec{x}(L_G(\vec{x}) \vee L_H(\vec{y}) \vee L_I(\vec{z})) \wedge \forall \vec{x}(L_G(\vec{x}) \vee \neg L_H(\vec{y}) \vee \neg L_I(\vec{z}))$ $C_G^- : \forall \vec{x}(\neg L_G(\vec{x}) \vee L_H(\vec{y}) \vee \neg L_I(\vec{z})) \wedge \forall \vec{x}(\neg L_G(\vec{x}) \vee \neg L_H(\vec{y}) \vee L_I(\vec{z}))$
$G = \forall x K$	$C_G^+ : \forall \vec{x}(\neg L_K(\vec{x}, g(\vec{x})) \vee L_G(\vec{x}))$ $C_G^- : \forall \vec{x} \forall x(\neg L_G(\vec{x}) \vee L_K(\vec{x}, x))$
$G = \exists x K$	$C_G^+ : \forall \vec{x} \forall x(\neg L_K(\vec{x}, x) \vee L_G(\vec{x}))$ $C_G^- : \forall \vec{x}(\neg L_G(\vec{x}) \vee L_K(\vec{x}, g(\vec{x})))$

Figure 2. The translation rules for classical logic.

$\{\vec{x}\}, \{\vec{z}\} \subseteq \{\vec{x}\}$, and $\{\vec{y}\} \cup \{\vec{z}\} = \{\vec{x}\}$. *Moreover, $\vec{x}, x$ are the free variables of K and $\vec{x}$ are the free variables of M. In $L_K(\vec{x}, g(\vec{x}))$, g is a globally new function symbol neither occurring in F nor being introduced in the transformation of any other subformula. The translation rules are presented in Figure 2. The definitional form of F is*

$$\delta(F) \quad = \quad \bigwedge_{G \in \Sigma(F)} (C_G^+ \wedge C_G^-).$$

The p-definitional form *(the definitional form obeying polarities) is*

$$\delta_p(F) \quad = \quad \left(\bigwedge_{G \in \Sigma^+(F)} C_G^+ \right) \wedge \left(\bigwedge_{G \in \Sigma^-(F)} C_G^- \right).$$

Roughly speaking, Plaisted and Greenbaum's variant is $\delta_p(\cdot)$, whereas Eder's variant is $\delta(\cdot)$.

Let us come back to the translation rules in Figure 2. We explain the case $G = (H \leftrightarrow I)$ in the following; the other cases are similar. Recall that the basic idea underlying (p-)definitional transformations is the introduction of a new label for any subformula occurrence of the given input formula. Suppose that we have already translated H and I where $L_H(\vec{y})$ is the label for the subformula occurrence H and $L_I(\vec{z})$ is the label for the subformula occurrence I.

For the subformula occurrence G, a label of the form $L_G(\vec{x})$ is introduced together with a closed equivalence of the form $\forall \vec{x}(L_G(\vec{x}) \leftrightarrow (L_H(\vec{y}) \leftrightarrow L_I(\vec{z})))$. A nonstructural transformation is applied to this equivalence resulting in the clauses depicted in the figure.

In the following, we do not distinguish between definitional forms as formulae and clause sets, where clauses are denoted as quantifier-free disjunctions of literals. It is well known (see (Eder, 1992; Plaisted and Greenbaum, 1986)) that $\delta(F) \cup \{\neg L_F\}$ as well as $\delta_p(F) \cup \{\neg L_F\}$ are unsatisfiable iff F is valid. Moreover, the time and space complexity of the transformation of a formula F is at most quadratic in the length of F.

Readers familiar with different approaches to skolemization may have observed that the technique used in the definitional transformation does not coincide with the usual skolemization techniques introduced in most textbooks. Indeed, the skolemization technique (Andrews, 1981) used above introduces Skolem terms depending on all free variables in the quantified subformula. Using this skolemization technique with the usual transformation to normal form yields considerably shorter (resolution) proofs for some classes of formulae (Egly, 1994a; Egly, 1994b).

EXAMPLE 3. *Let* $F = ((\forall x)(P(x) \wedge Q(x))) \rightarrow (\exists y)(P(y) \wedge Q(y))$. *Nine labels are introduced by the p-definitional transformation.* $\delta_p(F)$ *consists of the following clauses:*

$$
\begin{aligned}
C_1 &= \neg L_{P(x)}(x) \vee P(x) \\
C_2 &= L_{P(y)}(y) \vee \neg P(y) \\
C_3 &= \neg L_{Q(x)}(x) \vee Q(x) \\
C_4 &= L_{Q(y)}(y) \vee \neg Q(y) \\
C_5 &= \neg L_{P(x) \wedge Q(x)}(x) \vee L_{P(x)}(x) \\
C_6 &= \neg L_{P(x) \wedge Q(x)}(x) \vee L_{Q(x)}(x) \\
C_7 &= L_{P(y) \wedge Q(y)}(y) \vee \neg L_{P(y)}(y) \vee \neg L_{Q(y)}(y) \\
C_8 &= \neg L_{\forall x(P(x) \wedge Q(x))} \vee L_{P(x) \wedge Q(x)}(x) \\
C_9 &= L_{\exists y(P(y) \wedge Q(y))} \vee \neg L_{P(y) \wedge Q(y)}(y) \\
C_{10} &= L_F \vee L_{\forall x(P(x) \wedge Q(x))} \\
C_{11} &= L_F \vee \neg L_{\exists y(P(y) \wedge Q(y))}
\end{aligned}
$$

The clause set $\delta_p(F) \cup \{\neg L_F\}$ *can be refuted by resolution.*

What is the point in using a definitional transformation instead of the nonstructural transformation? The most obvious advantages of definitional transformations are the compact representation of the input formula by the definitional normal form and the preservation of the input formula's structure.

As already mentioned above, this is not the case for nonstructural transformations. A more subtle aspect of a comparison between such transformations is the minimal length of refutations of the resulting normal forms. In the remainder of this section, we investigate the behavior of minimal refutations if, for some class $(F_k)_{k \in \mathbf{N}}$ of first-order formulae, (i) $\gamma(F_k)$ and $\delta(F_k)$ are applied, and (ii) $\delta(F_k)$ and $\delta_p(F_k)$ are applied. We do not present a formal treatment but explain the intuition and basic ideas. A fully formal treatment with all proofs can be found in the cited literature.

Assume we have a sequence of first-order formulae of the form

$$(12.1) \quad (C_n \vee P \vee \neg P) \to F_n,$$

where F_n has only rather long cut-free proofs (e.g., in analytic tableau or cut-free sequent calculi). Orevkov (Orevkov, 1979) and Statman[1] (Statman, 1979) showed independently that there are classes of formulae $(F_n)_{n \in \mathbf{N}}$ which possess only nonelementary cut-free proofs, but which have short proofs with cuts. The reason for the nonelementary lower bound on proof length in such cut-free calculi is the nonelementary Herbrand complexity of F_n. Although the formulae introduced by Orevkov and Statman are quite different, they are essentially in clause form (up to skolemization). Since the length of a resolution refutation of a clause set C is, in the best case, logarithmic in the Herbrand complexity of C (Baaz and Leitsch, 1992; Eder, 1992), resolution cannot provide short proofs of F_n.

Let us discuss the details of the formula C_n, which is constructed from a short sequent proof ϕ_n of F_n with m cuts. It is of the form

$$(12.2) \quad \bigwedge_{i=1}^{m} \forall \, (C_n^i \to C_n^i)$$

where C_n^i is the cut formula of the i-th cut in ϕ_n.

Let us estimate the Herbrand complexity of a nonstructural transformation of (12.1). By condition 3. in Definition 6, the clauses obtained from $C_n \vee P \vee \neg P$ are all tautological and can be eliminated *without* increasing Herbrand complexity. But then, the answer is "It is the Herbrand complexity of F_n" which implies that, even for the extended formulae, any cut-free proof or any resolution proof remains nonelementary.

Before we consider the definitional form of (12.1), we explain how cuts and definitional forms are related. The discussion is based on Eder's extension

---

[1] Statman did not provide first-order formulae, but formulae in combinatory logic which were translated into first-order formulae with equality in (Baaz and Leitsch, 1992).

concept (Eder, 1992), which itself is a generalization of Tseitin's concept. Eder showed that, for any cut-free sequent proof of a first-order sentence F, there exists a (tree) resolution derivation of L_F from $\delta(F)$ and the length of the latter is polynomial in the length of the former. If the sequent proof of F contains cuts then applying $\delta(\cdot)$ to F is not sufficient for a preservation of proof length up to a polynomial. What is additionally necessary is the *definitional transformation of all cut formulae*. Hence, there is a (tree) resolution derivation of L_F from $\delta(F) \cup \delta(C^1) \cup \ldots \cup \delta(C^m)$ with essentially the same length as the corresponding sequent proof of F with m cuts, where the m cut formulae are C^i $(1 \leq i \leq m)$.

Let us come back to (12.1) and its definitional transformation. The formula C_n consists of formulae built from the cut formulae in a short sequent derivation of (12.1) with cut. If $\delta(\cdot)$ is applied, the definitional transformation yields all such formulae which enable a short simulation of all necessary instances of the cut rule in tree resolution, because all cut formulae are subformulae of (12.1). Since tree resolution is a Herbrand calculus, i.e., Herbrand complexity is a lower bound on proof complexity, the Herbrand complexity of the definitional transformation of (12.1) is also low. In the following theorem, taken from (Baaz et al., 1994), $RC(C)$ denotes the resolution complexity of C, i.e., the length of a shortest resolution refutation of C, and $|A|$ denotes the number of symbol occurrences in the string representation of A.

THEOREM 2. *There exists a sequence of valid sentences $(G_n)_{n \in \mathbf{N}}$ and constants a, b, c, such that, for $n > 1$, the following holds:*

1. $|\neg G_n| \leq 2^{an}$.
2. $RC(\delta(G_n) \cup \{\neg L_{G_n}\}) \leq 2^{2^{2^{bn}}}$.
3. $RC(\gamma(\neg G_n)) \geq c \cdot \mathsf{s}(n-1)$ *for any nonstructural transformation γ (see Theorem 1 for a definition of s).*

This theorem says that one can utilize "redundant information" in order to get a drastically shorter and more readable proof. The reason is that the cut formulae are used as lemmata, resulting in desirable proofs with more structure. These lemmata are obtained by the definitional transformation but not by nonstructural transformations, mainly because any C_n^i $(1 \leq i \leq m)$ is considered independent from $P \vee \neg P$. Hence, introducing some redundancy may enable speed-ups in proof length which cannot compensated by clever strategies. The reason is that, if the length of any minimal proof is nonelementary, then clearly the size of the search space is nonelementary too. However, if there are short (elementary) proofs then the size of the search space remains elementary. Hence, different normal forms of the same first-order formula, differing only linearly in size, can have Herbrand complexities (or minimal proofs) which are differing nonelementarily.

The most obvious difference between $\delta(F)$ and $\delta_p(F)$ is the shorter normal form for the latter transformation. In the best case, the saving is $2/3$, because one clause is needed in the latter case instead of (at most) three clauses in the former case. For this reason, it is often recommended to use a p-definitional transformation and to completely disregard a definitional transformation. One important aspect is neglected in the discussion focusing exclusively on the length of the resulting normal form. If $\delta(\cdot)$ is applied instead of $\delta_p(\cdot)$ then both polarities of the cut formulae are available in the resulting normal form, even if the cut formulae occur in one polarity only in the input formula. This observation will become important in the next comparison between definitional and p-definitional transformations.

Let us reconsider (12.1) and the short sequent proof ϕ_n with m cuts, where C_n is changed to

$$(12.3) \quad \bigwedge_{i=1}^{m} \forall C_n^i$$

and C_n^i is the cut formula of the i-th cut in ϕ_n. In contrast to (12.2), the cut formulae occur in one polarity only in (12.3). This has significant consequences if $\delta_p(\cdot)$ is applied instead of $\delta(\cdot)$. In the former case, only C_G^- for any $G = C_n^i$ ($1 \leq i \leq m$) is introduced, and, therefore, the abbreviated subformula C_n^i occurs in one polarity in the resulting normal form. In order to simulate a cut with cut formula C, both polarities of C are required in the normal form. Both polarities of C are provided by the definitional transformation because clauses in C_G^+ and C_G^- are available in the resulting normal form. The following result is taken from (Egly, 1996).

THEOREM 3. *There exists a sequence of valid formulae* $(H_n)_{n \in \mathbb{N}}$ *and constants a, b, c, such that, for* $n > 1$,

1. $|\neg H_n| \leq 2^{an}$,
2. $HC(\delta(H_n) \cup \{\neg L_{H_n}\}) \leq 2^{bn}$,
3. $HC(\delta_p(H_n) \cup \{\neg L_{H_n}\}) \geq c \cdot s(n-1)$.

Let us conclude this section with some remarks on the usefulness of structural transformations. It is often argued that structural transformations are nice theoretical tools without any practical relevance. Supporters of this opinion remark that the introduction of new labels (and new clauses) enlarges the search space and complicates proof search. What is often overlooked and neglected is the preservation of the input formula's structure. This preservation can often simplify proof search, especially in the case when nested implications or equivalences occur in the input formula. For instance, well-known examples like the proof of the unsolvability of the halting problem (Burkholder, 1987; Bruschi, 1991; Dafa, 1993; Dafa, 1994; Egly and Rath, 1996) can only be successfully tackled if the structure of the input formula is preserved.

If we give up the possibility to simulate some instances of cuts then even more compact representations are possible. One can generate a p-definitional transformation of an input formula F and apply preprocessing reductions to $\delta_p(F)$ in order to reduce the number of introduced labels (Egly and Rath, 1996). Alternatively, one can apply a mixed transformation scheme (Boy de la Tour, 1990) where subformulae are translated in such a way that the length of the resulting normal form is minimized. In both cases, the number of labels in decreased *without* giving up the preservation of the structure of F. In (Egly and Rath, 1996), it is shown that structural transformations can compete excellently with nonstructural transformations by presenting detailed run-time measurements for problems from an α-version of the (first-order nonclausal) TPTP library.

A more general remark concerns the comparison and test of automated deduction systems. It is a wide-spread practice to use hundreds or even thousands of problems to show the advantage of one system over the other(s). But what does the input look like? The usual test problems are presented as clause sets, sometimes optimized by hand in order to enable the systems to find a proof. If only clause sets are used then essentially the efficiency of the underlying deduction system is measured. From a methodological point of view, such a procedure is questionable because advantages from a clever reformulation of the given problem, from optimized transformations into the input representation of the system etc. are completely neglected. However, as the close relationship between definitional transformations, extensions and cuts indicates, these activities belong to the deductive solution of a problem.

4. EXTENSION AND INFERENCE

All computationally appealing calculi rest (more or less) on Herbrand's theorem. Any proof generated in such a calculus can be directly transformed into a Herbrand disjunction which is propositionally valid. Herbrand complexity is a lower bound for many calculi (like various connection calculi, some refinements of resolution) and the logarithm of the Herbrand complexity is a lower bound for many resolution refinements allowing for renaming of derived clauses. By such a renaming and applying self-resolution in an iterative manner, one can prove some clause sets C_i in polynomial length (with respect to the length of C_i) using, e.g., unrestricted resolution or linear resolution, but the Herbrand complexity of C_i is exponential (Baaz and Leitsch, 1992; Eder, 1992).

The problem now is that, for some classes of formulae (Statman, 1979; Orevkov, 1979), the Herbrand complexity is nonelementary in the length of a shortest proof with cuts. This disappointing result shows that a price has to be paid for computationally appealing calculi having desirable properties like local finiteness, minimality of substitutions etc. One "solution" for this problem might be giving up these properties and using a full logical calculus like Gentzen's LK *with* cut. From a computational point of view, however, this is not a solution, because the cut formula has to be mechanically determined, a problem which prevents the use of the (unrestricted) cut rule in automated deduction systems.

What can be done without giving up local finiteness (and minimality of substitutions) is to allow only restricted variants of the cut rule, where the cut formulae are computed from subformulae of the given input formula. We have already mentioned examples of such computations, namely skolemization, antiprenexing, and transformations to normal form. All these operations can be considered as applications of the restricted cut rule. Another technique, *functional extension* is discussed in the sequel of this section.

4.1. *Functional Extension Principles*

As already demonstrated in Section 2, functional extension principles can be used to change the quantificational structure of a formula. In this section, we introduce different variants and explain speed-up results. These results indicate a close relationship between such extension principles and the cut rule.

The basic idea of all functional extension principles is the introduction of a new $\exists$-quantifier. In case of restricted normal forms not containing existential quantifiers, skolemization is applied for their removal. We restrict our attention to clause form because any (closed) first-order formula can be translated into a set of clauses (either by structural or traditional transformations).

In the very beginning, functional extension principles were used as a problem reduction technique. Consider a clause set C and a clause $C = P(x) \lor Q(x)$ for illustration. Recall that the clause C represents a closed formula of the form $\forall(P(x) \lor Q(x))$. The same variable x occurring in both literals prevents a decomposition of the clause (and, therefore, the splitting of the clause set $C \cup \{C\}$) into two clause sets $C \cup \{C_1\}$ and $C \cup \{C_2\}$. Assume we would have a clause $D = P(x) \lor Q(a)$ instead of C. Then such a composition would be possible (with $C_1 = P(x)$ and $C_2 = Q(a)$, or vice versa) resulting in two *independent, smaller* subproblems. By an introduction of an $\exists$-quantifier and skolemization, D can be obtained from C. Since D replaces C, and a is a new Skolem constant, $C \cup \{C\}$ is unsatisfiable if $C \cup \{D\}$ is unsatisfiable, but the converse is not true in general. In other words, what we get is a correct but incomplete problem reduction technique. Functional extension principles of this kind were investigated in (Baaz and Leitsch, 1985; Baaz and Leitsch, 1990; Egly, 1990).

Although such a decomposition of clause sets strongly influences proof search, the effect on proof length is rather limited in unrestricted resolution. Some surprising results can be obtained if restricted resolution concepts are considered. For instance, in linear resolution, a decomposition of clauses can yield exponentially shorter proofs (Egly, 1994a). The same effect can be obtained if some tautological clauses of the form $A \vee \neg A \vee E$ (for an atom A and a clause E) are added, which correspond to the introduction of an atomic cut. On the other hand, tautology deletion can significantly lengthen proofs in such calculi.

DEFINITION 10. *Let S be a set of clauses. For a clause $C = C_1 \vee C_2$ from S, let $A = (\forall x_1) \ldots (\forall x_m)[(\forall \vec{u})C_1 \vee (\forall \vec{v})C_2]$. Let $\vec{u}$ denote the variables occurring only in C_1 and let $\vec{v}$ denote the variables occurring only in C_2. Let $\{y_1, \ldots, y_k\} \subseteq \{x_1, \ldots, x_m\}$, $\{z_1, \ldots, z_l\} = \{x_1, \ldots, x_m\} \setminus \{y_1, \ldots, y_k\}$, $Q_i \in \{\forall, \exists\}$, and Q_i^d is the quantifier dual to Q_i ($1 \le k \le m, 1 \le i \le k$). C_2' is obtained from C_2 by renaming y_i by y_i'. With $\vec{z} = z_1, \ldots, z_l$, the skolemized form of*

$$(\forall \vec{z})[(Q_1 y_1) \ldots (Q_k y_k)(\forall \vec{u})C_1 \vee (Q_1^d y_1') \ldots (Q_k^d y_k')(\forall \vec{v})C_2'] \wedge \bigwedge_{C \in S} C$$

is called an F-extension of S. If $k = 1$, we have a 1-F-extension. If $k = m$, we have an SF-extension (split F-extension).

As an example for F-extension, reconsider clause set C from Section 2.2. The clause C_1' is obtained by (split) F-extension. In contrast to the problem reduction techniques above, the clause derived by F-extension is added to the clause set. Hence, completeness is preserved.

PROPOSITION 1. *Let C' be an F-extension of C. Then C' is unsatisfiable iff C is unsatisfiable.*

Calculi can be extended by functional extension principles like F-extension. We use resolution as the underlying calculus but we remark that any other calculi like clausal tableaux, model elimination or connection calculi can also be used. We give the definition of FR-deduction.

DEFINITION 11. *Let S be a set of input clauses. A sequence $C_0, \ldots, C_n$ is called* FR-deduction *(resolution deduction with function introduction) of a clause C from S if the following conditions hold.*

1. *$C_n = C$.*
2. *for all $i = 0, \ldots, n$ and $j, k \geq 0$*

    a) C_i is a variant of an input clause, or
    b) C_i is a variant of a C_j for $j < i$, or
    c) C_i is a factor of a C_j for $j < i$, or
    d) C_i is a resolvent of C_j, C_k for $j, k < i$, or
    e) $\{C_0, \ldots, C_i\} \cup S$ is an F-extension of $\{C_0, \ldots, C_{i-1}\} \cup S$.

An FR-refutation *is an FR-deduction of $\square$ from S. If no F-extension appears in $C_0, \ldots, C_n$, we have an* R-deduction *of C from S. An R-refutation is an R-deduction of $\square$ from S.*

Technical remark. In the definition of FR-deduction, the input clauses as well as the derived clauses have to be considered in the F-extension step, because the newly introduced Skolem function symbols must differ from all function symbols introduced so far.

The following theorem is Theorem 4.6 in (Baaz and Leitsch, 1992).

THEOREM 4. *There is an FR-refutation of a clause set C iff there is an R-refutation of C.*

COROLLARY 1. *FR-deduction is sound and complete.*

We compare R-deduction and FR-deduction with respect to proof length. The discussion is based on (Baaz and Leitsch, 1992), where a translation of Statman's (sequence of) formulae (from combinatory logic) into first-order logic with equality is used. The resulting formula F is then translated into clause form (including the necessary equality axioms) resulting in a clause set S. Assume that $P_1, \ldots, P_m$ are the predicate symbols occurring in S. Then S is coded via the the predicate coding axioms

$$P = \{\neg P_i(\vec{x}_i) \vee S(p_i(\vec{x}_i)), P_i(\vec{x}_i) \vee \neg S(p_i(\vec{x}_i)) \mid 1 \leq i \leq m\}$$

($\vec{x}_i$ is the vector of all variables of P_i, S is a new predicate symbol not occurring in S, and any p_i is a new function symbol which corresponds to P_i and which does not occur in S) and the connectivity axioms

$$C = \{\neg S(neg(x)) \vee \neg S(x), S(neg(x)) \vee S(x), \neg S(x) \vee S(or(x, y)),$$

$$\neg S(y) \lor S(or(x,y))), \neg S(or(x,y)) \lor S(x) \lor S(y)\}.$$

Intuitively, the function symbol *neg* represents negation and the function symbol *or* represents disjunction.

Any pair of clauses for P_i (from P) can be considered as introduced by an extension of the form

(12.4) $\quad (\forall \vec{x}_i)(\exists y)(S(y) \leftrightarrow P_i(\vec{x}_i))$.

If we translate (12.4) into clause form then the newly introduced Skolem function is $p_i(\vec{x}_i)$. Similarly,

$$(\forall x)(\forall y)(\exists z)(S(z) \leftrightarrow (S(x) \lor S(y))) \quad \text{and} \quad (\forall x)(\exists y)(S(y) \leftrightarrow \neg S(x))$$

yield the connectivity axioms if $or(x,y)$ and $neg(x)$ are the Skolem functions introduced for $(\exists z)$ and $(\exists y)$, respectively. Hence, the coding scheme is based on extensions. Using the clauses resulting from Statman's sequence together with the predicate coding axioms and the connectivity axioms, the following theorem (Theorem 5.3 in (Baaz and Leitsch, 1992)) can be proved.

THEOREM 5. *There exists a sequence of clause sets* $(S_n)_{n \in \mathbf{N}}$ *and constants* $a, b,$ *such that, for* $n > 1,$

1. *any resolution deduction of* $\square$ *from* S_n *has length* $\geq a \cdot \mathrm{s}(n-1),$
2. *there exists an FR-deduction of* $\square$ *from* S_n *with length* $\leq 2^{bn}.$

Recall the case where the definitional transformation allows for the simulation of the cut rule by deriving an implication of the form $\forall(C \to C)$. Here, we have a clause of the form

$$D = \neg S(t(x_1, \ldots, x_l)) \lor S(t(x_1, \ldots, x_l))$$

which itself is a tautological clause. This clause D represents a formula of the form

$$(\forall x_1) \ldots (\forall x_l)(T(x_1, \ldots, x_l) \to T(x_1, \ldots, x_l)).$$

F-extension applied to D yields a clause which can be regarded as the skolemized form of

$$(\forall \vec{x})(((Q_{k+1}x_{k+1}) \ldots (Q_l x_l)T(x_1, \ldots, x_l)) \to$$
$$(Q_{k+1}^d x_{k+1}) \ldots (Q_l^d x_l)T(x_1, \ldots, x_l))$$

where $\vec{x}$ consists of $x_1, \ldots, x_k$ and $k < l$. Hence, with the coding scheme and F-extension, we can construct (the code of) a formula simulating the cut from a tautological clause.

The nonelementary speed-up in proof length does not depend on the underlying calculus; the result can also be obtained with other calculi like connection calculi or restricted forms of resolution.

4.2. *Generalizations of Functional Extension Principles*

Generalizations of F-extension are discussed in the sequel. One of these generalizations, defined by Eder, simulates the cut rule by two extensions, namely *junctorial extension* and *quantorial extension* (Eder, 1992).

DEFINITION 12. *A set of clauses S' is obtained from a set of clauses S by* junctorial extension *if $S' = S \cup S_0{}^2$ where*

$$S_0 = \{J \vee K^d \vee L^d, J^d \vee K, J^d \vee L\},$$

J has the form $p(x_1,\dots,x_k)$, M^d is the literal dual to M, and the following three conditions hold.

1. *The predicate symbol p does not occur in S and is possibly preceded by a negation sign.*
2. *The variables $x_1,\dots,x_k$ are pairwise distinct.*
3. *Every variable occurring in K or in L is in $\{x_1,\dots,x_k\}$.*

DEFINITION 13. *A set of clauses S' is obtained from a set of clauses S by* quantorial extension *if $S' = S \cup S_0{}^3$ where*

$$S_0 = \{J \vee K^d, J^d \vee L\},$$

and J has the form $p(x_1,\dots,x_k)$. The predicate symbol p does not occur in S and is possibly preceded by a negation sign. There is a variable x such that the following three conditions hold.

1. *The variables $x_1,\dots,x_k$ and the variable x are pairwise distinct.*
2. *Every variable occurring in K is in $\{x_1,\dots,x_k,x\}$.*
3. *There is a function symbol f of arity k not occurring in S such that L is obtained from K by replacing all occurrences of x in K by $f(x_1,\dots,x_k)$.*

The function f is called the Skolem function of the quantorial extension step.

---

[2] The clause set S_0 is the result of a transformation of $(\forall x_1)\dots(\forall x_k)\,(J \leftrightarrow (K \wedge L))$ into clause form (without negating the formula). Observe that $A \leftrightarrow B$ is an abbreviation for $(A \rightarrow B) \wedge (B \rightarrow A)$.

[3] The clause set S_0 is the result of a transformation of $(\forall x_1)\dots(\forall x_k)\,(J \leftrightarrow (\exists x)\,K)$ into clause form (without negating the formula).

Obviously, quantorial extension introduces a new function symbol. Observe that the structure of K is totally unrestricted. Hence, if we consider the branching degree of a node n in the search space generated by quantorial extension, then the branching degree of n is infinite. In contrast to quantorial extension, the branching degree of a node m generated by F-extension is finite, because the clause to which F-extension is applied must occur in the clause set under consideration.

The casual reader may have observed a strong similarity between the rules for the definitional transformations and the two extensions. Indeed, the above extension rules can be used to transform an arbitrary first-order formula into its definitional form in a structure-preserving way because the definitional transformation of a formula F can be obtained by extensions (see (Eder, 1992) for a detailed discussion).

Let us restrict our attention to quantorial extension in the sequel. Let us call the resolution principle extended by quantorial extension *QER-deduction*. From (Egly, 1993; Egly, 1994a) we know that QER-deduction is stronger than FR-deduction in the sense that there are clause sets for which the length of a shortest FR-refutation is exponential in the length of a shortest QER-refutation. In other words, FR-deduction cannot *polynomially simulate* QER-deduction. The reason for the nonexistence of a polynomial simulation is the possibility to instantiate variables in case QER-deduction is applied.

EXAMPLE 4. *Let S be a set of clauses and let $C = P(x,y) \vee Q(x,y)$ be a member of S. The equivalence $J \leftrightarrow (\exists z)\, Q(z,a)$ yields the clause set $S_0 = \{C_1, C_2\}$ with*

$$C_1 = \neg J \vee Q(c,a) \qquad\qquad C_2 = J \vee \neg Q(x,a)$$

where c is a globally new Skolem constant and a is a constant occurring in other clauses of S. S_0 is introduced by quantorial extension. *From C_1 and C_2, $Q(c,a) \vee \neg Q(x,a)$ is derived by one resolution step. By a further resolution step, $P(x,a) \vee Q(c,a)$ is derived. Such a clause cannot be derived by FR-deduction. The reason is the additional substitution of y by the constant a which is impossible in FR-deduction because resolution keeps substitutions minimal and F-extension cannot generate this substitution for y.*

If FR-deduction is extended by *tautology introduction*, i.e., clauses of the form $A \vee \neg A$ (for an atom A) can be introduced, then this extended resolution principle (called *FRTI-deduction*) can polynomially simulate QER-deduction. Instances of clauses can be obtained by introducing a proper tautological clause of the form $A \vee \neg A$ for an atom A and one resolution step. The length decrease enabled by the introduction of tautological clauses is surprising because these clauses are superfluous if unrestricted resolution is used and their deletion does not have any effect on the length of a minimal refutation. Recall that a similar behavior can be observed if restricted resolution principles are applied; additional tautological clauses can shorten refutations (see our exposition about splitting above).

A further property of QER-deduction and FRTI-deduction is the "unification" of the same argument positions in literals with the same predicate symbol and the same sign. Consider the clause $C = P(a) \vee P(b)$ for illustration. Since a and b are distinct constant symbols, there is no factor of C. If we introduce $J \leftrightarrow (\exists x) P(x)$, transform the result to clause form, derive $P(c) \vee \neg P(x)$ from the two resulting clauses by resolution and $D = P(c) \vee P(c)$ by two additional resolutions with C, then a factor of D, namely $P(c)$, is possible (see also Section 2.3 above). Such "strong" factors (of C) cannot be obtained by FR-deduction. The effects of strong factoring on the length of refutations have been investigated in (Egly, 1991; Egly, 1992; Egly, 1993; Egly, 1994a) where it is shown that FR-deduction cannot polynomially simulate extension principles enabling strong factors.

As already noted, local finiteness is an important property in automated deduction systems. Historically, the aim was even more enthusiastic: one wanted to get rid of quantifiers and to transform any formula into an equivalent quantifier-free formula which can be proved (or refuted) by pure propositional (finite) means. Although we know today that such a transformation is impossible because of undecidability of first-order logic, the historical approaches remain a good source for improvements (see Chapter 2 in (Hilbert and Bernays, 1939) for the details of the ε-calculus, and (Baaz and Leitsch, 1994a) and (Baaz et al., 1997) for two examples how induction principles can be obtained from this calculus).

One approach motivated by finding the aforementioned transformation is the ε-calculus. Quantifiers are represented as ε-*terms* (underlined in the following "definitions").

$$(\exists x)A(x,\vec{y}) \quad \text{is represented by} \quad A(\underline{\varepsilon_x(A(x,\vec{y}))},\vec{y}).$$
$$(\forall x)A(x,\vec{y}) \quad \text{is represented by} \quad A(\underline{\varepsilon_x(\neg A(x,\vec{y}))},\vec{y}).$$

The index variable x in the ε-terms above is a bound variable and can be renamed. Semantically, the ε-term $\varepsilon_x(A(x,\vec{y}))$ represents an object (if one exists) for which $A(x,\vec{y})$ holds. In contrast to skolemization, where a *satisfiability-equivalent* formula without strong quantifiers is obtained, the representation of quantifiers by ε-terms preserves logical equivalence.

There are some advantages using ε-terms instead of quantifiers and bound-ed variables.

(i) The axiomatization of first-order logic is simple; the axioms and rules for propositional logic together with the schema $A(t) \rightarrow A(\varepsilon_x(A(x)))$ suffices.

(ii) In contrast to bound variables, ε-terms have a meaning and can be evaluated (under some circumstances).

Why don't we use ε-terms instead of skolemization? The answer is that there are two drawbacks concerning the formalism and unification of ε-terms.

(i) The use of ε-terms in case of nested quantifiers is awkward. Con-sider the formula $(\exists x)(\exists y)(\exists z)P(x,y,z)$ and a step-wise elimination of quantifiers.

1. $(\exists x)(\exists y)\, P(x,y,\varepsilon_z(P(x,y,z)))$
2. $(\exists x)\, P(x,\varepsilon_y(P(x,y,\varepsilon_z(P(x,y,z)))),$
$\varepsilon_z(P(x,\varepsilon_y(P(x,y,\varepsilon_z(P(x,y,z))))),z)))$
3. The interested reader is invited to generate the formula for the final quantifier $(\exists x)$.

(ii) Unification of two ε-terms is *undecidable* in general. Hence, the local finiteness property is *not* retained if ε-terms are used.

The second drawback prevents the use of ε-terms in automated deduc-tion systems. What had been proved being computationally superior is the re-striction of "quantifier-elimination" to strong quantifiers and the correspond-ing generation of Skolem functions. Since the language of the formula is extended by applying skolemization, the resulting formula is satisfiability-equivalent (but not logically equivalent) to the given formula.

5. Conclusion

In the Sections 3 and 4, we have seen that an appropriate use of extension can lead to a spectacular (nonelementary) speed-up of proof complexity. The proof-theoretical principle behind all those speed-up results is cut-elimination and its nonelementary complexity. In case of skolemization and CNF-trans-formation, the speed-up effects are caused by the *preservation* of logical structure, which may be destroyed by inadequate methods in the normaliza-tion process. If there is not much structure in the original problem itself, it may be the case that all proofs using only the given "syntactic material" are

very long. Then we face the (much more difficult) problem to *create* structure. We have defined extension calculi capable of creating additional structure within clause logic; but the problem remains to find appropriate lemmata, i.e., to encode the corresponding cut formulae. By proving the inherent
nonelementary complexity of cut-elimination, Statman (Statman, 1979) and
Orevkov (Orevkov, 1979) (independently) defined lemmata leading to dramatic proof compressions. They proved that there exist sequences of problems admitting short (even linear) proofs with cuts, but the minimal length of
cut-free proofs cannot be bounded by an elementary function. Although the
problems of Statman and Orevkov are different (Statman uses equality while
Orevkov does not), the syntax of cut formulae (giving the short proofs) is of
striking similarity: they are inductively defined by implication and generalization. In particular, Statman's cut formulae are of the form

$$H_1(y) \quad = \quad (\forall x)f(p,x) = f(p,f(y,x)),$$
$$H_{m+1}(y) \quad = \quad (\forall x)(H_m(x) \to H_m(f(y,x))).$$

We may ask whether there are other types of formulae which may serve as
"essential" cut formulae. This question is of major importance to proof search
in automated deduction: assume that an analytic (or quasi-analytic) calculus is
applied to a problem but does not give a proof within acceptable time bounds.
Most probably, the minimal length of a refutation is too high. To overcome
this problem, an extension method is applied to introduce lemmata in order
to find a shorter proof. Trying all derivable formulae as lemmata may be as
destructive to proof search as the original method itself. Thus the problem
remains, which type of lemma is potentially useful in the case under consideration. This leads to the following crucial question:

*To what extent is it possible to restrict the syntax of formulae and — at the same time
— keep their power as cut formulae in a proof?*

In (Baaz and Leitsch, 1996) this problem was investigated for several normal forms and syntax types of formulae. It turned out that restricting cuts to
negation normal form (NNF) is always beneficial, while prenex form may de-
teriorate the situation. Moreover it is shown in (Baaz and Leitsch, 1996) that
elimination of monotone cuts (i.e., cut formulae over $\land, \lor, \forall, \exists$) over Horn
theories is of (exactly) exponential complexity. As proof search is exponential
in the length of a shortest proof, this result implies that a search for monotone
cuts over Horn theories does not make sense. Instead, substantial cut formulae must contain negations or implications; if they contain negation then the
occurrences of $\neg$ may be restricted to signed atom formulae (NNF).

   The results mentioned above can be considered as a first step toward a
theory of cut-introduction. Such a theory would not only deepen our under-

standing of mathematical proofs, but could also yield stronger tools (like e.g., *cut-refinements*) for automated deduction.

Acknowledgements: We thank H. Tompits and the second reader for comments on earlier versions of this chapter.

REFERENCES

Andrews, P. B.: 1981, 'Theorem Proving via General Matings'. *J. ACM* **28**(2), 193–214.

Baaz, M., U. Egly, and C. G. Fermüller: 1997, 'Lean Induction Principles for Tableaux'. In: D. Galmiche (ed.): *Proceedings of the Sixth Workshop on Theorem Proving with Analytic Tableaux and Related Methods.* pp. 62–75, Springer Verlag. LNAI 1227.

Baaz, M., U. Egly, and A. Leitsch: 1998, 'Normal Form Transformations'. In: J. A. Robinson and A. Voronkov (eds.): *Handbook of Automated Reasoning.* Elsevier Science.

Baaz, M., C. Fermüller, and A. Leitsch: 1994, 'A Non-Elementary Speed Up in Proof Length by Structural Clause Form Transformation'. In: *LICS'94.* Los Alamitos, California, pp. 213–219, IEEE Computer Society Press.

Baaz, M. and A. Leitsch: 1985, 'Die Anwendung starker Reduktionsregeln in automatischen Beweisen'. *Proc. of the Austrian Acad. of Science II* **194**, 287–307. In German.

Baaz, M. and A. Leitsch: 1990, 'A Strong Problem Reduction Method Based on Function Introduction'. In: *Proceedings of the International Symposium on Symbolic and Algebraic Computation (ISSAC).* pp. 30–37, ACM Press.

Baaz, M. and A. Leitsch: 1992, 'Complexity of Resolution Proofs and Function Introduction'. *Annals of Pure and Applied Logic* **57**, 181–215.

Baaz, M. and A. Leitsch: 1994a, 'Methods of Functional Extension'. In: *Collegium Logicum: Annals of the Kurt Gödel Society.* Springer Wien New York, pp. 87–122.

Baaz, M. and A. Leitsch: 1994b, 'On Skolemization and Proof Complexity'. *Fundamenta Informaticae* **20**, 353–379.

Baaz, M. and A. Leitsch: 1996, 'Cut Normal Forms and Proof Complexity'. Technical Report TR-CS-BL-96-1, Institut für Computersprachen, TU Wien.

Boolos, G.: 1984, 'Don't Eliminate Cut'. *Journal of Philosophical Logic* **13**, 373–378.

Boolos, G.: 1987, 'A Curious Inference'. *Journal of Philosophical Logic* **16**, 1–12.

Boy de la Tour, T.: 1990, 'Minimizing the Number of Clauses by Renaming'. In: M. E. Stickel (ed.): *Proceedings of the 10th International Conference on Automated Deduction: Kaiserslautern, 24.–27. Juli 1990.* Berlin, pp. 558–572, Springer Verlag. LNAI 449.

Bruschi, M.: 1991, 'The Halting Problem'. *AAR Newsletter* pp. 7–12.

Burkholder, L.: 1987, 'The Halting Problem'. *SIGACT News* **18**(3), 48–60.

Dafa, L.: 1993, 'A Mechanical Proof of the Halting Problem in Natural Deduction Style'. *AAR Newsletter* pp. 4–9.

Dafa, L.: 1994, 'The Formulation of the Halting Problem is Not Suitable for Describing the Halting Problem'. *AAR Newsletter* pp. 1–7.

Eder, E.: 1984, 'An Implementation of a Theorem Prover Based on the Connection Method'. In: W. Bibel and B. Petkoff (eds.): *AIMSA 84, Artificial Intelligence - Methodology, Systems, Applications, Varna, Bulgaria.* North-Holland Publishing Company.

Eder, E.: 1992, *Relative Complexities of First Order Calculi.* Braunschweig: Vieweg.

Egly, U.: 1990, 'Problem-Reduction Methods and Clause Splitting in Automated Theorem Proving'. Master's thesis, Technische Universität Wien, Institut für Computersprachen, Abteilung für Anwendungen der Formalen Logik, Resselgasse 3/1, A–1040 Wien.

Egly, U.: 1991, 'A Generalized Factorization Rule Based on the Introduction of Skolem Terms'. In: H. Kaindl (ed.): *Österreichische Artificial Intelligence Tagung.* Berlin, Heidelberg, New York, pp. 116–125, Springer Verlag.

Egly, U.: 1992, 'Shortening Proofs by Quantifier Introduction'. In: A. Voronkov (ed.): *Proceedings of the International Conference on Logic Programming and Automated Reasoning.* pp. 148–159, Springer Verlag. LNAI 624.

Egly, U.: 1993, 'On Different Concepts of Function Introduction'. In: G. Gottlob, A. Leitsch, and D. Mundici (eds.): *Proceedings of the Kurt Gödel Colloquium.* pp. 172–183, Springer Verlag. LNCS 713.

Egly, U.: 1994a, 'On Methods of Function Introduction and Related Concepts'. Ph.D. thesis, TH Darmstadt, Alexanderstr. 10, D–64283 Darmstadt.

Egly, U.: 1994b, 'On the Value of Antiprenexing'. In: F. Pfenning (ed.): *Proceedings of the International Conference on Logic Programming and Automated Reasoning.* pp. 69–83, Springer Verlag. LNAI 822.

Egly, U.: 1996, 'On Different Structure-preserving Translations to Normal Form'. *J. Symbolic Computation* **22**, 121–142.

Egly, U. and T. Rath: 1996, 'On the Practical Value of Different Definitional Translations to Normal Form'. In: M. McRobbie and J. K. Slaney (eds.): *Proceedings of the Conference on Automated Deduction.* pp. 403–417, Springer Verlag. LNAI 1104.

Hilbert, D. and P. Bernays: 1939, *Grundlagen der Mathematik II.* Springer Verlag.

Lee, S. and D. A. Plaisted: 1994, 'Use of Replace Rules in Theorem Proving'. *Methods of Logic in Computer Science* **1**, 217–240.

Orevkov, V. P.: 1979, 'Lower Bounds for Increasing Complexity of Derivations after Cut Elimination'. *Zapiski Nauchnykh Seminarov Leningradskogo Otdeleniya Matematicheskogo Instituta im V. A. Steklova AN SSSR* **88**, 137–161. English translation in *J. Soviet Mathematics,* 2337–2350, 1982.

Plaisted, D. A. and S. Greenbaum: 1986, 'A Structure-Preserving Clause Form Translation'. *J. Symbolic Computation* **2**, 293–304.

Reckhow, R. A.: 1976, 'On the Length of Proofs in the Propositional Calculus'. Ph.D. thesis, Department of Computer Science, University of Toronto.

Robinson, J.: 1965, 'A Machine-Oriented Logic Based on the Resolution Principle'. *J. ACM* **12**(1), 23–41.

Statman, R.: 1979, 'Lower Bounds on Herbrand's Theorem'. In: *Proc. AMS 75*. pp. 104–107.

Tseitin, G. S.: 1968, 'On the Complexity of Derivation in Propositional Calculus'. In: A. O. Slisenko (ed.): *Studies in Constructive Mathematics and Mathematical Logic, Part II*. Leningrad: Seminars in Mathematics, V.A. Steklov Mathematical Institute, vol. 8, pp. 234–259. English translation: Consultants Bureau, New York, 1970, pp. 115–125.

JÖRG DENZINGER AND MATTHIAS FUCHS

CHAPTER 13

A COMPARISON OF EQUALITY REASONING HEURISTICS

1. INTRODUCTION

Most logics of practical interest (e.g., first order, equality) in general are undecidable. Theorem proving in the context of an undecidable logic means exploring infinite search spaces. To this end, search-guiding heuristics must be employed. The use of so-called *fair* heuristics basically guarantees that each point in the search space will be reached after a finite period of time. Under these conditions, a proof will eventually be found if it exists (i.e., we have a semi-decision procedure). Note, however, that the notion "fairness" is mainly of theoretic interest. In practice, it is important to find a proof in an "acceptable period of time" rather than "eventually". Apart from this, resource limitations, in particular memory space, often make it impossible to conduct a fair search. Hence, in practice search procedures that can be unfair for several reasons are quite common.

From a different point of view, search-guiding heuristics control the application of inference rules. The quality of the control determines how long it will take to find a proof of a theorem presented by a user and thus critically influences the acceptance of the prover. Ideally, the control always chooses the inference rule as the next step of the proof that finally results in a shortest proof for the theorem at hand. One way to achieve this "optimal" control would be to generate all proofs of the problem, determine the (or a) shortest proof and then ground the selection of the next inference step on this proof.

Although this method for achieving an optimal control is impractical, there are two lessons to be learned here. Firstly, the time spent on deciding which inference step to do must be set off against the gains these computations provide. If one control algorithm allows us to prove a problem with 10,000 inference steps requiring 0.1 seconds for each of its 10,000 control decisions, while another algorithm results in only 101 inference steps with 10 seconds per decision, then the "bad" control presents a proof faster to the user than the "good" one. Secondly, control seems to require knowledge. The better the knowledge is, the better the control can be.

In summary, good control requires quickly processable, adequate and optimally selected knowledge. In reality, we have to construct heuristics that very often represent general and inadequate knowledge and produce proof runs in

W. Bibel, P. H. Schmitt (eds.), Automated Deduction. A basis for applications. Vol. II
© 1998 *Kluwer Academic Publishers. Printed in the Netherlands*

which less than one percent of the inferences made contribute to the proof found.

In this chapter we present and empirically evaluate concepts for developing control heuristics for theorem provers. The main focus is on demonstrating how and what knowledge can be used by heuristics and what gains this knowledge provides in the search process (and what additional obstacles are involved in using such knowledge). Starting with simple statistical criteria we will then present functional interpretations for symbols that have to be chosen by using knowledge about the given problem (and often also require some experiments to get them right). The third type of heuristics involves knowledge about the calculus used and about similarity of formulas with respect to this calculus. With this knowledge a goal-oriented control is possible.

The fourth type of heuristics tries to get rid of the involvement of the user in the encoding of knowledge into control heuristics by employing *learning*. By re-enacting proofs to problems that are similar to the problem at hand while still flexibly searching along the path given by this leading proof, a learning heuristic can be used to solve harder and harder problems in a domain. By first using easy problems (solved by one of the other heuristics) to solve problems that are a little more difficult, then using these to solve problems that are even a little more difficult and so on the prover can, in a *bootstrapping* manner, learn enough to finally solve very hard problem without the need for the user to try out parameter settings.

In our experiments we will base our comparison of these four types of control heuristics on instantiations for pure equational deduction using the completion method. We will use domains from the TPTP library to evaluate and compare the individual heuristics. We will also characterize the situations in which certain heuristics should be used. We will, however, not address the fairness of the heuristics introduced here. As mentioned above, fairness is a property that is mainly of theoretic interest and bears little or no importance in practice. Since the purpose of this chapter is of experimental and hence practical nature, a treatment of fairness would be out of place. Nonetheless we would like to point out that most of the heuristics to be presented are indeed fair when using them with "reasonable" parameter settings (cp. Fuchs (1997a)).

This chapter is organized as follows: After this introduction we will briefly present equational deduction and completion in Section 2. In Section 3 we will describe the four types of heuristics. An empirical comparison of these heuristics is the topic of Section 4. Finally, in Section 5 we close with some remarks about possibilities for more use of learning and knowledge-based techniques in control.

2. EQUATIONAL REASONING AND COMPLETION

Equational reasoning deals with solving the following problem:

Input: E, a set of equations over a fixed signature sig;

   $s = t$, a goal equation over sig

Question: Does $s = t$ hold in every model of E ?

Let $Th(E)$ denote the set of equations over sig that hold in every model of E. By Birkhoff's theorem we have $s = t \in Th(E)$ iff s can be transformed into t by *replacing equals by equals*. It is well-known that provers based on rewriting and completion techniques developed by Knuth and Bendix (1970) are efficient for this problem. In order to avoid failure of the completion procedure due to the fact that some equations cannot be oriented and therefore cannot be used as rules, we employ an *unfailing completion procedure* as our basic proof procedure (see Hsiang and Rusinowitch (1987), Bachmair *et al.* (1989)).

We assume that the reader is familiar with rewriting and completion techniques. For an overview see Avenhaus and Madlener (1990) and Dershowitz and Jouannaud (1990). We use the standard notation.

A signature $sig = (S, F, \tau)$ consists of a set S of sorts, a set F of operators and a function $\tau : F \to S^+$ that determines the input and output sorts of the operators. Let $T(F, V)$ denote the set of terms over F and a set V of variables. We write $t[s]_p$ to denote that $s \equiv t/p$, i.e. s is the subterm of t at position p. $T(F) = T(F, \emptyset)$ denotes the set of *ground terms* over F. Let K be a set of new constants (i.e., $F \cap K = \emptyset$). A *reduction ordering* $\succ$ is a well-founded ordering on $T(F \cup K, V)$ that is compatible with substitution and the term structure, i.e., $t_1 \succ t_2$ implies $\sigma(t_1) \succ \sigma(t_2)$ and $t[t_1]_p \succ t[t_2]_p$. If $\succ$ is total on $T(F \cup K)$ then $\succ$ is called a *ground reduction ordering*.

A *rule* is an oriented equation, written $l \to r$ such that $Var(r) \subseteq Var(l)$. A set R of rules is *compatible* with $\succ$ if $l \succ r$ for every $l \to r$ in R. If E is a set of equations then $R_E = \{\sigma(u) \to \sigma(v) \mid u \doteq v \text{ in } E, \sigma \text{ a substitution}, \sigma(u) \succ \sigma(v)\}$ is the set of orientable instances of equations in E. (We use $u \doteq v$ to denote $u = v$ or $v = u$.) Finally, we have $R(E) = R \cup R_E$.

Let $u \doteq v$ and $s \doteq t$ be equations in $E \cup R$. Let u/p be a non-variable subterm of u that is unifiable with s, say with most general unifier $\sigma = mgu(u/p, s)$. Then $\sigma(u[t]_p) = \sigma(v)$ is in $Th(R \cup E)$. If $\sigma(u[t]_p) \not\succ \sigma(u)$ and $\sigma(v) \not\succ \sigma(u)$ then $\sigma(u[t]_p) = \sigma(v)$ is a *critical pair* of R, E. $CP(R, E)$ denotes the set of all critical pairs of R, E.

We are now ready to define the unfailing completion procedure used by our prover DISCOUNT (see Avenhaus *et al.* (1995) and Denzinger *et al.* (1997b)).

The set E and g, a ground equation over $F \cup K$ (the skolemized goal $s = t$) and a reduction ordering $\succ$ are given. The procedure uses sets R, E and CP. The input equations are put into CP (therefore $E = \emptyset$). Then the following loop is repeated until the normal forms of the terms of g are identical or subsumed by an equation of E or until the set CP is empty: Select an equation $u = v$ from CP. Let u' and v' be normal forms of u and v with respect to $R(E)$. If neither $u' = v'$ is subsumed by an equation in E nor $u' \equiv v'$, then all critical pairs between $u' = v'$ and E and R are added to CP. If u' and v' are comparable with respect to $\succ$, then the respective rule is added to R. Otherwise $u' = v'$ is added to E. All elements of E and R that can be reduced with the new rule or equation are removed from R or E and their normal forms are added to CP.

Analyzing the completion procedure reveals that the important indeterminisms that require control are the computation of a normal form of a term and the selection of the next critical pair. Although it is possible that different functions for computing a normal form can effect the search for proofs of certain problems, these effects are essentially due to critical pairs that are not selected early enough to avoid this problem. Therefore the selection of the next critical pair is the crucial point in a completion procedure.

3. SELECTION HEURISTICS

As already stated, controlling the search of a completion-based theorem prover amounts to the selection of the next critical pair to be processed. In contrast to, for example, many resolution-based theorem provers, our completion procedure allows the control mechanism a relatively good anticipation of what the equation or rule that is going to be added looks like. We do not have to work with ancestors, but can rate (explicitly represented) critical pairs. However, most of the following criteria can also be employed if we are dealing with ancestors (in the absence of explicitly represented descendants).

The general idea for controlling the selection of a critical pair is to associate an integer value (or value for short) with each critical pair and to always select a critical pair with a minimal value. If there are several pairs with a minimal value, then the selection is made according to the FIFO strategy. Note that the more pairs are given the same value the less the control remains in the hand of the developer of the system. Therefore one goal when developing functions that compute values for critical pairs is to avoid assigning the same value to many (or even more than a few) critical pairs. It would be ideal to have an injective function, but unfortunately this is often not possible. This is due to the (often vague) knowledge that is used by the functions. Since these functions provide a focus for the search of the prover we call them *focus functions*.

The main problem when developing focus functions is how to encode knowledge the developer has about

- mathematical expertise
- the proof calculus
- several domains
- already found proofs for other problems

and much more into the functions. In the following subsections we will provide examples for this encoding. As we will see in section 4, the different functions enable our system to solve many problems, but there are unfortunately many problems that can be solved by only a few of the functions (and often these functions differ).

Note that it is very rarely possible to combine several (different) focus functions to form a single one (in order to obtain a function for solving more problems than the functions that are combined). The knowledge represented by different focus functions is often contradictory. Therefore each combination must either favor one of the functions (thus rendering the others nearly useless) or the same value will be assigned to many critical pairs (with all the problems already described). So, in general the combination of several (different) focus functions has to be achieved by other means, as for example by cooperation between them as in the TEAMWORK method (see Denzinger (1995) and Chapter 3 of this part of this volume).

3.1. *Statistical Criteria*

Focus functions employing statistical criteria have the following guideline: *"Small (size) is beautiful"*. As already stated, we always select those critical pairs to which a focus function assigned minimal values. This takes care of the attribute small. The focus functions we present in this subsection have to characterize the size of a critical pair.

Obviously, counting the symbols in a term measures the size of the term. But if we look at the definition of terms, we can observe that they are constructed using two sets, function symbols and variables. It makes sense to take this into account when counting symbols in a term, since terms can be substituted for variables (during the reasoning process) while the other symbols will persist. By giving function symbols and variables different base weights, we already have several possible term weights.

DEFINITION 3.1. (Weight). *Let t be a term and w_V and w_F parameters. Then the term weight* Weight *of t is recursively defined by*

$$\text{Weight}(t) = \begin{cases} w_V, & t \in V \\ w_F + \sum_{i=1}^{n} \text{Weight}(t_i), & t \equiv f(t_1, \ldots, t_n). \end{cases}$$

Since a critical pair consists of two terms, there are several ways to combine the term weights to obtain a weight for the whole critical pair. In fact, these possible combinations will also exist if we compute term weights according to other criteria. Therefore we make the following general definition which also gives us a naming convention.

DEFINITION 3.2. (Add, Max). *Let W be a term weight and $u = v$ an equation. Then the focus functions* AddW *and* MaxW *are defined by*

$$\begin{aligned} \text{AddW}(u = v) &= W(u) + W(v) \\ \text{MaxW}(u = v) &= \max\{W(u), W(v)\} \end{aligned}$$

So, by AddWeight we characterize the focus function that combines the term weight Weight of the two terms of a critical pair by adding them.

Please note that even small variations of the parameters w_V and w_F result in quite different focus functions (see Section 4.2).

3.2. *Semantical Criteria*

In theory function symbols are just symbols without any deeper meaning (semantics). In practice, however, a user of a theorem prover associates some meaning at least with a few of these symbols. Therefore ways should be provided for the user to communicate these meanings to the prover. Naturally, the result must be a syntactical procedure, an *interpretation* of the term resulting in a term weight.

A first idea to provide such interpretations is to refrain from treating all function symbols equally (as in the case of Weight), but to permit different values for different elements of F. This leads us to the term weight FWeight.

DEFINITION 3.3. (FWeight). *Let t be a term, w_V a parameter and $\nu : F \to N$ a numerical interpretation of the function symbols. Then the term weight* FWeight *of t is recursively defined by*

$$\text{FWeight}(t) = \begin{cases} w_V, & t \in V \\ \nu(f) + \sum_{i=1}^{n} \text{FWeight}(t_i), & t \equiv f(t_1, \ldots, t_n). \end{cases}$$

Naturally, there are also more complex possibilities for an interpretation of function symbols. Stickel (1984) used polynomials as interpretations. This defines the term weight Poly.

DEFINITION 3.4. (Poly). *Let t be a term, w_V a parameter and ρ a function associating with each n-ary element of F a polynomial in n unknowns, its polynomial interpretation. Then the term weight Poly of t is recursively defined by*

$$\text{Poly}(t) = \begin{cases} w_V, & t \in V \\ \rho(f)(\text{Poly}(t_1), \ldots, \text{Poly}(t_n)), & t \equiv f(t_1, \ldots, t_n). \end{cases}$$

Polynomial interpretations already permit a very fine tuning of the control of a theorem prover, since they can be used to assign very small weights to only a few selected terms, thus forcing an early selection of critical pairs containing these terms. However, much experience is required to be able to choose the right polynomials.

Allowing for arbitrary functional interpretations provides even more fine tuning possibilities, but almost always requires several tries until a suitable interpretation is found. According to our experience FWeight is satisfactory for most of the problems for expressing further knowledge of a user.

Another kind of semantical information are "defining" equations. Such equations have the form $f(x_1, \ldots, x_n) = t$, where t is typically a large term (without an occurrence of the function symbol f). The term weights presented so far will in most cases result in focus functions that select such a defining equation very late in the search process due to the size of t. Nevertheless, a defining equation allows for the elimination of a function symbol (provided that an appropriate reduction ordering is used) and very often advances the search process. In order to select defining equations and variants of it early we use the following focus function.

DEFINITION 3.5. (GTW). *Let W be a term weight and $u = v$ be an equation. The focus function GTW is defined by*

$$\text{GTW}(u = v) = \begin{cases} W(u), & u \succ v \\ W(v), & v \succ u \\ W(u) + W(v), & \text{else.} \end{cases}$$

The definition of GTW has the side effect that critical pairs that cannot be oriented (and that therefore cause more computational effort, since equations typically produce more new critical pairs than rules) are in a way penalized, since both terms contribute to the value of a pair.

3.3. *Goal-oriented Criteria*

The problem we want to solve is to show the validity of a goal equation. If we look at the completion procedure and the focus functions presented so far, we observe that this goal equation does not influence the search process at all (except by providing a termination condition for the main loop). Since the computation of normal forms is an important part of the completion procedure that increases its efficiency (by pruning the search space), strong rules are welcome, even if they are not necessary for the proof of a given goal (but in fact such strong rules are very often needed in a proof). Therefore it is advisable to avoid an exclusive concentration on the goal equation during the search.

But nevertheless the goal should somehow be taken into account when selecting the next critical pair. Focus functions that consider the goal in their computations provide a goal orientation (instead of goal direction) of the search process.

Goal-oriented focus functions include aspects of the goal into the computation of a value of a critical pair. Typically, these aspects are the difference between goal and critical pair with respect to one or several features or *measures*. Our focus function OccNest (Denzinger and Fuchs, 1994) concentrates on structural differences between goal and critical pair, namely on the occurrences of function symbols and the nesting of them.

DEFINITION 3.6. (occ, nest). *Let t be a term. The number of occurrences* $\mathrm{occ}(f,t)$ *is recursively defined by*

$$\mathrm{occ}(f,t) = \begin{cases} 0, & t \in V \\ \sum_{i=1}^n \mathrm{occ}(f,t_i), & t \equiv g(t_1,\ldots,t_n), f \not\equiv g \\ 1 + \sum_{i=1}^n \mathrm{occ}(f,t_i), & t \equiv f(t_1,\ldots,t_n). \end{cases}$$

The nesting $\mathrm{nest}(f,t)$ *computes the maximal number of consecutive occurrences of f in t and is recursively defined by*

$$\mathrm{nest}(f,t) = \Upsilon(f,t,0,0)$$

where

$$\Upsilon(f,t,c,a) =$$

$$\begin{cases} \max(\{c,a\}), & t \in V \text{ or } t \text{ is a constant} \\ \max(\{\Upsilon(f,t_i,0,\max(\{c,a\})) \mid 1 \le i \le n\}), & t \equiv g(t_1,\ldots,t_n), f \not\equiv g \\ \max(\{\Upsilon(f,t_i,c+1,a) \mid 1 \le i \le n\}), & t \equiv f(t_1,\ldots,t_n) \end{cases}$$

We use Weight as the basic value of a term t and $\mathrm{occ}(f,t)$ and $\mathrm{nest}(f,t)$ for refinements to express the structure of t. Next we extend occ and nest to equations $u = v$ by

$$\begin{aligned} \mathrm{occ}(f,(u,v)) &= \max(\{\mathrm{occ}(f,u),\mathrm{occ}(f,v)\}) \\ \mathrm{nest}(f,(u,v)) &= \max(\{\mathrm{nest}(f,u),\mathrm{nest}(f,v)\}) \end{aligned}$$

Finally we define the value $\mathsf{OccNest}(u = v)$ according to the following idea: We start with the basic weight $\mathsf{AddWeight}(u = v)$ of the equation $u = v$ and modify it with a penalty to describe the difference between the structural complexity of $u = v$ and the goal $s = t$. There are several ways to combine these values. We chose to apply the penalty in the form of a factor. This factor has to be at least 1; this is ensured by using the function $\psi(x) = 1$, if $x \leq 0$; otherwise $\psi(x) = x + 1$. In order to be flexible, we make use of a set $D \subseteq F$ to describe which operators $f \in F$ should contribute to the value $\mathsf{OccNest}(u = v)$.

DEFINITION 3.7. ($\mathsf{OccNest}$). *Let $u = v$ be a critical pair and $s = t$ the goal. Let furthermore $D \subseteq F$. For all $f \in F$ we define:*

$$\Delta_{\mathsf{occ}}(f) \;=\; \mathsf{occ}(f, (u, v)) - \mathsf{occ}(f, (s, t))$$
$$\Delta_{\mathsf{nest}}(f) \;=\; \mathsf{nest}(f, (u, v)) - \mathsf{nest}(f, (s, t))$$

$$m_f = \begin{cases} 1, & f \notin D \\ \psi(\Delta_{\mathsf{occ}}(f)) \cdot \psi(\Delta_{\mathsf{nest}}(f)), & else. \end{cases}$$

Then we have :

$$\mathsf{OccNest}(u = v) = \mathsf{AddWeight}(u = v) \cdot \prod_{f \in F} m_f.$$

Naturally, it depends on the given problem whether OccNest will provide a good or bad focus. One can expect OccNest to work fine, if the goal to prove has some minimal structure to compare critical pairs with. But if there is too much structure, which means more structure than in most of the critical pairs, then OccNest will have problems. If there are very different axioms (with respect to the structure measured by occ and nest) that result in many very different critical pairs, then OccNest has a good chance of selecting useful pairs. Then the set of axioms may even be large without causing trouble for OccNest (which unfortunately is not the case for functions using statistical criteria).

3.4. *Focus Functions Using Learned Knowledge*

The focus functions presented so far are either very general, which means that they can be used for many problems, but will fail for the hard ones, or they require help from the user of the system who has to have experience both in the problem domain and the system in order to help it in solving hard problems. Obviously, this does not favor the use of the system by users who want to have solutions of their problems and who do not want to spend several months (or more) to learn the (right) usage of the system.

An alternative to letting the user learn is to allow the prover to learn. In the past few years several methods have been suggested how a prover can learn from previous successful proof attempts and how it can use the learned knowledge to solve other (harder) problems (see Kolbe and Walther (1994), Denzinger and Schulz (1996)).

Our focus function FlexRe uses a so-called *source problem* with a source proof to search for the proof of a given *(target) problem*. The general idea is to re-enact the source proof, but also to allow for deviations that give the approach some flexibility (which means here that FlexRe not only allows the prover to find the source proof again, but also proofs of other problems that contain many steps of the source proof and a few new steps).

DEFINITION 3.8. (FlexRe). *Let P be a set of equations (usually the critical pairs that have to be selected in order to prove some source problem). For any critical pair u = v (with u′ = v′ or $u_1 = v_1$ and $u_2 = v_2$ denoting ancestors of u = v) we have*

$$D(u = v) = \begin{cases} 0, & \text{if } u = v \text{ subsumes an equation in } P \\ 100, & \text{otherwise.} \end{cases}$$

$$d(u = v) = \begin{cases} \psi\left(q, D(u = v)\right), & \text{no ancestors} \\ \psi\left(d(u' = v'), D(u = v)\right), & \text{one ancestor} \\ \psi\left(\gamma(d(u_1 = v_1), d(u_2 = v_2)), D(u = v)\right), & \text{two ancestors} \end{cases}$$

where $q \in \{0, \ldots, 100\}$ and

$$\psi(x, y) = \begin{cases} 0, & y = 0 \\ \min(x, y) + \lfloor q_2 \cdot (\max(x, y) - \min(x, y)) \rfloor, & \text{otherwise.} \end{cases}$$

$$\gamma(x, y) = \min(x, y) + \lfloor q_1 \cdot (\max(x, y) - \min(x, y)) \rfloor, \ q_1, q_2 \in [0; 1].$$

The focus function FlexRe is defined by

$$\text{FlexRe}(u = v) = (d(u = v) + p) \cdot H(u = v), \quad p \in \mathbf{N}.$$

where H can be one of the previously introduced (or any other) focus functions.

FlexRe attempts to re-enact the source proof given by P by giving a small distance value $d(u = v)$ to critical pairs that "agree" with the source proof. The distance increases as critical pairs (and their ancestors) are "farther away" from the source proof. (See Fuchs (1996a), Fuchs (1997a), Fuchs (1997b) or Denzinger *et al.* (1997a) for more details.)

The parameter p controls the effect of the distance measure $d(u = v)$ on the final weight FlexRe$(u = v)$. $d(u = v)$ will be dominant if $p = 0$. In this case, if $d(u = v) = 0$, FlexRe$(u = v)$ will also be 0 regardless of $H(u = v)$. As p grows, H increasingly influences the final weight, thus mitigating the inflexibility of the underlying method, namely using $d(u = v)$ *alone* as a measure for the suitability of a fact $u = v$. For very large p, the influence of $d(u = v)$ becomes negligible, and FlexRe basically degenerates into H.

FlexRe is very profitable when source and target proof (i.e., a proof of a problem that is to be found) share many facts or in other words profit from almost the same selection of critical pairs. If, however, the source is inappropriate, then focusing on the source proof as it is done by FlexRe will be counterproductive.

4. EXPERIMENTAL COMPARISON

A sensible and meaningful comparison of search-guiding heuristics, i.e. focus functions, is always a difficult task. Most focus functions (and all focus functions presented here) are generic in the sense that they are parameterized; specific values must be chosen for the respective parameters in order to obtain an operational focus function. The choice of parameters, however, critically influences performance. Given a certain problem to be solved, one parameter setting may enable a focus function to succeed very quickly, whereas a perhaps only slightly different setting causes it to fail.[1] Therefore, strictly speaking the following comparison merely compares certain instances of (generic) focus functions and does as such not necessarily give a general picture of the potential of each (generic) focus function. But from a practical point of view, such a restricted use of focus functions is inevitable. Nonetheless, even their restricted use can give us some ideas of possible tendencies and "phenomena".

4.1. *First Experimental Study: "General-purpose" Focus Functions*

The first part of our experimental studies compares instances of AddWeight, MaxWeight, GTWeight, and OccNest by choosing $w_F = 2$ and $w_V = 1$. For OccNest the additional parameter $D = F$. Furthermore, we employ the instance of AddWeight obtained by letting $w_V = w_F = 1$. (Hence, this instance

---

[1] The notion "failure" is to be interpreted as "failure to succeed within some given period of time".

Table I. Summary of Experiments with "General-Purpose" Focus Functions

Domain	AddWeight$_1^2$	MaxWeight	GTWeight	AddWeight$_1^1$	OccNest
GRP	87 (69.6%)	85 (68%)	89 (71.2%)	91 (72.8%)	92 (73.6%)
LCL	22 (68.8%)	22 (68.8%)	22 (68.8%)	22 (68.8%)	15 (46.9%)
ROB	5 (26.3%)	0 (0%)	8 (42.1%)	8 (42.1%)	6 (31.6%)

simply counts symbols.) AddWeight$_1^2$ and AddWeight$_1^1$ denote these two instances of AddWeight, respectively.

This first batch of experiments essentially deals with testing "general-purpose" focus functions, i.e. focus functions employing statistical and goal-oriented criteria, on a large number of problems in order to obtain a first taste of and some insight into their strengths and weaknesses. Note that FWeight is a generalization of Weight; we have FWeight = Weight if $v(f) = w_F$ for all $f \in F$. Note also that Poly is a generalization of FWeight (and thus of Weight). In other words, FWeight is a special case of Poly that admits only linear polynomials. We will concentrate on FWeight and its potential in Section 4.2. We refrain here from any further, in a way more specialized use of Poly, because a reasonable and beneficial application of non-linear polynomials is very problem-dependent and therefore questionable for large-scale comparisons based on a variety of problems.

The problems selected for our experimental comparison are taken from the TPTP problem library (Sutcliffe *et al.*, 1994). Three domains were considered: GRP (groups), LCL (logic calculi), and ROB (Robbins algebra). Naturally, only the problems specified solely by positive unit-equality literals were eligible. There are 125, 32, and 19 such problems in the domains GRP, LCL, and ROB (TPTP version 1.2.0), respectively. Table I summarizes the experiments with DISCOUNT (using the instances of the focus functions outlined above) in terms of number of problems solved (and success rate). DISCOUNT was run on a SPARCstation 10 and was granted at most 600 seconds per problem.

Table I gives us a rough idea of the performance of the tested focus functions with respect to the three domains. It reveals that, for instance, OccNest is well suited for problems from GRP (92 of 125 problems solved), but less suitable for problems from LCL (only 15 of 32 problems solved). Likewise it shows that MaxWeight performs very poorly in the domain ROB (0% success rate).

Table I, however, does not provide enough information to declare one focus function the "winner" and another one the "loser"—not even for the domains under investigation. OccNest has the highest success rate for GRP (73.6%), but this does not mean that OccNest "subsumes" any of the other focus functions in the sense that any problem which can be solved by the other focus functions can also be solved by OccNest. That is, the (solvable) problems are not necessarily in a subset relation. Moreover, the fact that all tested focus functions (except for OccNest) have the same success rate for LCL does not imply that they can be interchanged arbitrarily: The problems solved by these focus functions are not necessarily the same.

Furthermore, there occasionally are significant differences in run-time. Hence, for a certain (sub-) domain, a certain focus function might be more profitable than another one although the success rate does not indicate this circumstance.

All these possibilities could actually be observed in the experiments as the detailed tables II through IV demonstrate. The entries in the tables are approximate run-times in seconds. We restricted the presentation of detailed results to those results which contribute to a meaningful comparison. Thus comprehensiveness of the tables could be improved, while the size of the tables could be reduced reasonably. To this end, we generally omitted all problems that none of the tested focus functions could solve before the time-out of 10 minutes. (23, 7, and 11 problems did not pass this filter in the domains GRP, LCL, and ROB, respectively.) In the domain LCL we also ignored 11 problems that all focus functions could solve in less than 20 seconds. (The difference in performance was rather insignificant for these problems; the ratio of run-times of slowest and fastest focus function was always less than 6.) For the same reason 51 problems were ignored in connection with GRP. Furthermore, for some in a way related problems in the domain GRP (e.g., problems GRP154-1 through GRP159-1 and GRP170-1 through GRP170-4) the focus functions behaved almost identically. Table II shows only one "representative" of those "classes" of problems, respectively.

The results displayed in tables II–IV sustain the hypothesis that there is no (single) "universal heuristic" which performs well on a wide range of problems. (See also McCune and Wos (1992) in this context.) Table II, for instance, shows that while OccNest is the only focus function (among the tested ones) which solves, e.g., GRP168-1 and GRP175-1, it nonetheless fails for GRP002-2 and GRP121-1 where all the other focus functions succeed. The same goes for MaxWeight which solves GRP184-1 alone, but is also the only one to fail for GRP014-1.

Similar observations can be made in Table III: Although OccNest has the worst success rate in the domain LCL, it is the only focus function capable of solving LCL111-2 and LCL138-1. All focus functions except for MaxWeight fail for LCL160-1. Both OccNest and MaxWeight, however, fail for LCL164-1

Table II. Excerpt from Detailed Results for Domain GRP

Problem	AddWeight$_1^2$	MaxWeight	GTWeight	AddWeight$_1^1$	OccNest
GRP002-2	3s	164s	181s	1s	—
GRP002-3	3s	1s	4s	2s	434s
GRP014-1	15s	—	14s	9s	278s
GRP119-1	7s	450s	418s	342s	368s
GRP121-1	8s	300s	414s	225s	—
GRP138-1	34s	273s	4s	5s	1s
GRP154-1	25s	255s	1s	7s	1s
GRP166-3	33s	2s	1s	278s	1s
GRP167-1	—	—	272s	255s	—
GRP168-1	—	—	—	—	1s
GRP169-1	—	255s	181s	27s	—
GRP170-1	—	—	267s	94s	1s
GRP171-1	19s	65s	2s	9s	1s
GRP171-2	5s	53s	2s	8s	1s
GRP173-1	5s	1s	1s	6s	87s
GRP175-1	—	—	—	—	1s
GRP175-2	93s	—	—	177s	1s
GRP178-1	131s	—	—	—	10s
GRP178-2	131s	—	—	—	28s
GRP184-1	—	250s	—	—	—
GRP184-2	—	250s	—	—	10s
GRP190-1	24s	309s	134s	7s	3s

which can be solved by any of the other three focus functions within just one second.

The results in tables II and III indicate that the "right" choice of focus function is crucial for success: The right one may produce a proof within a couple of seconds, whereas the "wrong" one may still be searching on after hours. For certain (sub-) domains, "clear winners" can be identified: In the domain ROB (cp. Table IV), AddWeight$_1^1$ is unarguably the superior focus function. It performs *always* at least as well as the best of the other four focus functions.

Unfortunately, such a "clear winner" situation is an exception. The situa-

Table III. Excerpt from Detailed Results for Domain LCL

Problem	AddWeight$_1^2$	MaxWeight	GTWeight	AddWeight$_1^1$	OccNest
LCL111-2	—	—	—	—	407s
LCL116-2	10s	47s	14s	5s	3s
LCL133-1	1s	326s	1s	1s	330s
LCL138-1	—	—	—	—	126s
LCL153-1	2s	2s	1s	2s	—
LCL154-1	2s	2s	1s	2s	—
LCL155-1	2s	2s	1s	2s	—
LCL156-1	1s	2s	1s	1s	—
LCL157-1	1s	2s	1s	1s	—
LCL158-1	2s	2s	1s	2s	—
LCL159-1	2s	2s	5s	2s	—
LCL160-1	—	2s	—	—	—
LCL161-1	1s	17s	1s	1s	—
LCL164-1	1s	—	1s	1s	—

tion as in tables II and III is much more common. In such a situation, statistics or a posteriori analyses have a very limited value when it comes to suggesting a focus function for tackling a new problem. Under these circumstances, more sophisticated approaches have to be taken (see section 5).

Table IV. Excerpt from Detailed Results for Domain ROB

Problem	AddWeight$_1^2$	MaxWeight	GTWeight	AddWeight$_1^1$	OccNest
ROB002-1	1s	—	1s	1s	1s
ROB003-1	23s	—	40s	9s	30s
ROB004-1	26s	—	48s	11s	65s
ROB008-1	—	—	55s	51s	—
ROB009-1	—	—	29s	19s	—
ROB010-1	1s	—	1s	1s	1s
ROB013-1	1s	—	1s	1s	1s
ROB023-1	—	—	392s	66s	76s

Nevertheless there are a few leads that can be followed in special cases in order to determine a focus function or to narrow down the choices. For instance, the goal-oriented OccNest essentially penalizes the fact that a critical pair deviates from the structure of the goal in terms of exceeding certain measures (here occurrences and nesting of function symbols). If the goal has very little structure (e.g., a goal $a = b$ where a and b are two constants), then almost all critical pairs exceed the measures of the goal. If, conversely, the goal has a lot of structure ("large" terms in at least one side of the goal equation), then many critical pairs do not exceed any measure of the goal. In both cases, the purpose of OccNest to distinguish "exceeding" from "non-exceeding" critical pairs is diluted and even comes to nothing. Hence, an a priori analysis of the goal as to its structure can give us hints (but no guarantees) regarding the profitableness of OccNest.

As already stated in Section 3.2, GTWeight often achieves good results for examples with defining equations. MaxWeight sometimes also has this capability.

4.2. *Second Experimental Study: Using domain-specific knowledge*

The focus functions examined in the first empirical study use fairly general knowledge (e.g., "small is beautiful"). With this policy, they perform satisfactorily well on a wide range of, say, moderately difficult problems. When it comes to dealing with "challenging" problems, however, the lack of available domain-specific knowledge becomes a notable impediment.

In this second experimental study we investigate the performance of focus functions that are able to utilize domain-specific knowledge. We examine the focus function AddFWeight which allows us to integrate knowledge in an admittedly coarse way in form of different weights for different function symbols. By integrating more specific knowledge, a focus function can be adapted to the needs of certain problems of certain domains. Thus, the focus function becomes specialized in such problems. In a way we exchange the flexibility—the potential to deal satisfactorily well with a wider range of problems—of general-purpose focus functions for specialization—the ability to deal with certain "hard" problems.

Naturally, specialization comes at the expense of losing the ability to perform well on a wider range of problems. Hence, it does not make sense to compare AddFWeight with general-purpose focus functions on the basis of large-scale experiments. This would be completely besides the point. The point is that AddFWeight can be tuned to solve problems general-purpose focus functions have difficulties with. This is what Table V is to illustrate. Table V shows that even using knowledge simply by choosing appropriate pa-

Table V. Results of AddFWeight

Problem	w_V	Function Symbol : Weight	Run-time	"General Purpose"
GRP169-1	33	*inv* : 44, *mul* : 12, *glb* : 245, *lub* : 37, *id* : 54, *a* : 24, *b* : 2	1s	27s (AddWeight$_1^1$)
GRP169-2	23	*inv* : 28, *mul* : 4, *glb* : 26, *lub* : 247, *id* : 39, *a* : 7, *b* : 30	1s	28s (AddWeight$_1^1$)
LCL111-2	1	*not* : 6, *implies* : 1, *true* : 1	49s	407s (OccNest)
ROB005-1	3	*negate* : 1, *add* : 7, *c* : 1	26s	—
ROB022-1	2	*negate* : 1, *add* : 3, *c* : 1	20s	—
ROB023-1	3	*negate* : 1, *add* : 6	4s	66s (AddWeight$_1^1$)

rameter settings, i.e., weights for function symbols, can give rise to significant improvements. These improvements range from speed-ups to truly acceptable run-times where general-purpose functions fail. For problems LCL111-2, ROB005-1, ROB022-1, and ROB023-1 the parameter settings were provided by a user. (These settings are displayed in Table V using the function-symbol names as given in the TPTP.) This user utilized intuition, experience, and brief analyses of a few trial runs in order to produce these settings. Such a proceeding is unsatisfactory because it is time-consuming and requires an *experienced* user. In particular the latter prerequisite severely hampers the use of automated deduction systems by the "average user" who might be interested in such a system, but is put off by its intricacies.

There are, however, automatable alternatives to the trial-and-error procedure conducted by an experienced user. Approaches to parameter adaptation fall into the area of learning (see also the following subsection). Basically, when given a source problem that has already been proven, the parameter setting is "optimized" for, or adapted to this problem employing some machine-learning technique (e.g., a genetic algorithm as in Fuchs (1995) or Fuchs (1997a)). The parameter setting obtained this way can then be made use of for target problems that are "similar" (see again the following subsection) to the source problem. The parameter settings for GRP169-1 and GRP169-2 stem from such an automated process as described in Fuchs (1995) or Fuchs (1997a) using the source problems GRP190-1 and GRP191-1, respectively. (The function-symbol names in Table V for problems GRP169-1 and GRP169-2 are obvious abbreviations of the names used in the TPTP.) But note that in particular the parameter settings for problems GRP169-1 and GRP169-2 exhibit decisive differences, which demonstrates that within a domain defining the similarity of problems is a difficult and crucial task.

4.3. *Third Experimental Study: Learning*

The preceding subsection pointed out that the integration of domain-specific knowledge is often indispensable to cope with difficult problems. In the case of AddFWeight knowledge corresponds to certain settings of its parameters, namely the weights of function symbols. Regardless of how knowledge is utilized and represented, one of the main questions is where it comes from. Some of the parameter settings for AddFWeight were provided by the user. Others, however, were found by a machine-learning approach for parameter adaptation (Fuchs, 1995).

In general, learning plays a key role for improving problem-solving behavior by providing (relevant) knowledge. Essentially, learning can be characterized as analyzing problem-solving processes and creating knowledge based on these analyses (e.g., specific parameter settings). The acquired knowledge can then be employed to support the search for a solution of a "similar" problem.[2] Hence, the knowledge that can be used when attempting to solve a problem A stems from a learning process applied to the solution of a problem B found in the past. (Problem A and B have to be "sufficiently similar in a certain sense".) Problems A and B are referred to as target and source (problem), respectively (see also 3.4).

FlexRe (cp. definition 3.8) exploits (learned) knowledge in the form of a set of facts (here critical pairs). The selection of these critical pairs allows for proving a source problem. FlexRe attempts to prove a similar target problem by giving preference to these critical pairs as described in 3.4. The focus function H required by FlexRe is $AddWeight_1^2$ (which is of course only one of several possible choices). The remaining parameters were set as follows: $p = 20$, $q = 0$, $q_1 = 0.75$, $q_2 = 0.25$. (See Fuchs (1997a) for details.)

Table VI demonstrates that the knowledge which FlexRe can utilize makes it possible to prove a number of problems that are out of reach for all our general-purpose focus functions (and also pose serious problems for powerful and renowned systems like OTTER, cp. Denzinger *et al.* (1997a)). As a matter of fact, with the help of FlexRe 113 problems (90%) of GRP can be solved.

It is interesting to note that once FlexRe "gets started" with a number of source problems (here GRP179-1, GRP190-1, and GRP191-1) solved in some "conventional" way (for instance by employing one of the general-purpose focus functions), it can solve more and more difficult problems in a kind

---

[2] We shall here not discuss the crucial issue concerning similarity. See Fuchs and Fuchs (1997), Denzinger *et al.* (1997a) or Chapter 3, Section 4.3 of this part of this volume.

Table VI. Results of FlexRe

Target	Source	FlexRe	AddWeight$_1^2$	"General Purpose"
GRP169-1	GRP191-1	36s	—	27s
GRP169-2	GRP190-1	38s	—	28s
GRP179-2	GRP179-1	37s	—	—
GRP179-3	GRP179-1	38s	—	—
GRP186-1	GRP179-1	41s	—	—
GRP186-2	GRP179-1	40s	—	—
GRP183-1	GRP179-2	40s	—	—
GRP183-2	GRP183-1	42s	—	—
GRP183-3	GRP183-1	40s	—	—
GRP183-4	GRP183-1	42s	—	—
GRP167-3	GRP183-1	129s	—	—
GRP167-4	GRP183-1	130s	—	—
GRP167-1	GRP167-3	32s	—	255s
GRP167-2	GRP167-4	35s	—	251s

of bootstrapping fashion. Consider, for instance, target problem GRP167-1. FlexRe proves this problem using the source GRP167-3, which in turn can be proven with source GRP183-1. GRP183-1 can be proven with GRP179-2, and the latter with GRP179-1. Note that this chain constitutes a "genuine" bootstrapping process in the sense that, for instance, FlexRe fails to prove target GRP167-3 immediately with source GRP179-1. Thus, identifying a suitable source problem when given a certain target problem is a pivotal and difficult task (possibly requiring "a little trial and error"). Methods for automating this task are introduced in Fuchs and Fuchs (1997) and Denzinger *et al.* (1997a).

In connection with learning, the (pool of) problems to be solved must meet a requirement that could be characterized as "didactically arranged". In other words, there must be a pool of problems which more or less continuously range from *related* simple to moderately difficult and eventually to challenging problems. This requirement is not a limitation of machine-learning approaches. On the contrary, it is actually natural. A human student (tutored by a reasonable teacher) also starts out with the simpler problems of a chosen domain and then gradually works his way up to the more difficult ones.

It is worth noting that the method in Fuchs and Fuchs (1997) or Denzinger *et al.* (1997a) does not require a (human) teacher presenting the problems in a suitable (didactic) order. It merely needs a pool of "didactically arrangeable" problems. Arranging them in an appropriate order (implicitly) is part of the method.

Unfortunately, very few domains of the TPTP meet the requirement. The domain GRP is essentially the only domain that is accessible with learning methods. The main reason for this is that a large part of GRP was provided by the mathematician B.-I. Dahn. His objective was not to provide a collection of problems that are hard for equational reasoning systems, but to present a domain (lattice-ordered groups in this case) as it would be presented to a human, beginning with almost obvious theorems and closing with the challenging ones. From this fact machine-learning approaches can profit, too.

5. CONCLUSION

The ability of automated deduction systems (including equational reasoning systems) to control the application of inference rules is crucial for performance, success, and acceptability. The availability of a complete (and correct) set of inference rules is only the beginning. In a semi-decidable area like automated deduction in general and equational reasoning in particular search effort is inevitable. Without appropriate means to control the search a set of inference rules is –from a practical point of view– next to worthless.

Recent advances in implementation techniques (e.g., term indexing as described by Graf (1995)) have produced remarkably fast and efficient implementations of reasoning systems such as OTTER (McCune, 1994), EQP, and WALDMEISTER (Hillenbrand et al., 1996). In particular these state-of-the-art systems have corroborated the fact that sheer speed alone does not and cannot solve intricate search problems—it merely fills the available and limited memory a lot faster. As a matter of fact, a "conventional" (slow) implementation can beat a sophisticated one by several orders of magnitude if it uses a superior control of the search (see Fuchs (1996c)). In other words, speed cannot compensate for deficiencies in search control. Note, however, that speed and search control can supplement each other because they hardly interfere with each other (at least in the case of equational reasoning systems).

While it is definitely worth having efficient implementations, research in search control is the key for significant improvements. Our experiments here have demonstrated the well-known fact that there is no single universal search-guiding heuristic (focus function) performing well on a wide range of problems. Choosing an appropriate (generic) focus function and a suitable parameter setting is a difficult and important task that determines success and failure. Mostly, this task is left to the user who must be fairly experienced in order to make judicious choices. This circumstance is quite unsatisfactory and severely limits the usefulness of a system for an unexperienced user. Recent approaches that deal with this issue make use of machine-learning techniques (e.g., instance-based learning and case-based reasoning) to relieve the user from this burden. These approaches have produced encouraging results (see Fuchs (1996b), Fuchs (1997c), Fuchs and Fuchs (1997), and Denzinger *et al.* (1997a)).

REFERENCES

Avenhaus, J. and Madlener, K. (1990): Term Rewriting and Equational Reasoning, in *Formal Techniques in Artificial Intelligence*, R.B. Banerji (ed.), North Holland, Amsterdam, pp. 1–43.

Avenhaus, J., Denzinger, J., and Fuchs, M. (1995): DISCOUNT: A system for distributed equational deduction, Proc. 6th Conference on Rewriting Techniques and Applications (RTA-95), Kaiserslautern, GER, LNCS 914, pp. 397–402.

Bachmair, L., Dershowitz, N., and Plaisted, D.A. (1989): Completion without Failure, *Colloquium on the Resolution of Equations in Algebraic Structures*, Austin, USA (1987), Academic Press.

Denzinger, J. (1995): Knowledge-Based Distributed Search Using Teamwork, Proc. 1st International Conference on Multi-agent Systems (ICMAS-95), San Francisco, USA, AAAI-Press, pp. 81–88.

Denzinger, J. and Fuchs, M. (1994): Goal Oriented Equational Theorem Proving Using Teamwork, Proc. KI-94, Saarbrücken, GER, LNAI 861, pp. 343–354.

Denzinger, J. and Fuchs, Marc and Fuchs, M. (1997a): High Performance ATP Systems by Combining Several AI Methods, Proc. 15th International Joint Conference on Artificial Intelligence (IJCAI-97), Nagoya, JAP, Morgan Kaufmann, pp. 102–107.

Denzinger, J. and Kronenburg, M. and Schulz, S. (1997b): DISCOUNT - A distributed and learning equational prover, *Journal of Automated Reasoning* 18(2), pp. 189-198.

Denzinger, J. and Schulz, S. (1996): Learning Domain Knowledge to Improve Theorem Proving, Proc. 13th Conference on Automated Deduction (CADE-13), New Brunswick, USA, LNAI 1104, pp. 62–76.

Dershowitz, N. and Jouannaud, J.-P. (1990): Rewrite Systems, in *Handbook of Theoretical Computer Science*, J. van Leeuwen (ed.), Elsevier, Volume B, Chapter 6, pp. 243–320.

Fuchs, M. (1995): Learning Proof Heuristics By Adapting Parameters, in Machine Learning: Proceedings of the Twelfth International Conference (ICML-95), A. Prieditis & S. Russell (eds.), Morgan Kaufmann Publishers, San Francisco, USA, pp. 235–243.

Fuchs, M. (1996a): Experiments in the Heuristic Use of Past Proof Experience, Proc. 13th Conference on Automated Deduction (CADE-13), New Brunswick, USA, LNAI 1104, pp. 523–537.

Fuchs, M. (1996b): Experiments in the Automatic Selection of Problem-solving Strategies, LSA Report LSA-96-09E, Center for Learning Systems and Applications (LSA), University of Kaiserslautern, URL http://www.uni-kl.de/ AG-AvenhausMadlener/fuchs.html.

Fuchs, M. (1996c): Powerful Search Heuristics Based on Weighted Symbols, Level and Features, Proc. 9th Florida Artificial Intelligence Research Symposium (FLAIRS-96), Key West, USA, ISBN 0-9620-1738-8, pp. 449–453.

Fuchs, M. (1997a): *Learning Search Heuristics for Automated Deduction*, Ph.D. thesis, Verlag Dr. Kovač, Hamburg, ISBN 3-86064-623-0.

Fuchs, M. (1997b): Flexible Re-enactment of Proofs, Proc. 8th Portuguese Conference on Artificial Intelligence (EPIA-97), Coimbra, POR, LNAI 1323, pp. 13–24.

Fuchs, M. (1997c): Automatic Selection of Search-guiding Heuristics, Proc. 10th Florida Artificial Intelligence Research Symposium (FLAIRS-97), Daytona Beach, USA, ISBN 0-9620-1739-6, pp. 1–5.

Fuchs, Marc and Fuchs, M. (1997): Applying Case-based Reasoning to Automated Deduction, Proc. 2nd International Conference on Case-based Reasoning (ICCBR-97), Providence, Rhode Island, USA, LNAI 1266, pp. 23–32.

Graf, P. (1995): *Term Indexing*, Springer LNAI 1053.

Hillenbrand, T. and Buch, A. and Fettig, R. (1996): On Gaining Efficiency in Completion-Based Theorem Proving, Proc. 7th Conference on Rewriting Techniques and Applications (RTA-96), New Brunswick, USA, LNCS 1103, pp. 432-435.

Hsiang, J. and Rusinowitch, M. (1987): On word problems in equational theories, Proc. 14th International Colloquium on Automata, Languages and Programming (ICALP-87), Karlsruhe, GER, LNCS 267, pp. 54–71.

Knuth, D.E. and Bendix, P.B. (1970): Simple Word Problems in Universal Algebra, *Computational Algebra*, J. Leech, Pergamon Press, pp. 263–297.

Kolbe, T. and Walther, C. (1994): Reusing Proofs, Proc. European Conference on Artificial Intelligence (ECAI '94), Amsterdam, HOL, pp. 80–84.

McCune, W. and Wos, L. (1992): Experiments in Automated Deduction with Condensed Detachment, Proc. 11th Conference on Automated Deduction (CADE-11), Saratoga Springs, USA, LNAI 607, pp. 209–223.

McCune, W. (1994): OTTER 3.0 Reference Manual and Guide, Techn. Report ANL-94/6, Argonne Natl. Laboratory.

Stickel, M. (1984): A Case Study of Theorem Proving by the Knuth-Bendix Method: Discovering that $x^3 = x$ Implies Ring Commutativity, Proc. 7th Conference on Automated Deduction (CADE-7), Napa, USA, LNCS 170, pp. 248–258.

Sutcliffe, G., Suttner, C., and Yemenis, T. (1994): The TPTP Problem Library, Proc. 12th Conference on Automated Deduction (CADE-12), Nancy, FRA, LNAI 814, pp. 252–266.

CHAPTER 14

COOPERATING THEOREM PROVERS

1. INTRODUCTION

Cooperation is one of the foundations of civilization. Today, nearly everything around us is the result of cooperative efforts. For example, the production of a car is only possible, because many different factories produce the necessary parts that are then assembled by different specialists. But not only production is a cooperative task. The development of a new product involves joint efforts of specialists of many areas of expertise combining their knowledge in order to find solutions to all problems involved in such a development process.

For all these tasks cooperation achieves a combination of different abilities, different knowledge and different actors, thus producing *a whole that is more than the sum of its parts*. Naturally, cooperation can be achieved by many different interactions. Cooperating agents can tackle a problem *all at the same time*, communicating in real-time and hence resulting in a parallel "implementation" of the problem solving effort. This is often the case in development teams.

But cooperation is also achievable over a longer period of time with only *one agent working on a problem at each point of time*. The latest stages of the assembly of a car are such an example, but also many scientific developments mirror such a behaviour: over the time scientists add more and more pieces to the solution of a problem until a correct solution is found.

The history of Fermat's last theorem that there are no non-zero integers x, y, z such that $x^n + y^n = z^n$, for $n \geq 3$, illustrates this way of cooperative and often competitive research. Arithmetic methods were applied by several authors to confirm Fermat's conjecture for many values of n. Further progress was made by relating the conjecture to a problem of the factorization of polynomials with complex coefficients. This was a basis for the use of computers to push the bound for n up to 4,000,000 until 1993. Faltings proved in 1983 that for each n there are at most finitely many different counterexamples to Fermat's conjecture. Then it was shown that these counterexamples would have implications for a conjecture about elliptic curves. In 1993 Wiles announced a proof of Fermat's conjecture. Later, a gap was found in the proof. In 1994, in cooperation with Taylor, Wiles realized, that the reasons for the

W. Bibel, P. H. Schmitt (eds.), Automated Deduction. A basis for applications. Vol. II
© 1998 *Kluwer Academic Publishers. Printed in the Netherlands*

failure of one method in the proof enabled the use of another method in order to complete the proof successfully.

In computer science both kinds of cooperation can help designers of computer program systems. Having tasks done in parallel results in more efficiency, can add safety to a system, and allows for more structure. Having series of programs where each performs parts of a task allows for better maintenance of each program, better error recovery and better cooperation of the system developers. The price to pay is the development of an interaction and communication structure for the agents in the system with potential bottlenecks and some overhead. If there is not enough communication, then either agents cannot do their work or they have to compute the necessary information by themselves. The latter results in much redundancy and therefore less efficiency. If there is too much communication necessary agents may be overwhelmed by unimportant information and are not able to continue their work.

Automated deduction is a very complex task and therefore it can be expected that cooperation is very helpful. Even more, there are many different theorem proving techniques (calculi) and also many different theorem provers each of which has different strengths and weaknesses. Again, this is a typical situation in which cooperation should be applied. Unfortunately, so far not many cooperating theorem proving systems have been developed, especially not many that include provers with different search paradigms.

In this chapter we will present cooperation concepts for theorem provers with three different purposes: the TEAMWORK method for homogeneous prover networks, the TECHS approach to form an automated heterogeneous prover network, and its extension to an interactive prover network in combination with ILF. Our TEAMWORK method represents a concept for building a cooperating prover based on a single basis prover. It lets several incarnations of this prover employ different control strategies to form a *homogeneous prover network*. In order to achieve a fully *automated heterogeneous prover network* we propose the cooperation of several different basis provers employing quite different search paradigms. Also we want to use already existing theorem proving systems in this network which resulted in our TECHS approach.

Some broad definitions of cooperation would call each interactive theorem prover a cooperative one (due to the interaction of prover and the human user). In this chapter we will present a concept for *cooperative interactive prover networks* that include at least several different incarnations of a basis prover or several basis provers and that can make optimal use of our TECHS concept. In fact, it is an interactive extension of it. This allows to choose the appropriate level of interaction for all proof tasks.

2. Cooperation: General Concepts and Problems

Cooperation of computing agents is the main topic of the field *multi agent systems*, which mainly lies between the fields distributed systems and artificial intelligence. Various concepts have been developed in order to enable agents to solve a given problem together. Naturally, questions like what an agent is, what communication is, or which goals agents should share greatly influence which concept should be chosen in a particular situation.

In this section we will present some concepts developed for multi agent systems that can be used to develop systems of cooperating *search* agents and we will outline the basic assumptions that are made by these concepts. We will briefly discuss how to implement these concepts and which problems one has to face when using such a concept for conducting a cooperative search.

But first we will present a concept for distributed search that is not cooperative and that sets a surveyor's wooden rod on which cooperation should improve: *competition* of agents.

2.1. *Competition of Agents*

If there are several methods for solving a problem and it is not clear which method is best for a given problem instance, then one can make use of several agents by assigning each method to an agent and then let the agents work independently and in parallel. The first agent solving the problem determines the solution time for the whole group of agents.

This concept, which belongs to the parallelization approaches (see II.3), does not use any cooperation between agents at all. Because of this fact it can be easily implemented (often a simple shell script starting programs using the different methods is sufficient). If the problem instances one wants to solve are quite different and there is no good classification of the problems that allows to select at least a not so bad method for solving it, then such a primitive multi agent approach performs at the average better than each single method, even if we count as effort of the multi agent system the product of runtime of the best agent for an instance and the number of agents.

In Ertel (1992) this approach was first used for automated deduction and some other examples can be found in II.3.10.

2.2. *Blackboard Approaches*

One of the first approaches for achieving cooperation between different agents was the blackboard approach (see Erman *et al.* (1980), Hayes-Roth (1985)). Different agents with different abilities tackle a problem by picking problems from a central storage device, the *blackboard*, partially solving them and putting the solved parts and the remaining unsolved parts back on the device.

Although the blackboard architecture can be implemented on computer networks, the best implementation of a blackboard system is on a shared-memory machine or a single-processor machine that allows to have more than one process. Problems that have to be solved when using a blackboard approach for a multi agent system include finding methods to divide problems into reasonable subproblems (that should then be solved by other agents), providing a control that assigns problems to agents, and extracting the final solution from the blackboard.

Some parallelization methods for search problems can be seen as blackboard approaches, as for example the ROO system (Slaney and Lusk, 1990). But in these methods all agents have the same abilities and the blackboard is mainly a synchronization and control tool.

2.3. *Cooperation by Negotiation*

Negotiation is a very important concept in many multi agent system approaches. A cooperative effort for solving a given problem is achieved by one agent that offers work to agents that are idle and determines those (available) agents that are best suited for this work. This task is modeled as a negotiation between the agent that has work to distribute and the agents that can do this work.

This kind of negotiation is the basis of the so-called *contract-net protocol* for multi agent systems (Smith, 1980). The problem to solve is given to one agent. Each agent is capable to either solve a problem or to divide it into several subproblems. It is also able to decide or anticipate to a given problem how much effort for solving it is necessary. If an agent decides to solve a problem by dividing it, it sends messages describing the subproblems to all other agents. Agents without work compute the effort they need to solve these subproblems and send these measures as bid to the distributing agent. This agent selects the best bids and assigns the subproblems to the respective agents. Naturally, each agent can divide a subproblem into other subproblems, again. If an agent has completely solved a problem it reports its results to the agent that assigned the problem to it.

The contract-net protocol requires that a problem can be divided into independent subproblems. Furthermore, these problems have to be described on a high-level, for the biding process, and on a concrete level, for the agent that wins the biding. Also it must be possible to compute the bids. Negotiation based approaches can be implemented both on multi processor machines and networks of computers. But notice that the necessary communication among agents can be time consuming.

The communication behaviour of the DARES system (see Conry *et al.* (1990), Section 3.3) is very similar to the behaviour of systems using negotiation.

3. COOPERATION OF AND IN THEOREM PROVERS

Before we present the different concepts for cooperation that are used in automated deduction, we present two general paradigms for theorem proving that influence what cooperation concepts can be chosen in a particular situation. The first paradigm is the analytical view on the problem. According to this paradigm theorem proving is performed by case analysis. Therefore the representation of the search state in a prover uses trees or graphs. Examples of calculi following this paradigm are analytic tableaux or model elimination (see I.1).

The second paradigm is the saturation or generating view on the problem. According to this paradigm theorem proving is the generation of new knowledge until the required knowledge is found. A search state is represented as a set of "facts" (formulas, clauses, equations, and so on). Here, examples of calculi are resolution and paramodulation or the Knuth-Bendix completion (see Chapter 2 of this part of this volume). Note that there are provers that have features from both paradigms.

Regardless of the paradigm a prover follows, there are several possibilities on the conceptual level to present proof systems as a group of cooperating agents. For example, each prover that employs decision procedures for certain tasks that are implemented as reusable modules could, in theory, also be implemented as several agents: one agent representing the main search mechanism of the prover, the other agents representing its decision procedures. A concrete example would be unification and matching as decision procedures that are called each time an inference rule requires their use. In practice, a realization as multi agent system would be much too expensive (in terms of runtime). For the rest of this chapter we are not interested in systems where the cooperating agents exist only on the conceptual level. Even with this restriction there are still several different types of approaches to cooperating theorem proving.

3.1. *Cooperation Pipelines*

We have seen in the introduction that one kind of cooperation between agents is a kind of processing pipeline. One agent takes a problem, uses its expertise to perform some steps towards the solution, and passes it on to the next agent in line. In several automated deduction systems such a *cooperation pipeline* is realized as one or several "preprocessors" and one or several "postprocessors".

Examples for preprocessors are the generation of clausal normal form, as realized by FLOTTER for SPASS (Weidenbach *et al.*, 1996), or a limited saturation of the input, as realized by the delta iterator for SETHEO (Schumann, 1994). Postprocessors may analyze proof statistics or give a readable presentation of the proofs generated by a prover. Examples are the pipeline of analyzing programs for DISCOUNT (Denzinger and Schulz, 1994) or the transformation into block calculus done in the ILF system (Dahn and Wolf, 1994) or by ILF-SETHEO for the SETHEO system (Wolf and Schumann, 1997).

In a cooperation pipeline, each time only one agent works on a problem, so that pipelines offer neither control nor communication problems. However, many researchers require that in a multi agent system several agents can work in parallel (as indicated in the concepts of Section 2). In the following, we will concentrate on cooperation concepts for automated theorem provers that fulfill this condition.

3.2. *Cooperation by Splitting Problems*

If we look at the cooperation approaches of Section 2 then one necessary condition is the ability to divide a problem into several (independent) subproblems. Naturally, this results in the ability of a system to use several agents in parallel. Unfortunately, in automated deduction systems it is not easy to fulfill this condition.

Provers that employ the analytical paradigm provide, in theory, the ideal means to divide problems into subproblems. But practice has shown that the application of such an inference rule requires not much computational effort (so that as single task of an agent the ratio computation to communication would be disastrous) and there is no way to decide when to stop assigning subproblems to other agents and solve the problem locally. Also, variable bindings have to be communicated. The same reason, too little computational effort versus high communication costs, also makes distribution concepts for other calculi on the level of single inference rule applications in most cases infeasible.

Many interactive provers allow for the use of *meta-inference rules* (often also called tactics or methods, see Constable *et al.* (1985) or Benzmüller et al. (1997)) whose application requires more computational effort and that result in several subproblems that need to be solved (and that may not be independent). In interactive provers there are also often steps that require some guessing (which is usually provided by the user). If this guessing process can be limited to choosing a guess out of a small set of possibilities then different agents can try out different guesses.

3.3. *Cooperation by Partitioning the Search Space*

Often it is not possible to partition problems automatically into appropriate subproblems that can be solved by different agents. Therefore many cooperation approaches aim at a partitioning not of the problem to solve but of the search space generated by this problem. Then the different sectors are traversed by the different agents (that are of the same basic type using the same inferences). While static partitioning belongs to the domain of parallelization approaches, dynamic partitioning requires a cooperation of agents and naturally some interchange of information.

A *static partitioning*, as for example developed in Suttner (1995), determines a priori which agent traverses which parts of the search space. No more communication is necessary. In a *dynamic partitioning* information is passed between the agents and due to this information agents traverse other parts of the search space or they traverse their parts in a different order as it would be the case without this information.

There are two general principles for cooperation of search systems based on the interchange of information:

- demand-driven cooperation and
- success-driven cooperation.

Demand-driven data interchange is characterized by agents that determine information they need and send requests for this information to the other agents. The DARES system, see Conry *et al.* (1990), is an example for this kind of cooperation. Each agent gets a part of the input facts and generates for some time period consequences of its facts. Then the agent tries to determine facts that would complete the proof or would support its current work but that it is not able to infer. Generalized descriptions of such facts are then sent to the other agents. An agent receiving a description interrupts its work, examines its set of facts for facts that match the description, and sends the facts found to the requesting agent. After receiving all the matching facts of its colleagues the requesting agent continues to search for a proof using the received facts until it is time for new requests.

Obviously, demand driven cooperation is not easy to achieve since it is difficult to detect which facts a prover needs to succeed. This was already the reason why a splitting of the proof problem is not easy to achieve. In fact, if we could determine the facts a prover needs to solve a proof problem without already knowing the proof, the whole task of automated theorem proving would be easy because a very good control could be designed.

Success-driven interchange of data means that an agent determines facts among its own results that it found useful so far. These facts are communicated to the other agents. An example of this kind of cooperation is the TEAMWORK method that will be the subject of Section 4 or the clause-diffusion method of Bonacina and Hsiang (1995) and Bonacina (1997).

The idea of the latter method is to assign each generated fact to one agent that is responsible for doing all necessary inferences with it. Therefore each agent must also know all the facts of the other agents but it is guaranteed that each inference with a particular fact is only made by one agent. The crucial part of this approach is the assignment of facts to agents since an agent prefers inferences between facts that it is responsible for. In Bonacina (1997) assignment strategies based on common ancestors are presented that were quite successful.

Since the selection of the data to interchange can be based on a retrospective view on the work of an agent, there are several criteria that can be successfully used. The criterion of the clause-diffusion method is rather weak: a newly generated fact must be not redundant. In the TEAMWORK method stronger criteria are used, see Section 4.4. In both methods there is no guarantee that a selected fact will really be needed in a proof.

If we look at the cooperation concepts of Section 2 then we can classify blackboard systems and the contract-net protocol as demand driven. It should be noted that there are concepts that include aspects of both principles. An example for the area of automated deduction will be presented in Section 5.

Dynamic partitioning of the search space, both demand- and success-driven, is the result of the interchange of data. There are several types of data that can be interchanged:

–   Descriptions of parts of the search space that seem to be interesting. Such descriptions may be realized as

  • good facts that are known to be valid with respect to the actual problem
  • decision sequences that lead to the interesting parts
  • paths in trees or graphs.

– Descriptions of parts of the search space that should be avoided. Such descriptions may be realized as

- facts that should not be used in inferences
- facts that are known not to be valid
- decision sequences that should be avoided
- paths in trees or graphs.

– Control strategies

It depends on both the calculus and the control strategy used by an agent which type of data can be used. It should be noted that, in most known co-operating provers, valid data, with respect to the given problem, is the (only) chosen type.

3.4. *Cooperation between Different Theorem Provers*

The approaches of Section 3.3 employed agents that used the same basic inference mechanism, so that the behaviour of the resulting multi agent systems can be described with respect to the behaviour of a single agent using the same mechanism. But interchange of data is also possible between agents that use different inference mechanisms, although then it is not clear whether they have a common search space and how this space can be partitioned.

Although an analysis of the resulting search process is difficult, the different calculi that are used in today's theorem provers have different strengths and weaknesses so that cooperation between such provers may result in very powerful multi agent theorem provers. That cooperating theorem provers can be more powerful than their single agents has been empirically shown in Sutcliffe (1992). In this work different inference mechanisms, in form of agents, are coupled by a so-called tuple-space that provides a kind of communication via shared memory. This success-driven approach communicates all new facts generated by an agent to all other agents that can work with these results.

The results presented there have shown that the interchange of all new facts generates too much overhead. In addition, no already existing theorem provers were used as agents. While in general the use of existing systems definitely is no requirement, it must be noted that the performance of the individual agents of Sutcliffe (1992) was very poor so that the performance of the multi agent system was no match for state-of-the-art, non-cooperating systems. Naturally, this derates the empirical results as a basis for determining the success of cooperation.

The main reason for this poor performance is the developing effort that is necessary to design, implement and maintain such systems. Very efficient implementations of the basic inferences, using indexing techniques, sophisticated management of the search state and the potential successor states, and heuristical control based on a long experience of the developers are the key results of this effort. Since it is a full-time job to improve on these results, it is very difficult to also include cooperation aspects into these provers. Therefore, cooperation concepts that allow to use state-of-the-art provers with only small modifications are very interesting in order to show the usefulness of cooperation and in order to interest people with applications for automated deduction. We will present such a cooperation concept in Section 5. For now, just note that the following problems have to be solved by all cooperation concepts using existing high-performance provers.

- The changes of the provers must be held to a minimum. Changes are necessary because usually the automated provers are not prepared to read and use data provided after a proof run has been started and is still running (an aspect which is necessary for making use of cooperation).
- Automated theorem provers usually perform much more inferences that do not contribute to a proof for a given problem than inferences that do contribute (a ratio of 1000:1 or worse has to be expected). Communicating useless results to other provers usually results in additional obstacles that may even result in not finding a proof. Therefore a good selection of the data to be interchanged is essential.
- Theorem provers like to claim as many resources (memory, CPU time) as possible. Having several such agents requires a resource control component for the whole system or an own computer for each agent (or even both).

4. AUTOMATIC COOPERATION IN HOMOGENEOUS PROVER NETWORKS: THE TEAMWORK METHOD

The TEAMWORK method (Denzinger, 1995) is a cooperation concept that realizes a success-driven dynamic partitioning of the search space. The partitioning is accomplished by using different control heuristics for a basic prover and the dynamic is achieved by interchanging selected facts of the agents.

The TEAMWORK method is aimed at the use in a network of computers which requires limitations to the amount of communication. As an additional feature the use of unfair and very specialized control heuristics (that very seldom find proofs alone) is possible. This allows for gains in efficiency. Therefore the multi agent system has to be able to "rescue" agents that got lost in regions of the search space that are unlikely to lead to a proof. TEAMWORK also includes the use of planning techniques (Denzinger and Kronenburg,

1996) that enables a system based on it to employ agents that are based on learned knowledge (see Denzinger and Schulz (1996a), Chapter 2 of this part of this volume and Denzinger *et al.* (1997)).

4.1. *Tasks and Interaction*

In a TEAMWORK-based system an agent plays alternately different roles: expert/specialist, referee and one agent has the role of supervisor (but this agent may change which results in a so-called *floating control*). In the role of an expert or specialist an agent works on solving the given problem. As expert it employs the basic inference mechanism to generate new facts and delete redundant ones while as specialist it may either use other inference mechanisms to solve the problem or performs tasks that help the supervisor, as for example classifying the given problem. An agent is either expert or specialist during a so-called *working period*. The agents with an expert role use different control heuristics for the basic inference mechanism.

After the working period all agents assume their referee role. As referees they compute a *measure of success* for the work they have done and they select outstanding facts they have generated. After this first phase of a *team meeting* one agent assumes the supervisor role (the same agent that was supervisor at the end of the last meeting). This agent receives the measures of success from the other agents, determines the agent with the best measure –the winner– and resigns the supervisor role while the winner assumes this role.

The new supervisor receives the selected facts of the other agents, integrates them into its own search state, and broadcasts this improved search state to all other agents. Based on the measures of success of all working periods and a long-term memory about different experts, specialists, and referees the supervisor also selects for all agents the particular expert, specialist and referee roles for the next working period. It uses the same data to determine the length of this period and this completes the cycle.

4.2. *Experts*

As already stated, an agent in the role of an expert is a generating theorem prover. The behavior of such a prover is mostly determined by the selection of a new fact out of the set of all facts that potentially could be generated using the appropriate inference rules and the actual set of facts. Each agent uses another selection heuristic in order to explore a different part of the search space.

In Chapter 2 of this part of this volume we already presented several selection heuristics for equational deduction based on completion. In the following we will briefly describe two examples for heuristics that are not intended to solve many proof problems alone but are often able to finish proof attempts much faster than other heuristics, when provided with an appropriate starting point. So, they are only useful as agents in a team. Both heuristics can be classified as using goal-oriented criteria because their main idea is to measure the similarity of a potential fact and the goal to prove (Denzinger and Fuchs, 1994). In the following we will sketch these two heuristics for the case of the Knuth-Bendix completion (but these ideas can also be transfered to other calculi).

The first selection heuristic, CP-in-Goal, tries to match or unify a potential fact with a subterm of the goal that must have a certain minimal structure. A statistical measure for the potential fact is used as basis for CP-in-Goal and then the weight of the rest of the goal, according to this statistical measure, multiplied with a penalty is added. The penalty is bigger, if there is only a unifier and not a match, and it is even more big, if neither a match nor a unifier is found. Since the potential fact and the goal are equations, the penalty is also based on whether one substitution achieves the similarity of both sides of the equations or two substitutions are needed (one for each side) or similarity can only be achieved for one side. Naturally, the less the similarity the higher is the penalty. The second heuristic, Goal-in-CP, is similar to CP-in-Goal but goal and potential fact have to be exchanged. Therefore we search for a subterm of the potential fact that matches or unifies with the goal.

The basic idea of CP-in-Goal is to find and prefer potential equations that reduce the difference between the two terms of the goal or at least can be applied to one of the goal terms. Goal-in-CP tries to find equations that contain the goal and some context that has to be eliminated by further inferences. Typically, both situations are not often found at the beginning of proof attempts. But later on, after other experts generated a search state with such situations, the two heuristics direct the search very fast to a successful end state. So, both heuristics require cooperation with other heuristics.

4.3. *Specialists*

Specialists either generate new facts by other means than the basic proof mechanism or they help the supervisor in classifying the given proof problem. These other means to generate new facts include to look them up in a *long-term memory* that contains sets of axioms and known consequences of these axioms. If the axioms of the given proof problem are a superset of one of the sets in the long-term memory then all the known consequences of the set are also consequences of the problem. This resembles a human agent that searches in a library for books that help him with his problems. Naturally, there may be many known consequences that cannot all be added as facts because then the search space would "explode". Selecting useful consequences is the task of referees, see the next subsection.

Classifying a problem is very useful if further knowledge concerning such a classification is available. For example, to the before mentioned sets of axioms there may also be some information available that suggests good selection heuristics for proof problems that contain them. This is used for the planning task of the supervisor (see Section 4.5 and Denzinger and Kronenburg (1996)).

A specialization of this is to find solved proof problems that are somewhat similar to the problem at hand. If such a problem can be found then the supervisor can use the selection heuristic FlexRE (see Chapter 2 of this part of this volume) in order to re-enact the proof of this problem. Finding similar problems for FlexRE is the task of specialist PES (Proof Experience Specialist, see Denzinger *et al.* (1997)).

PES is based on the similarity of single facts: A fact s is 100% similar to a fact t, if s subsumes t. It is less similar (say, only 80%), if s subsumes t modulo AC and it is even less similar (say, only 20%) if s is homeomorphically embedded into t. Note that this is not a symmetrical definition. In order to measure the similarity of a source problem (Ax_S, g_S) to the target problem (Ax_T, g_T) (after the source is transformed by an appropriate signature-match), PES computes three measures (s_1, s_2, s_3).

s_1 measures the coverage of the source by the target by searching for each axiom in Ax_S the most similar axiom in Ax_T and then computing the mean value of these similarity grades. s_2 measures the coverage of the target by the source, similar to s_1. The main reason for a low s_2-value are many "superfluous" axioms in the target. Finally, s_3 measures the similarity of the source goal g_S to the target goal g_T. Then PES returns the most similar source problem to the problem at hand that it was able to find during the working period, provided that a certain minimal similarity, computed as weighted sum of s_1, s_2 and s_3, is given.

4.4. Referees

An agent in its role as referee judges the work the agent did during the last working period as expert or specialist. Besides a general measure of success that indicates how good the whole work has been, a referee also selects a few of the facts that the agent has generated and that have proven to be necessary or important. Both tasks of a referee should be based on retrospective criteria that allow a success-driven cooperation.

Judging a single fact is nearly the same task as selecting the next potential fact that is performed by experts. So, in principle, one could use experts also as referees, but the main problem of experts is the speculative aspect of their selection heuristic, an aspect that we do not want to have in referees. Therefore we may use some criteria that are also used by experts but we must also include retrospective criteria as

— the number of potential facts that can be generated from a given fact, partitioned with respect to the responsible inference rules
— the number of redundant facts that have been eliminated using a given fact, partitioned with respect to the responsible inference rules
— the success of the ancestors of a fact
— the success of the descendants of a fact
— domain specific criteria

The whole success then can be computed as weighted sum of the different criteria. Note that some of the retrospective criteria may be positive, as for example the elimination of many facts, while others sometimes may be negative, as for example a large number of facts to be generated. More about possible criteria and their combination can be found in Denzinger and Fuchs (1996).

Naturally, the criteria a certain referee uses also depend on the expert or specialist whose new facts have to be judged. The specialist that simply finds facts in the long-term memory needs a referee that uses criteria like

— similarity to the goal
— eliminations of facts possible with a given fact or
— judgements already included in the long-term memory.

Also learning experts may need specialized referees.

The measure of success for an expert/specialist should be based on

— the number of new facts
— the number of eliminated facts
— the number of potential facts
— the judgements of the individual new facts
— the trend of the latest selected potential facts

Again, these criteria have to be combined and may be weighted differently for different experts/specialists. For example, a bad trend indicates for expert FlexRE that no new focus facts have been found (see Chapter 2, Section 3.4 of this part of the volume). This can indicate that a re-enactment of the chosen proof does not lead to a proof for the actual problem.

Agents use their referee role to allow for both, competition and cooperation between them. Since all not-selected facts of the experts that are not the winner in a cycle are *forgotten*, not only the amount of communication is held low but also an explosion of the search space is avoided.

4.5. *The Supervisor*

The agent that plays the role of the supervisor is the central control of a TEAMWORK-based system. Although the time that there is an agent playing the supervisor role is small compared to the length of a working period, the supervisor greatly influences the behavior of the whole multi agent system. By generating a new start search state for *all* experts and the specialists, no expert will get lost in unpromising regions of the search space. By selecting the particular expert, specialist and referee roles the agents will assume in the next cycle, the whole team can be adapted to the problem at hand and also different phases in a proof attempt can be supported optimally.

Since all agents are capable to assume the role of an expert, the generation of a new start state is very easy. The selected facts of the other agents are simply treated as newly generated facts and the appropriate procedures of the experts can be used for integrating them.

For the selection of the next team the supervisor uses a reactive planning process based on a long-term memory and a short-term memory. The long-term memory is organized as sets of frames describing domains of interest, experts, specialists, and referees. These frames are mainly based on previous proof experience, either generated automatically or abstracted by a human expert. The short-term memory contains information on the actual proof attempt like detected domains, similar problems or the reports of the referees of all cycles and for all agents. Based on this information plan skeletons for the use of experts and specialists are selected, filled with additional agents, checked against the observed behavior, and, if necessary, corrected or chosen completely new (Denzinger and Kronenburg, 1996).

A plan skeleton consists of several small teams that are intended to be used in different phases of a proof attempt. As we have seen, in the beginning of an attempt it is necessary to have specialists for the classification of the given problem and standard experts capable of quickly solving easy problems in many domains. Later on in such an attempt, depending on the domain of a problem, learning experts and specialists for the domain should be employed. If enough work is done, experts like CP-in-Goal and Goal-in-CP can be used to finish a proof. Since it is almost never obvious whether a new phase has been reached, the supervisor has to try out experts of a new phase and then has to decide if they perform better than the ones of the old phase. Included in a plan skeleton are also recommendations for the length of the working period.

Having a supervisor allows the user of a TEAMWORK-based system to get rid of the task of configuring the usually large sets of parameters both multi agent systems and deduction systems have. The experiments in Denzinger and Kronenburg (1996) have shown that the system is capable of adapting itself to a given problem: using the same start team and the same set of experts, specialists, and referees the system is able to solve even hard problems of many different domains without any interaction or configuration by the user. While there are often teams, selected by experienced users, which perform better, nevertheless the team with a planning supervisor still showed synergetic effects and could solve the same problems.

4.6. *Analysis*

It has been empirically shown that TEAMWORK produces synergetic effects, so that not only proofs for given problems are found much faster than by single experts (faster than the time needed by the best single expert even if the team's time is multiplied by the number of computers used), but also problems can be solved that are beyond the capabilities of all our single experts (Denzinger, 1995). This can also be observed in other applications involving search in very large search spaces, e.g. optimization problems (Denzinger and Scholz, 1997). In Table I we show the results of our DISCOUNT system for two examples from the TPTP problem library (see Sutcliffe *et al.* (1994)). There are two different observable synergetic effects that may occur together:

- One agent is able to generate most of the facts necessary for a proof but is not able to find a few also needed ones (in acceptable time). These needed facts are provided by other agents as a result of cooperation. This effect is the reason for the speed-up achieved by DISCOUNT for example LCL111-2 of Table I.
- The agents are only able to work well for parts of the proof. Similar to a cooperation pipeline one agent using the search state of another one can proceed with the proof. Then another one takes over the search and so on. Since in contrast to a cooperation pipeline it is not clearly defined when one agent takes over from another one, the competition aspect of TEAMWORK is needed. This effect, in form of a goal-oriented expert taking over from a standard one, is responsible for solving example BOO008-2 that none of our experts could solve working alone.

Naturally, in order to achieve these effects, the referees must do their job. As demonstrated in Denzinger and Schulz (1996b), 10-50% of the facts selected by referees contribute to the found proof. Compared to the success of experts in selecting the next potential fact this is better by a factor of 100 and more.

Table I. Examples for TEAMWORK

Example	Teamexpert 1	Teamexpert 2	best known expert	TEAMWORK
LCL111-2	-	428.9s	406.7s	127.9s
BOO008-2	-	-	-	33.6s

This shows the usefulness of success-driven cooperation in general and of our retrospective criteria in particular.

TEAMWORK also has proven to be the ideal basis for learning theorem provers, i.e. provers that use experiences from previous proof attempts. Several problems of such provers, like appropriate storage and retrieval of experiences, detection of misleading knowledge, or combination of knowledge from different sources, are already solved by TEAMWORK: long-term memory, classifying specialists, referees and proof planning, and cooperation are these solutions.

Nevertheless, there are also some weak points of the method that deserve attention. The agent interaction scheme with the different roles is rather complicated and, to be effective on a network, requires the realization of a secure broadcast. Therefore we developed a library of objects, methods and skeleton methods to save someone interested in an application of TEAMWORK from working on the level of system software (Denzinger and Lind, 1996). Nevertheless, usually the knowledge used by referees, specialists, and the supervisor has to be acquired for each such application anew.

Without good referees and a good control it is possible to have teams that perform worse than their members. Also it is not possible to use different search paradigms with equal rights and some people do not like to have a central control in a multi agent system. We do not see the latter as a problem while the former will be subject of the next section.

5. AUTOMATIC COOPERATION IN HETEROGENEOUS PROVER NETWORKS: THE TECHS CONCEPT

As TEAMWORK has shown, success-driven cooperation of whole theorem provers as agents can increase the abilities of an automatic theorem proving system. Naturally, a multi agent system using agents employing deduction mechanisms stemming from different philosophies, using state-of-the-art techniques, and mixing both, success-driven and demand-driven cooperation, should be even more successful.

The development of such a system from scratch with an appropriate communication and control architecture is not impossible, but would require the cooperative effort of many human experts from leading research groups not only in the field of automated deduction. In addition to an enormous effort such a project would also have to deal with software engineering problems like the continuous improvement of the known provers (as demonstrated by the CADE prover competitions CASC-13 and CASC-14) or the necessity of developing "open systems" in which new provers can be easily integrated.

While such a venture definitely is interesting, nevertheless it is possible to achieve cooperation between already existing provers with much less effort involved but still showing an increase in the abilities of the cooperating system. In the following, we will present such a cooperation concept, the TECHS approach (TEams for Cooperative Heterogeneous Search), that includes success- and demand-driven cooperation between state-of-the-art provers without the need for many changes in these provers. Naturally, these changes can be integrated into all new versions of these provers and the addition of more provers into the developing heterogeneous prover network is easy.

5.1. *Architecture and Interaction*

Our TECHS concept is based on the cooperation of several fully automated provers or different incarnations of such provers that preferably will run on different computers in a local area network. So, the provers on their machines will be agents in our multi agent system. The problem to be solved has to be given to each agent. If an agent is not capable to fully understand the problem then it may only get the parts it understands.

For a theorem prover, not fully understanding a problem means that the problem either contains interpreted symbols it is not able to deal with or parts of the problem are not given in the logic the prover is developed for. As an example, the problem may be given as a set of formulas of first-order logic with equality and one prover may only be able to deal with first-order logic without equality and another one is a pure unit-equality prover. Then for the first prover an axiomatization of the equality predicate has to be added, while the second prover only gets the formulas that are unit equalities. Obviously, all provers must either be able to work with the logic the proof problem is formulated in or there must be a translation from this logic into their logic, or they must work in a more specialized logic (naturally, in the latter case we cannot expect that such a prover finds a proof, but it may help the others). E.g. a prover using intuitionistic logic may well support a prover working with classical logic. On the other hand, when a theorem is to be justified in intuitionistic logic, a prover for classical logic may be useful only when its results are interpreted in intuitionistic logic - e.g. by double-negating the proved formulas. Note that in this case no proof in intuitionistic logic would be obtained. Instead, only the *provability* of the translated result in intuitionistic logic is established. For some of the theoretical issues, please refer to Denzinger and Fuchs (1997).

Naturally, before we can take a closer look on the interaction of the provers, we have to establish what are the means the provers can use to communicate. The most basic, low-level abilities we require from a prover are that it has to be able to

— write formulas into a file during a proof attempt and to
— read formulas from a file during a proof attempt and to integrate these formulas into its search state for use in the further search.

The latter task has to be done at some predefined points of time. Note that the first task is realized in nearly every prover while the second one in some provers requires some modifications.

The tasks indicate how the cooperation of provers is achieved: by interchanging formulas. As we have already seen, not each newly generated formula should be communicated to the other agents. Therefore, as in the case of TEAMWORK, referees filter the formulas passing between the agents and their selection methods are both demand- and success-driven. In fact, there is not only one referee for each prover. Instead, each agent has 1 to $n-1$ send-referees (if there are n agents in the system) and one receive-referee.

A *send-referee* judges all information generated by its agent and selects such information that has proven to be useful, thus realizing a success-driven interchange. If a send-referee has to do its task for one agent or a group of agents that have certain requirements, then it can already try to meet these requirements (very weak form of demand-driven interchange). If an agent has only one send-referee the task of this referee is similar to the selection of good facts by referees in TEAMWORK.

A *receive-referee* collects the data sent by the send-referees of all other agents and tries to select that data that helps its agent at the moment the most (real demand-driven interchange). With the send-referees as a first filter the amount of communication between agents is held low, but still an agent receives quite some amount of data, because all other agents send their selected results. Then the receive-referee cuts down on this amount of data, again.

Figure 1 shows our TECHS architecture for three agents with the full number of referees. The referees are depicted half within their provers and half outside of them. This indicates that it is either possible to include them fully into the provers or they can be additional agents that read the files produced

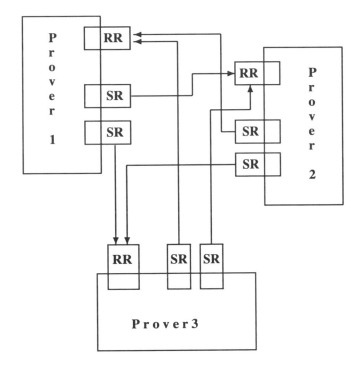

Figure 1.   TECHS architecture for 3 provers

by a prover or a send-referee and write their results in files, again. Naturally, an integration of the referees into the provers has many advantages: they can access all the data of the provers, as for example statistical data in case of a send-referee or actual goals in case of a receive-referee. Also they do not have to use two files but only one. But for first versions of referees the realization as agents is easier, because it is not necessary to do large modifications in the provers.

Note that, depending on the provers that are part of the network, not all referees may be needed. A send-referee can select formulas for several or all provers and, if the send-referees select only a small number of formulas, receive referees may be unnecessary. While it may be very useful to have for two provers send-referees that already try to make their selection with the other prover in mind, nevertheless it is possible to use a standard send-referee in first versions. This allows for an easy integration of new provers into the network.

5.2. *Concepts for Send-Referees*

There are two different types of send-referees: those that try to select formulas only based on their success for the search of the prover that generated them (pure success-driven) and those that also use criteria that are based on knowledge about receiving provers (weak demand-driven). The first type can use the same criteria as the referees in TEAMWORK (see Section 4.4). In analytical provers they select formulas out of the set of lemmata, i.e. formulas describing the path to a closed subtree in the case-analysis. As already stated, pure success-driven send-referees can be used for all possible provers as receivers.

Weak demand-driven send-referees are much more limited in their use, because they can only be used for one prover, sometimes only for one specific incarnation of it. In addition to the criteria used by pure success-driven send-referees they use criteria that make use of knowledge about the receiving prover. Note that this knowledge has to be of general nature, because the actual demands of the receiver cannot be known without a large additional amount of communication. In addition, the used criteria should be able to suggest formulas that are useful for the receiver but that the receiver is not able to generate by itself. Otherwise the probability is high that the receiver already has the selected formulas. So far, we have two criteria that we use for this:

— For analytical provers working on clauses, atomic formulas (or their negation) are very interesting, because they are needed to close branches. They are not often generated, since analytical provers have only very limited generating capabilities. Especially atoms with long derivations are interesting, since analytical provers very seldom do such long derivations.

— For generating provers, formulas are of interest that are rated high by the particular control strategy of the receiving prover, but that have ancestors that are rated very low. Since the rating of some ancestors is low also the probability that the receiving prover already has generated the formula is low, because it will not generate these ancestors.

Note that the second criteria requires the implementation of the rating mechanism of the control of the receiving prover into the send-referee. Therefore we have tested it only for the cooperation of different incarnations of a prover, but there with quite some success (see Fuchs and Denzinger (1997)).

5.3. *Concepts for Receive-Referees*

Receive-referees allow to filter data with respect to the *actual* requirements of a prover that receives this data. Unfortunately, often it is not possible to define the actual requirements. Therefore we have to rely heavily on heuristics and even more on several heuristics each of which selects a few formulas to be passed to the prover.

Since the send-referees already provide a success-driven view on all the formulas a receive-referee obtains, we found it a good idea to let the receive-referee choose the best formula of each send-referee. For choosing from the remaining formulas our receive-referees employ criteria that can be best described as "what-if" criteria. This means that these criteria try to measure the consequences the addition of a formula has for the actual search state of the prover.

What-if criteria for generating provers are

– number of formulas of the actual state that will be subsumed or simplified by a formula
– number of consequences of a formula with the actual state
– similarity of a formula to goals

A criterion for analytical provers is the number of paths that are closed by a formula.

Even if a receive-referee uses only approximations of these criteria measuring one formula can entail a high computational effort. Therefore it is important that the send-referees already made a selection, thus ruling out most of the formulas generated by the other provers. Naturally, for each prover there will be additional criteria that are based on the particular calculus of the prover, its implementation, and its control. In contrast to earlier concepts, for receive-referees the combination of criteria is not a weighted sum but selecting all those formulas that are for one or several criteria above given thresholds.

5.4. *Analysis*

Our TECHS concept for cooperation in heterogeneous prover networks has proven to cause synergetic effects. If we use it to let different incarnations of the same prover cooperate (Fuchs and Denzinger, 1997), it performs for some examples even better than TEAMWORK although the communication via files is much slower than the communication TEAMWORK uses (provided that we use only well suited control heuristics in all incarnations, otherwise the supervisor allows for good TEAMWORK results but the TECHS concept cannot adapt).

Table II. Examples for TECHS

Example	Prover 1	time	Prover 2	time	TECHS
BOO007-4	SPASS	403.4s	DISCOUNT	-	19.3s
LDA010-2	SPASS	-	DISCOUNT	-	682.7s
LAT001-1	SETHEO	-	SPASS	-	8.9s
LAT001-3	SETHEO	31.7s	SPASS	134.4s	11.2s

By making necessary modifications in the systems SPASS, SETHEO, and DISCOUNT we showed that the TECHS concept is easy to integrate into existing provers. Cooperation between a "general" prover, as SETHEO or SPASS, and a "specialized" one, as DISCOUNT specialized in unit equality, led to huge speed-ups and the solution of problems none of the provers could solve alone in an acceptable amount of time. Examples BOO007-4 and LDA010-2 of Table II (again taken from TPTP) are an excerpt of Denzinger and Fuchs (1997) and give an example for these effects. Many more examples are in the cited paper.

The cooperation of provers that are "general" produced also improvements, although it seems that so far the analytical provers profit more from the cooperation than the generating ones. The examples LAT001-1 and LAT001-3 of Table II demonstrate the possible speed-ups and the improvement in the number of solved problems, again. For both examples SPASS only acted as provider of (obviously very useful) data while SETHEO found a proof (that it was not able to find alone in case of LAT001-1).

The following improvements of our concept are possible. The communication via files is definitely not an efficient way. The use of packages like PVM would be an improvement but our experiments have shown that this is not urgent.

Most analytical provers perform their search based on depth bounds and backtracking. Therefore the actual search state when integrating formulas from other provers does not give a good view on the actual requirements of such a prover. Either requirements of earlier states have to be stored or requests of such provers to other provers have to be allowed at all times to improve on this. Requests to other provers require massive modifications in these other provers, but storage of earlier requirements and redoing this earlier search is possible (by modifications of the analytical prover only).

Finally, our concept does not make any use of dividing a given problem into subproblems (only, if one of the prover does this for itself). As already stated, an automatic division of problems into subproblems and finding the best provers for such subproblems is very difficult and often impossible. Even if it were possible, then either for each new prover we would have to include appropriate information into all other provers to allow them to find the subproblems it is good at solving, or negotiations are necessary in which each prover must anticipate not only if it is able to solve a problem but also in which time in order to find the best suited one (see Section 2.3). But these problems can be solved by integrating the user into the network, as a kind of supervisor, as we will demonstrate in the next section.

6. Interactive Cooperation in Heterogeneous Prover Networks: ILF and TECHS

Given the state of the art, really complex and large proof projects still require user guidance. Interactive theorem provers, e.g. Gordon and Melham (1993), Nipkow and Paulson (1992), Constable et al. (1985), II.1.1, Farmer et al. (1993), concentrate on this aspect of the theorem proving process. They presume a user which is editing a proof in a specific formal calculus. The user can call proof tactics supplied by the system. These tactics are small programs which automate routine parts of the proof editing process.

In order to increase the degree of automation, some interactive systems have been augmented by an automated theorem prover. This prover is called in order to solve specific proof problems supplied by the user. Therefore – according to our classification of cooperation – these systems realize a demand-driven cooperation of the user and an automated theorem prover.

ANALYTICA (Clarke and Zhao, 1992) and FAUST/MEPHISTO (Kumar et al., 1993) are examples of automated theorem provers which have been specifically designed to support interactive theorem proving. In fact, they have been implemented within existing interactive systems (MATHEMATICA, HOL) using the available tactic mechanisms.

Within the DFG-Schwerpunktprogramm "Deduktion" interactive theorem provers have been combined with some high performance automated theorem provers which had been developed separately. Thus the user of the Omega system can hand over proof problems to the automated prover Otter (Benzmüller et al., 1997) and the KIV user can refer to the prover 3TAP (see II.1.4). Then, proofs found by the automated theorem prover are used to generate proof tactics for the interactive prover which produce similar proofs within the interactive system. The final result is a formal proof within a single calculus.

Using a system which combines interactive and automated theorem proving as described, requires a user who

- knows the calculus of the interactive system
- knows the proof tactics of the interactive system
- knows the abilities of the automated theorem prover in order to decide which problems it should handle and how the prover should be configured.

The last point becomes most crucial when the concept is extended to the use of a variety of different automated theorem provers.

Having a human user as an agent in a distributed system does not require a change in the TECHS architecture as described in the last section. However, the position of the user in the flow of control is special. On one hand, his demands deserve the highest priority. On the other hand, he is not able to interfere directly with the fast communication between the automated provers in the TECHS architecture. Instead, his influence will be directed towards tuning provers and the referees of the automated provers prior to launching an automated proof search.

When the user in reality is as unaware of the details of the automated provers as the provers are of each other, this tuning has to be done by the system based on an analysis of the user demands. Conceptually, in the TECHS approach, it is the users send-referee which generates and outputs the tuning commands to the other systems through their corresponding receive-referees.

The users receive-referee has to extract from the provers exactly the information which is relevant to the users proof editing process. For the user, the most important thing to know is a pure confirmation of the correctness of his arguments, i. e. a confirmation that a certain goal formula is a semantic consequence of a given set of axioms. Each automated theorem prover will supply such confirmations and some of them are able to print a formal proof, but the internal structure of the proof delivered by the automated prover is of minor importance. It is the aforementioned justification which matters. For later reorganization of the proof, e. g. deleting some intermediate formulas, it may be useful to know, which formulas have been exactly used in a formal proof. Therefore, the user must be informed by his/her receive-referee

- that a proof was found,
- what was proved
- which assumptions have been used.

The display of the proof or its translation into a specific calculus can be useful, but are conceptually less relevant. Moreover, translation and integration of proofs from various different sources can be time consuming and lead to very detailed formal proofs which are hard to analyze by the human. Therefore proof integration should be postponed until it is really required by the user.

6.1. *Block Structured User Communication*

We mentioned that - according to the TECHS approach - the user communicates with a complex prover network through his/her referees. Then the network appears to the user as a monolithic system, which produces justifications of user arguments. Consequently, these justifications appear as building blocks of the large chain of arguments edited interactively by the user. One of the provers, who gave the final confirmation of the users argument, appears as the users direct partner, delivering the block. But this prover may have communicated again with other provers, which provided subblocks that it could use in its argument, etc. Thus, viewed over a period of time, each agent in the prover network produces a sequence of blocks justifying a certain inference. The atomic blocks in this system are the proofs, found by a single agent without interaction.

This suggests an overall proof structure which is a combination of interrelated blocks. Besides being a concise description of the combination of the arguments, it documents clearly the particular responsibility of each agent for the overall argument.

A formal definition of such a proof concept was given in Dahn and Wolf (1994). A *block* is a sequence of lines. Each line contains a formula and a status *untried, tried, proved, unproved, assumption* or *subproof* (Pf). Here, Pf is a reference to another block. A *block structured proof* is a set of blocks. Given such a proof, we consider *usability of lines* to be the least transitive relation such that

— axioms are usable for all lines,
— a line l in a block B is usable for all lines l' in B succeeding l and
— if such a line l' has status *subproof* (Pf), then l is also usable for all lines in the block referred to by Pf.

A block structured proof is *correct*, if the contents of each line with status *proved* is a semantic consequence of the contents of all lines which are usable for this line. Moreover, the contents of each line with status *subproof* (Pf) must be a consequence of the formulas in usable lines and the theory saying that each formula in a line of Pf is a consequence of these usable formulas and the assumptions in block Pf. A block structured proof is *complete* if it does not have any line with status *untried, tried* or *unproved*.

Note that the concept of a correct proof is strictly semantic and is not related to any specific collection of inference rules. There are many possibilities for sound inference rules for block structured proofs. E.g. – given a sound automated theorem prover P – an inference rule may permit to change the status of a line l to *proved*, if P can prove the contents of l within a given time limit from axioms and the contents of the lines usable for l. When P delivers a proof which can be translated into a block structured proof Pf, the proof containing l will remain correct if the status of l is changed to *subproof* (Pf) and references to axioms used by P are replaced by references to the original axioms and lines.

For many provers there are algorithms which transform the output files into block structured proofs. Wolf (1997) describes such an algorithm for model elimination proofs with factorization. Since different blocks in a block structured proof can use different rules of inference, in this way proofs from different automated theorem provers can be linked with a block structured proof edited interactively by the user. A sound and complete natural deduction style calculus, suitable for interactive proof editing, has been introduced in Dahn and Wolf (1994).

6.2. *Generation of Proof Problems*

The combination of automated theorem provers with an interactive theorem prover creates the perspective of editing a correct proof without knowledge of any logical calculus. This becomes extremely important, when deductive tools are to be used by human experts without special training in logics.

In fact – given a block structured proof – it is easy to extract automatically all proof problems which have to be solved: The contents of each line which does not have a status *assumption, proved* or *subproof* ($\_$) has to be proved from the axioms and from the contents of the usable lines.

When these problems have been extracted, automated theorem provers can be launched to solve them. If there remain unsolved problems, the user can insert additional lines into the proof. These new lines have initially the status *untried*. Then one or several provers can be launched demand driven to prove the contents of this new line. Moreover, subsequent lines can make use of this line. This modifies proof problems which have been treated before by automated provers. Either these already running provers are able to take this new line into account – which is one of the features that provers employing TECHS have – or new provers have to be launched success driven.

When an automated prover has been launched to verify a line, the status of that line changes to *tried*. When the prover succeeds, the status is changed to *proved*. When the prover fails or exceeds its time limit, the status becomes *unproved*.

So far, the only action of the user was to insert new lines. The only feedback he got consisted of changes of the status of proof lines. Thus, the system resembles an ordinary editor with a spell check facility.

Realistic proof problems come frequently with a fairly large set of axioms. But most lines will require for their justification only a small part of the theory and of the available lines. Since redundant theories can prevent automated theorem provers from finding a proof, the user can enhance the efficiency of the system by pointing to parts of the theory which should be used to prove a specific line. Human experts usually have a fairly good idea of the structure of the theory and the relative dependencies between its parts. Therefore, this kind of interaction is quite accessible to them. However, there is a danger that some parts of the theory which are obvious for the user are omitted. Such parts have to be added automatically.

It would be desirable to determine relevant parts of the theory automatically. III.2.9 gives a method to do this in some cases.

6.3. *Interactive Environment for Automated Provers*

The cooperation of automated theorem provers with an interactive theorem prover as described here creates an environment with specific properties. Normally, many proof problems have to be solved at the same time and at each time the user is able to continue the interactive editing of the proof. Therefore, in the ILF system, automated provers run asynchronously in the background.

When there are many automated theorem provers running on the same system, this can affect user interaction severely. Hence, the generation of problems described here is only feasible on parallel hardware or based on a network of computers.

The proof problems generated from a block structured proof are very similar to each other. They share many axioms. When a line l is usable for lines l_1 and l_2, all theorem provers that try to verify l_1 and l_2 have to treat the contents of l. This leads to a large amount of duplicated work if these provers are not able to reuse results of each other. Establishing a success-driven cooperation between the provers in the sense of our TECHS approach of Section 5 gave a considerable improvement. Especially the exchange of preprocessing results, like clauses generated from the contents of l, speeds up provers launched later. Therefore, preprocessing of proof problems has to be split into two phases - a phase based on single formulas giving reusable results and a second phase analyzing the complete proof problem.

Using automated theorem provers in cooperation with an interactive theorem prover changes the role of proof tactics essentially. Tactics are programs that extend partial proofs. In standard interactive theorem provers they apply inference rules in order to construct partial subproofs. But in our setting, their main purpose is to modify open proof problems in such a way that they can be treated by automated provers.

It is a major task for the tactics to analyze the concrete proof problems and to select relevant parts of the available theory. To do this, the tactic can analyze the contents of the line to be proved as well as its position in the entire proof. An example of such a tactic is given in Chapter III.1.1.

6.4. *Global Control*

Insertion of a line at position k in a block with n lines generates one new proof problem and modifies $n - k$ problems. In the worst case, none of these $n - k$ problems has been solved before.

Realizing a network with the current non-cooperating provers, enforces launching a new set of provers for each of the $n - k$ problems. Then the number of proof problems generated is quadratic in the number of insertions. Since each proof problem is handled by one or more automated theorem provers, the number of simultaneously running provers soon becomes a major problem (*prover plague*). It is especially severe, when the interactive theorem prover applies proof tactics which insert many lines automatically.

On the other hand, TECHS cooperative provers working on these problems will incorporate the contents of the new line into their database, review the proof searching process and continue their work. Only the verification of the new line requires to launch new provers.

To kill an uncooperative prover in favor of a new prover working for the justification of the same line can be inefficient when the newly available knowledge is not really necessary for this justification. Since it is undecidable, whether the contents of a given line is necessary to verify a subsequent line, we propose a simple heuristic to restrict the number of proof problems generated.

This heuristic is based on the observation that in interactively edited proofs it is unlikely that a formula in line l is necessary to justify a line which is far beyond l. Therefore, we propose to fix a number m such that the insertion of a line at a position launches new provers only for the new line and for the next m succeeding lines. Then, the number of running provers will be linear in the number of insertions. $m = 3$ has turned out to be a reasonable value. Applying this heuristic may require that the user checks failed proof attempts in order to see whether a distant line should be made available manually.

An efficient method for restricting the prover plague is proof structuring using blocks. Since the lines in one block b are not usable for lines in another block b' at the same or larger depth of the proof, insertion of a new line into b will not modify proof problems for lines in b'. Fortunately, users experienced with informal proofs, tend to structure proofs in order to concentrate on specific subproblems.

ILF uses *flexible killing* to handle the prover plague. According to this strategy, each automated theorem prover is guaranteed a minimal time to work. When the user generates new proof problems with the interactive prover, these are handed over to a background system which preprocesses the problems for the available automated theorem provers. The background system requests a number of processors from a load control system.

The load control system negotiates with load information systems on the available machines as described in subsection 2.3 to find the requested number of processors with a load below a specified threshold. In reality, the load control system has to take into account also the number of proof jobs launched but not yet reflected in the load information obtained as bid. If the requested number of processors is found, the automated provers are launched. Otherwise the load control system checks periodically whether there are provers which have already been working for their guaranteed time. Then these provers are killed and the waiting proof jobs are launched on the processors which become available. With this strategy, a certain load level of the whole system can be maintained in view of a discontinuous flow of requests from the interactive theorem prover.

In some cases, model checkers can be used to determine unsolvable proof problems and prevent the launching of theorem provers. Chapter III.1.1 describes the effect of this technique for proofs on lattice-ordered groups.

6.5. *The Necessity of Proof Presentation*

Since the automated theorem provers appear to the user as black boxes, there is a danger that their use covers errors. The main reason for this danger is not faulty behavior of the provers but unjustified self-confidence of the user. It may happen, that the user enters a line with a contents contradicting the contents of previously entered lines. Based on these lines, anything can be proven in a formally correct way.

Therefore, a critical look at the automatically found subproofs is necessary. In some cases, e.g. when a refutational prover succeeds without using the negated goal in the proof, ILF will issue a warning. But the user himself is much better in discovering errors when he finds unexpected intermediate results or unused axioms in the subproofs. Of course, this requires the conversion of the prover output into a human readable format.

6.6. *Results of Cooperation of Automated and Interactive Theorem Provers*

The use of automated theorem provers in an interactive environment enables the editing of proofs in a way similar to writing informal proofs. The user can concentrate on the problem and leave the generation of formal proofs in special logical calculi to the automated provers. He can proceed in larger steps and he has to be less precise in specifying the order in which arguments have to be applied. This makes formal verification of arguments accessible to a wider audience.

As described above, the cooperation of the interactive prover PROOFPAD with the automated provers DISCOUNT and SETHEO was realized within the ILF system. With such a configuration a proof of the intermediate value theorem for real functions was obtained with 16 user commands, based on ordinary axioms for continuous functions and real numbers. This approaches the level of detail found in mathematical text books. In this proof, 32 lines were generated by proof tactics and 23 subproofs were found automatically. Chapters III.1.1 and III.2.8 describe environments to use the cooperation of interactive and automated provers in other fields.

7. CONCLUSION

In theorem proving, as in many other fields using large computer program systems, there are many different systems helping human beings solving given problems. There is more and more need for concepts that allow to combine these different realistic state-of-the-art systems in order to use the accumulated expertise of many researchers and developers. Multi agent concepts that are based on the view of such program systems as agents and that provide a communication and control structure (and perhaps additional agents) seem to be an affordable way to achieve some cooperation between existing systems.

In this chapter we presented such multi agent concepts both for the fully automated case and the case involving interaction with the user. For the fully automated case we developed two concepts, TEAMWORK emphasizing one basic inference mechanism and TECHS concentrating on equal importance of all systems. Our experiments have shown that the achieved cooperation resulted in multi agent systems with abilities none of the single agents had.

The TECHS concept allows the integration of the user as (control) agent of the system. Since the essence of TECHS aims at achieving fully automated systems, the user's wishes decide how smooth the transition from interactive to fully automated really is. Since the TECHS concept is open, i.e. it allows to add more and more agents, and it also allows to improve on the cooperation quality by adding better referees, thus realizing a kind of getting to know new provers, problems that can at first only be solved with much interaction may come more and more into the reach of the fully automated prover network.

The requirements on a prover for being able to be part of a prover network are very low, so that most provers need only small modifications. On the other side, after the modifications a prover can cooperate with itself, either as different incarnations or by using earlier results of itself from previous proof attempts, which improves the acceptance of a prover by a user.

But not only provers can be part of a prover network. Data bases, mathematical libraries, even prover networks of other users can be integrated to enhance the abilities of the individual network of a user. As our experiments have shown, slow communication media still allow for a good cooperation. The latter may encourage the cooperation via the world wide web.

REFERENCES

Beckert, B. and Hähnle, R. *et al.* (1996): The Tableau-Based Theorem Prover 3TAP, Version 4.0, Proc. CADE-13, New Brunswick, LNAI 1104, pp. 303-307

Benzmüller, C. and Cheikhrouhou *et all.* (1997): ΩMEGA: Towards a Mathematical Assistant, Proc. CADE-14, Townsville, LNAI 1249, pp. 252–255.

Bonacina, M.P. and Hsiang, J. (1995): The Clause-Diffusion methodology for distributed deduction, *Fundamenta Informaticae* **24**, pp. 177–207.

Bonacina, M.P. (1997): The Clause-Diffusion Theorem Prover Peers-mcd (System Description), Proc. CADE-14, Townsville, LNAI 1249, pp. 53–56.

Clarke, E. and Zhao,X. (1992): Analytica – A Theorem Prover for Mathematica, Internal report CMU-CS-92-117.

Conry, S.E. and MacIntosh, D.J. and Meyer, R.A. (1990): DARES: A Distributed Automated Reasoning System, Proc. AAAI-90, pp. 78–85.

Constable, R. L. *et al.* (1985): Implementing Mathematics with the Nuprl Proof Development System, Prentice-Hall Inc.

Dahn, B. and Gehne, J. *et al.* (1997): Integration of Automated and Interactive Theorem Proving in ILF, Proc. CADE-14, Townsville, LNAI 1249, 1997, pp. 57–60.

Dahn, B.I. and Wolf, A. (1994): A Calculus Supporting Structured Proofs, *J. Inf. Proc. Cyb.*, pp. 261–276.

Dahn, B.I. and Wolf, A. (1996): Natural Language Presentation and Combination of Automatically Generated Proofs, Proc. FroCoS'96, München, Springer, pp. 175–192.

Denzinger, J. (1995): Knowledge-Based Distributed Search Using Teamwork, Proc. ICMAS-95, San Francisco, AAAI-Press, pp. 81-88.

Denzinger, J. and Fuchs, M. (1994): Goal Oriented Equational Theorem Proving Using Teamwork, Proc. KI-94, Saarbrücken, LNAI 861, pp. 343–354.

Denzinger, J. and Fuchs, D. (1996): Referees for Teamwork, Proc. FLAIRS '96, Key West, ISBN 0-9620-1738-8, pp. 454–458.

Denzinger, J. and Fuchs, D. (1997): Knowledge-based Cooperation between Theorem Provers by TECHS, SEKI-Report SR-97-11, University of Kaiserslautern.

Denzinger, J. and Fuchs, Marc and Fuchs, M. (1997): High Performance ATP Systems by Combining Several AI Methods, Proc. IJCAI-97, Nagoya, Morgan Kaufmann, pp. 102–107.

Denzinger, J. and Kronenburg, M. (1996): Planning for Distributed Theorem Proving: The Teamwork Approach, Proc. KI-96, Dresden, LNAI 1137, pp. 43–56.

Denzinger, J. and Lind, J. (1996): TWlib - a Library for Distributed Search Applications, Proc. ICS'96-AI, Kaohsiung, pp. 101-108.

Denzinger, J. and Scholz, S. (1997): Using Teamwork for the Distribution of Approximately Solving the Traveling Salesman Problem with Genetic Algorithms, SEKI-Report SR-97-04, University of Kaiserslautern.

Denzinger, J. and Schulz, S. (1994): Recording, Analyzing and Presenting Distributed Deduction Processes, Proc. PASCO'94, Linz, pp. 114–123.

Denzinger, J. and Schulz, S. (1996): Learning Domain Knowledge to Improve Theorem Proving, Proc. CADE-13, New Brunswick, LNAI 1104, pp. 62–76.

Denzinger, J. and Schulz, S. (1996): Recording and Analyzing Knowledge-Based Distributed Deduction Processes, *Journal of Symbolic Computation* **21**, pp. 523–541.

Ertel, W. (1992): OR-parallel theorem proving with random competition, Proc. LPAR'92, St. Petersburg, LNAI 624, pp. 226–237.

Erman, L.D. and Hayes-Roth, F. and Lesser, V.R. and Reddy, D.R. (1980): The Hearsay-II speech-understanding system: Integrating knowledge to resolve uncertainty, *Computing Surveys* **12**, pp. 213–253.

Farmer, William M. and Guttman and Joshua D. and Thayer, Javier F. (1993): IMPS: An Interactive Mathematical Proof System, The MIITRE Corporation.

Fuchs, D. and Denzinger, J. (1997): Cooperation in Theorem Proving by Loosely Coupled Heuristics, SEKI-Report SR-97-03, University of Kaiserslautern.

Gordon, M.J.C.; Melham, T.F. (eds.) (1993): Introduction to HOL: A Theorem Proving Environment for Higher Order Logic, Cambridge Univ. Press.

Hayes-Roth, B. (1985): A Blackboard Architecture for Control, *Artificial Intelligence* **26**, pp. 251–321.

Kumar, R. and Schneider K. and Kropf T. (1993): Structuring and Automating Hardware Proofs in a Higher-Order Theorem-Proving Environment, *Int. J. Formal System Design*, pp. 165–230.

Nipkow, T and Paulson, L. C. (1992): Isabelle-91, Proc. CADE-11, Saratoga Springs, LNAI 607, pp. 673–676.

Schumann, J. (1994): DELTA - A Bottom-Up Preprocessor for Top-Down Theorem Provers. System Abstract, Proc. CADE-12, Nancy, LNAI 814, pp. 774–777.

Slaney, J.K. and Lusk, E.W. (1990): Parallelizing the Closure Computation in Automated Deduction, Proc. CADE-10, Kaiserslautern, LNAI 449, pp. 28–39.

Smith, R.G. (1980): The Contract-Net Protocol: High Level Communication and Control in a Distributed Problem Solver, *IEEE Trans. Comp.*, **C-29**, pp. 1104–1113.

Sutcliffe, G. (1992): Heterogeneous Parallel Deduction System, Proc. of FGCS'92 Workshop W3, appeared as Technical Report ICOT TM-1184, TU Munich.

Sutcliffe, G., Suttner, C., and Yemenis, T. (1994): The TPTP Problem Library, Proc. CADE-12, Nancy, LNAI 814, pp. 252–266.

Suttner, C.B. (1995): Parallelization of Search-based Systems by Static Partitioning with Slackness, PhD-Thesis, Infix-Verlag, DISKI 101.

Weidenbach, C. and Gaede, B. and Rock, G. (1996): SPASS & FLOTTER Version 0.42, Proc. CADE-13, New Brunswick, LNAI 1104, pp. 141–145.

Wolf, A. (1997): A Translation of Model Elimination Proofs into a Structured Natural Deduction, Proc. FLAIRS-97, Daytona-Beach, ISBN 0-9620-1739-6, pp. 11–15.

Wolf, A. and Schumann, J. (1997): ILF-SETHEO: Processing Model Elimination Proofs for Natural Language Output, Proc. CADE-14, Townsville, LNAI 1249, pp. 61–64.

INDEX